DE NEUFVILLE, Richard and David H. Marks, eds. Systems planning and design; case studies in modeling, optimization, and evaluation. Prentice-Hall, 1974. 438p il tab bibl 73-17053. 18.95. ISBN 0-13-881599-2.　C.I.P.　no LC

The editors, professors at M.I.T., are well-known authorities in the field of engineering systems design. This well-indexed text contains actual case studies in the application of systems analysis techniques to actual problems in planning and design. The cases are grouped into three major sections that focus on modeling the system, search for optimal ranges of design, and evaluation and choice of alternatives. Coverage includes a large-scale public water supply network, airport and railroad transportation systems, as well as thermal energy supplies. Ten of the twenty-seven chapters are contributions (in part) of the editors. The work of 34 people is incorporated in these case studies. Technically oriented professionals, teachers, and students of engineering and systems theory will find this book useful. Although the price of $18.50 may deter individual purchases, libraries that serve the above clientele may wish to consider it.

Systems Planning and Design

CASE STUDIES IN MODELING, OPTIMIZATION, AND EVALUATION

CIVIL ENGINEERING AND ENGINEERING MECHANICS SERIES

N. M. Newmark and W. J. Hall, *editors*

Systems Planning and Design

CASE STUDIES IN MODELING, OPTIMIZATION, AND EVALUATION

Edited by

RICHARD DE NEUFVILLE

and

DAVID MARKS

106391

PRENTICE-HALL, INC., *Englewood Cliffs, New Jersey*

Library of Congress Cataloging in Publication Data

DE NEUFVILLE, RICHARD,
   Systems planning and design.

   Includes bibliographies.
   1. Systems engineering.   2. Engineering design.
3. Mathematical optimization.   I. Marks, David H.,
joint author.   II. Title.
TA168. D46   620'.72   73-17053
ISBN 0-13-881599-2

To

JANET AND JUDITH

© 1974 by
Prentice-Hall, Inc.
Englewood Cliffs, N.J.

Printed in the United States of America

PRENTICE-HALL INTERNATIONAL, INC., *London*
PRENTICE-HALL OF AUSTRALIA, PTY., LTD., *Sydney*
PRENTICE-HALL OF CANADA, LTD., *Toronto*
PRENTICE-HALL OF INDIA PRIVATE LIMITED, *New Delhi*
PRENTICE-HALL OF JAPAN, INC., *Tokyo*

Contents

Part III
OPTIMIZATION   153

Part **IV**
EVALUATION   289

Preface

This book is a reader of case studies of the application of systems analysis techniques to actual problems in planning and design. The studies are intended to complement the growing, and already extensive, theoretical literature on systems analysis. The cases have been selected to illustrate the relative strengths and the usefulness of alternative approaches to formulating and solving large-scale planning problems.

Though the theoretical desirability of using systems analysis is widely acknowledged, there is much uncertainty as to how these methods can and should be used in practice. The fact is that most expositions of the subject either discuss generalities or focus on analysis techniques. A large number of potential practitioners have been left somewhat confused about how the methods can and should be applied. This text intends to help solve that problem by providing good examples and guidelines for the use of systems analysis.

Most of the cases are grouped into three major sections that focus on: modeling the system; search for optimal ranges of design; and evaluation and choice of alternatives. Some additional studies refer to the future prospects of systems analysis and the problems of implementation. The cases were all chosen to provide a greater understanding of the modern methods of analysis that are now being introduced into the process of systems planning and design: econometric and nonexperimental modeling; large-scale mathematical programming and simulation, decision analysis, and multiobjective evaluation.

This text is not, by any means, a simple juxtaposition of material published elsewhere. The editors have extensively adapted and, often, completely rewritten the studies so that they will form a coherent whole. It is expected that readers will appreciate the way the case studies have been selected to complement each other by comparing the use of alternative techniques to the same kind of problems and, thus, to show the advantages and disadvantages of alternative formulations of a real problem.

Care was also taken to present the material at a level that will be accessible to technically oriented professionals with minimal training or orientation in systems methods. Yet many of the cases report on developments at the forefront of current research. Much of the material, furthermore, is now being published for the first time.

A broad range of students and professionals are expected to find this book useful. Since the cases emphasize practical problems and avoid tedious analytical formalisms, it is expected that practicing engineers and planners will be able to use the text extensively for reference and self-study. Teachers can expect to find the book useful as a supplementary text for their subjects in systems analysis. These subjects now must rely mostly on theoretical texts which do not have space to develop a spectrum of significant cases illustrating real-world applications [for example, *Introduction to Systems Engineering, Deterministic Models* by Au and Stelson (1970), *Systems Analysis for Engineers and Managers* by de Neufville and Stafford (1971), or *Mathematical Foundations for Design* by Stark and Nicholls (1972)]. The editors have, in fact, tested and validated the concept of using this material as a supplement to theoretical presentations by using the text in their graduate classes at the Massachusetts Institute of Technology.

After several years of collaboration, both the authors share extensively in the responsibility for the presentation of the text. To the extent that credits can be distinguished, the conception of Part III, on optimization, is primarily due to David Marks, and that of the introductory material and Parts II and IV, on modeling and evaluation, to Richard de Neufville. Essentially all the cases were redrafted by Richard de Neufville before being sent to press, and the entire manuscript was reviewed by both editors.

We are heavily indebted to many persons in the preparation of this manuscript. Over many years we have benefitted extensively from the counsel of our colleagues, both at MIT and elsewhere, many of whom are represented in this reader. Among the persons who have most particularly stimulated and encouraged us during our professional careers, and who are thus indirectly responsible for helping us complete the text, we would most gratefully like to acknowledge Professors Adolf May and Harmer Davis at the University of California at Berkeley; Professor Andrei Rogers at Northwestern; Professor Jon Liebman of the University of Illinois at Champaign-Urbana; Professor Lester Hoel at Carnegie-Mellon; Professor Peter Loucks at Cornell; Professor Anthony Tomazinis at the University of Pennsylvania; as well as Professors Peter Eagleson, John Schaake, David Major, and Ralph Keeney at MIT. We are also grateful to the Civil Engineering Department and the Civil Engineering Systems Laboratory at MIT for the financial support and unique intellectual opportunities they have provided us.

We owe a particular debt to Ms. Rosemary Carpenter for her invaluable assistance. She has been totally responsible for the actual preparation of the manuscript, for making sure that it was consistent from chapter to chapter, for securing all necessary releases, and for supervising the many detailed tasks that must be done to prepare a text.

RICHARD DE NEUFVILLE

DAVID H. MARKS

PART I

Introduction

1

Introduction

Substantial change has been taking place, since about 1960, in the procedures for planning, designing, and managing large-scale physical systems. That is, the systems approach is being adopted. More specifically, a variety of new techniques, such as linear programming and decision analysis, are being introduced. Many of these methods owe their existence as practical tools to the development of the computer. In a fundamental sense, the transformation of the planning and design process we are witnessing is the inevitable result of one of the really major technological changes of our time. In that general respect, the systems approach seems to have an assured and enviable future.

The particulars of the future development of the systems approach in planning and design are not, however, at all clear. So far, the number of applications is low and their success difficult to evaluate. There is much confusion among the experts as to what approaches are valid, let alone practical, for what kind of problems and for which situations. Perhaps this disarray is a natural characteristic of any rapidly developing field. The questions remain, nonetheless, and need to be addressed.

This text attempts to provide some guidelines as to how and when the systems approach should be applied for planning and design. Each case study has been carefully selected for its value in exploring particular methodological and pragmatic issues; each case is intended to provide evidence as to which methods and techniques are appropriate for which circumstances. It is hoped that they will jointly clarify our perception of how systems analysis can be used and practiced.

Because the nature of new strategies for planning and design is so broad a topic, it is not possible to cover it all. Only a portion of the question can be examined satisfactorily in a single text. Recognizing that it was not possible to be comprehensive, we chose to cover a limited number of topics in which the questions were best formulated and where the evidence appeared most solid. As a result, the cases do not address such questions as: how are goals defined? how is policy formulated? how are programs distorted through the implementation process? These are interesting issues and should eventually be confronted. Meanwhile, we have attempted to present some guidelines for the use of available techniques and procedures.

TOPICS COVERED

The cases are generally focused on three major areas: modeling, optimization, and evaluation. These are described below.

The modeling section illustrates how different analytic approaches should be used to attack the quite dissimilar problems of representing the way services can be provided—as described by the production and cost functions—and of representing the potential use of a service by the public, as described by the demand function. The cases on modeling are presented in Part II, and the issues in modeling are discussed in detail in Chapter 5.

The section on optimization deals principally with how mathematical programming techniques can be used best in practical analysis. An underlying issue throughout is the question of what simplifying assumptions can be made safely, without sacrificing essential elements of a problem, to achieve computational power and insights into the nature of the optimal design. Because of the assumptions that must be made before any optimization technique can be used, these approaches do not define the optimum; a parallel issue, therefore, revolves around the most appropriate blend of optimization and detailed simulation. The cases on optimization and simulation appear in Part III, and the attendant issues are discussed in detail in Chapter 12.

The evaluation section considers the various means for dealing with the multidimensional criteria for evaluation, whose importance generally varies between the groups interested in any significant public project. For systems of this sort, the traditional single-dimensional benefit-cost criteria are insufficient. The analyst must take a multiobjective approach. He must encode the nonlinearities in the way persons and groups sense values into a multi-attribute utility function, and he must take account of risk. Finally, the analyst should recognize the inevitable limits to the usefulness of any methodology in the evaluation process. The cases illustrating evaluation are presented in Part IV, and the issues are described in detail in Chapter 20.

In addition, the remainder of this section, Part I, includes some pieces on the future of systems analysis and on the prospects of its use in practice.

Chapter 2 surveys the current uses of systems analysis and its conceptual definition implicit in current university curricula. The authors then hypothesize where and how the several techniques of systems analysis will best be used. Chapter 3 reports on some of the recent studies of what factors influence the implementation of the results of any systems analysis. Because of the substantial methodological problems inherent in investigations of this sort, the work is still quite primitive and largely speculative. It does, however, suggest some very real issues which systems planners and designers ought to keep in mind. These considerations are, finally, reinforced in Chapter 4 by the discussion of the actual use of systems analysis in the design of New York City's water supply.

LEVEL OF PRESENTATION

The cases have been edited, often extensively rewritten, to ensure that they will be accessible to a technically oriented professional with a minimal background in systems approaches. Specifically, we have attempted to define the major theoretical issues in plain English as they appear so that a reader may understand the overall thrust of the argument, even if he is not yet fully versed in the analytic details. We have tried to eliminate all terminology that is particular to a special field and not widely understood. It is hoped, as a result, that professionals in different fields of planning and design will be able to use and appreciate the several cases even though they deal with many fields.

But the editorial process has not lowered the level of discussion of the cases. Many of them, in fact, involve techniques and issues at the forefront of current research. To make the material accessible to a broad spectrum of readers, however, these sophisticated cases are deliberately paired against more introductory presentations of similar material.

A reader desiring an introduction to the use of systems analysis in planning and design would probably wish to omit Chapters 9, 11, 14, 18, 21, and 24 on a first reading. Conversely, a more advanced reader might well choose to bypass Chapters 8, 10, 13, 19, 22, and 23. This would depend, of course, upon the reader's level of preparation in modeling, optimization, and evaluation. The introductions (Chapters 1, 5, 12, and 20), the more speculative pieces (Chapters 2, 3, and 26), and the descriptive material (Chapters 4 and 25) are recommended for any level of preparation.

The topics of these studies are mostly problems in transportation and water resources. This coverage is consistent with our hypothesis that systems analysis will be most usefully applied to large-scale regional planning problems whose elements are highly interconnected, as they are by a network of routes or rivers. This emphasis does not, however, represent an attempt to imply that systems approaches cannot be advantageous in other areas. Indeed, about one-fourth of the cases deal with a variety of other problems, including some of detailed design.

RELEVANT BASIC TEXTS

This reader of case studies is intended primarily as a supplement to theoretical presentations or to knowledge that the reader may have. Although particular theoretical points in each study have been explained in some detail, the underlying concepts of production functions, demand models, linear programming, and decision analysis are not given here. The reader may, consequently, wish to refer to one or more of the most current basic texts.

The following are particularly recommended for the reasons given. For an easy, comprehensive view of the basic elements of linear programming, the reader is referred to

AU, T., AND STELSON, T. E., 1969. *Introduction to Systems Engineering, Deterministic Models*, Reading, Mass.: Addison-Wesley Publishing Company, Inc.

A more intensive treatment of mathematical fundamentals, together with a long discussion of probability, is given by

STARK, R., AND NICHOLLS, R., 1972. *Mathematical Foundations for Design, Civil Engineering Systems*, New York: McGraw-Hill Book Company.

Finally, the important concepts associated with modeling and evaluation, together with the fundamentals of optimization, are given in

DE NEUFVILLE, R., AND STAFFORD, J. H., 1971. *Systems Analysis for Engineers and Managers*, New York: McGraw-Hill Book Company.

Naturally, more advanced presentations are available for each field. References to relevant major works are provided in the introductions to modeling, optimization, and evaluation.

2

Role of Systems Analysis in Engineering Design*

RICHARD DE NEUFVILLE AND FRANCIS Y. H. CHIN

The purpose of this chapter is to define the existing status of systems analysis in engineering, to explore where the field might go from here, and to suggest how universities and professional groups might go about developing a capability in this area.

INTRODUCTION

An already substantial and increasing number of professionals in industries, governments, and universities believe that the systems approach has a significant role in the planning and the design of large-scale public projects. The number of firms using these approaches, coupled with the number of universities offering or planning systems-oriented courses and programs is strong evidence of the support for the systems methodology.

In a general way, the systems approach implies a comprehensive attack on design problems. The current emphasis on this, the holistic approach, is almost certainly to some extent a reaction to the previous emphasis on

*A prior version of this chapter was presented at the Annual Meeting of the American Society of Engineering Education in Lubbock, Tex., June 1972. The material was also reprinted, by permission, in the *Proceedings of the International Symposium on Systems Engineering and Analysis*, Purdue University, Oct. 1972 and has been published in the *Highway Research Record* No. 462, Sept. 1973.

7

detailed analysis of particular issues. But, and most importantly from our point of view, this trend is reinforced and accelerated by our rapidly expanding capabilities for dealing with large-scale problems. Indeed, the development of the computer and of an extensive catalog of powerful computer-based techniques permits consideration, explicitly and analytically, of more alternatives and more possibilities than ever before. The opportunities offered by these methods, which are now inextricably attached to the systems approach and which it is convenient to refer to as elements of engineering systems analysis, appear to be very great.

Consequently, industry, government and universities have devoted considerable effort to the development of capabilities in systems analysis. Computer facilities, large computer-based models, mathematical programming approaches, and so on, are in evidence throughout engineering planning and design. All these investments of time and resources might, mistakenly, lead one to believe that there is a high level of confidence in the validity of the new methods of analysis now associated with the systems approach.

Actually, however, there are simply not many examples of particular cases in which systems analysis and systems techniques have been applied to real-world problems with especially beneficial results. Although we would hope for a significant body of evidence justifying the confidence and the resources dedicated to systems analysis, it is not yet available. Such evidence needs to be sought, not only to validate (or refute) the confidence for the directions that are being taken, but also to define, more precisely, what these directions should be.

With regard to the analytic portions of the systems approach, the basic issues can, perhaps, be stated in terms of three specific questions:

1. To what classes of problems can the various techniques of systems analysis be applied profitably?

2. Which of the many techniques available are appropriate to particular problems?

3. What techniques are appropriate and what new ones are needed?

It is important to be able to answer these questions accurately so that time and effort are not wasted. It is quite likely, and is suggested often by practicing professionals, that systems approaches are often mistakenly applied where they may not be especially useful or applied to improperly formulated problems. We should learn to avoid this. Likewise, a clearer understanding of what specific techniques were really useful would do a great deal to rationalize the variety of subjects that are, at present, offered as part of systems curricula at universities throughout the United States and elsewhere. More insight into the strengths and weaknesses of alternative approaches would also permit the universities to form more capable and resourceful analysts.

ASSESSMENT OF CURRENT VIEWS

Approach

The first step of our approach to the problem of determining what kind of role systems analysis should have was, then, a national assessment of what problems exist. Practicing professionals in both industry and in universities were polled to determine how the respondent felt about systems analysis or engineering, and his point of view. Maximum use was made of previous relevant surveys.

We sent a survey to major practitioners. Prospective university respondents were determined from the list of departments known to be using either of the two most recent texts on systems analysis and engineering, *Introduction to Systems Engineering, Deterministic Models*, by Au and Stelson (1); or *Systems Analysis for Engineers and Managers*, by de Neufville and Stafford (4). This list was expanded using the results of the survey of educational programs conducted by Vidale (11) and a list of universities offering a degree in systems analysis or engineering supplied by the American Society of Civil Engineers. Specific faculty members were identified by checking through the catalogs of the relevant colleges. The firms were chosen from a national listing of U.S. consulting firms which contained brief descriptions of their interests. The evidence obtained from the questionnaires was supplemented by the surveys of Vidale, already noted, and by Johnson (7). In addition, the articles by Wagner (12), Kavanagh (8), Eldin (5), Tabak (10), and Gross (6) were used.

Agreement on Overall Definition

In general, there was remarkably widespread agreement not only on what systems analysis and engineering were about, but also that it forms a meaningful field in itself. Although the range of replies varied in particular details, the essence of the responses was quite similar.

The systems approach is seen as a comprehensive attack on problems, as mentioned before. It is widely agreed that the skills and knowledge needed to carry out a systems approach are not now to be found in any one academic department or discipline, but must be taken from several. This suggests the need for multidisciplinary activities that somehow transcend the individual disciplines.

Systems analysis is the embodiment of the specific techniques of the systems approach. It is a process that applies appropriate economic and other theory, in a rational and systematic manner, to generate optimal plans and designs. This is to be done using whatever tools are appropriate but exploiting particularly the new computer-based methods of modeling, mathematical programming, simulation, and the like. It is also usually thought to imply the use of computer-based and other appropriate techniques.

Disagreement on Specifics

The overall agreement on the general definition of systems analysis and engineering does not imply a common understanding of how the methodology is to be used for actual application. Quite the contrary is actually true. There is, apparently, little specific agreement on the strengths, weaknesses, and relevance of the particular techniques or approaches available.

Although, logically speaking, it is possible for this disagreement to arise because there might really be little to choose from between the techniques, such does not seem to be the case. Individual experience appears to indicate, again and again, that many particular approaches are, in fact, much more applicable to one class of problems than others. The evident disagreement about which methods should be used arises because we do not, as a profession, have any clear answers to the problem of selecting the methods that are applicable to any problem.

The questionnaires confirmed that there is widespread disagreement about what techniques should be encouraged in a systems program. There was quite some disagreement about whether sufficient texts of high quality were available. Although it is conceivable that these differences stem from a lack of publicity or familiarity, it is more likely to reflect a fundamental question as to what should be in a good systems text.

It is relevant to note, too, that the widely held notion that adequate systems texts were not available is in the face of the well-known abundance of excellent texts on operations research and optimization techniques. It is therefore possible to infer that there is a consensus to the effect that optimization procedures—linear programming, for example—constitute only a limited portion of the methods required in systems analysis and engineering.

Optimization Versus Modeling, Evaluation, and Implementation

If one were to predict the future from the published evidence of the past, one would be forced to conclude that we were nearly universally agreed that optimization methods are at the core of systems approaches to engineering. Yet this is not the case.

Our questionnaires and the surveys of others suggest that respondents felt that as much emphasis needed to be placed on the modeling and evaluation techniques as on specific forms of optimization. Johnson's detailed examination of how professionals in the field of water resources used computer methods showed that simulation approaches were commonly preferred to optimization (7).

These results confirm an impression that the prevailing predominance of optimization approaches in academic circles is not due to their overwhelming importance—but to their mathematical elegance and tractability. Since such work is normally carried out within a theoretical framework rather than as part of a large-scale implementable study, it is denied the test of battle. Thus there is little real concern given to problems of implementation and validation of assumptions.

The other analytic elements that appear to be important to systems engineering, such as modeling and evaluation, are certainly much more subjective than optimization procedures. That they are, therefore, less satisfactory or less available to the purist explains, in part, why less attention is paid to them.

An argument can also be made for the need of a critical mass and project focus for the development of the systems approach. Many faculties, for example, appear to have a solitary "systems" man, who is forever searching without much success for easy problems to knock off to prove his worth. Many of these problems are continuously being rediscovered in the literature, are largely solved, and were not of much interest in the first place. When individual efforts can be combined, however, and preferably focused on a specific project and its implementation, it seems more likely that we can get beyond relatively trivial applications of optimization to the more cogent issues. The relative inattention to relevant aspects of systems analysis and design indicates areas in which efforts should be increased. Kleindorfer, for example, suggests that a systems design effort requires essentially four elements (9):

1. An analytic capability to operate on the problems as described.

2. A descriptive methodology so that a problem can be accurately described.

3. A normative point of view so that the relevant weights which would enter into any objective function or decision-making algorithm can be estimated.

4. A budgeting and control apparatus to ensure that optimal plans or designs, based upon the first three elements, do actually get executed.

According to him, and we would agree, any optimization or operations research methodology (mostly part 1) which does not effectively involve the other parts is destined to be sterile in practice. A comprehensive, multidisciplinary program is required.

Multidisciplinary Programs

Because of the particular orientation of the established departments, it is universally reported that multidisciplinary efforts are difficult to establish. Members of any discipline usually find that their rewards are oriented toward that discipline and that, if there is to be a choice, the multidisciplinary effort inevitably suffers. Worse, established departments often refuse to approve broader programs which would reduce their own influence. The question is, then, how should one go about implementing a systems program with its requisite multidisciplinary flavor?

In answering this question, it appears useful to define what, precisely, one means by an interdisciplinary effort. For example, suppose we take an interdisciplinary effort to be one in which all the skills needed to attack a problem are brought together. If we agree that this is reasonable, then we should recognize that engineering "disciplines," civil engineering in particular, have long been "multidisciplinary" efforts in that they combined mathematics, mechanics, geology, hydrology, and thermodynamics in amounts considered sufficient to address the civil engineering problems that the profession was confronting.

The point is, what may be seen as an interdisciplinary problem today may not appear to be one in a few years. On the other hand, the capabilities necessary to an effective systems approach to design can be brought together in an expanded concept of a department. This is the approach now being implemented by the MIT Civil Engineering Department. Alternatively, separate systems engineering departments can be established, as has been done in many instances, as described by Vidale (11).

It is not clear, therefore, that we should focus our attention on solving the problem of how to establish multidisciplinary efforts, since that need may disappear. The essential problems would appear to be that of learning (1) how to work together with other disciplines, and (2) to choose which elements of these disciplines are most useful to us. Because the solutions to these problems appear contingent on what kinds of problems one is working on, the MIT effort seeks to integrate the systems approach into departments rather than to establish separate systems departments.

HYPOTHESES ABOUT FUTURE DEVELOPMENTS

The overriding issues before those who wish to practice engineering systems analysis appear to be those of where the approach should best be used, and which techniques are best suited to which problems. Until these issues are better resolved, the field is bound to continue to be somewhat nebulous.

To advance our understanding it may be necessary to cast these issues into operational questions that can be addressed concretely. We propose that it is useful to advance specific hypotheses to be examined; their confirmation (or denial) would then at least partially resolve existing disagreements.

With regard to which classes of problems the systems approach is most useful, we suggest the following hypothesis:

Hypothesis I: The systems approach will make the greatest contribution in the large-scale engineering planning problems that must consider many substantial but interdependent projects and the system that links and manages them. The areas of transportation and water resources, where the projects (highway links or dams, as the case may be) are inevitably connected by the network of roads or rivers, typify the type of problem. By this hypothesis, areas such as structural design, in which individual buildings can stand almost totally independently, would be less suited for a systems approach.

To the extent that we should, indeed, be using systems approaches to deal with planning issues, it appears that we can further specify the kinds of skills that should be combined in an analysis and, by extension, in a college curriculum. An understanding of both the physical system and of societal preferences and behavior is required. Specifically:

Hypothesis II: The skills inherently necessary for an effective systems approach must include those necessary to the definition of the problem both deductively, through the use of engineering production functions, and inductively by means of systems modeling and econometrics. Actual knowledge of and experience with the particular system or problem is, of course, the key to an effective use of these techniques.

Hypothesis III: Skills helpful in the specification of criteria for the problem are essential for a useful optimization and analysis. Knowledge of how individual and societal preferences are developed, as through utility theory, welfare economics, and sociology, and how they are applied in specific cases, via decision analysis or game theoretic analysis of collective choice, is necessary.

The optimization and other available techniques which will be used suggest a further hypothesis. Indeed, since each of these methods inherently involves stringent assumptions (such as linearity, additivity, and decomposibility) which are contrary to physical reality, it would seem that the techniques are necessarily approximate and therefore useful principally for large-scale planning. So we propose:

Hypothesis IV: The systems approach is most useful in the planning of the overall configuration of programs and in the definitions of regions of optimality in which it is profitable to apply the more traditional detailed design. The analyst should, then, devote significant effort to sensitivity analyses, both of the physical parameters of the problem, to discover areas of potential redesign, and of the evaluation criteria, to indicate how different public groups may be satisfied.

This further suggests a hypothesis concerning the general relevance of specific computer-based programming techniques:

> *Hypothesis V:* Optimization and the more detailed simulation techniques should be used hierarchically and interactively, the one to define regions of optimality, the other to examine, in detail, specific alternatives within that region. Knowledge gained from building simulation models can help to improve optimization models. This implies that relatively simple optimization techniques may be most appropriate in an initial analysis.

Further, we can anticipate that the exploitation of computers will enable us to examine problems at greater levels of complexity and detail. As a consequence, we may expect a more realistic, multidimensional representation of problems:

> *Hypothesis VI:* The development of computer capabilities permits, and thus requires, explicit consideration of randomness and stochastic effects; of the dimensions of objectives that are, in fact, relevant to a problem; and of means, such as decision analysis, to choose alternatives within this complexity.

Finally, because it is reasonable to suppose that we are really concerned with problem solving rather than merely with problem analysis, it seems clear that we must be concerned with implementation. From the results of systems analyses available so far, it would appear that the profession has not been eminently successful in this regard. Therefore, we suggest a final hypothesis:

> *Hypothesis VII:* The effective implementation of systems analysis and design requires that we develop knowledge about the effects of different organizational structures (political, governmental) and management systems (such as program budgeting).

DEVELOPMENT OF A SYSTEMS PROGRAM

Policy Implications

The hypotheses about the nature of future developments in the use of systems analysis and design imply certain implementation policies. They define fairly specific kinds of questions for research. And they suggest a distinct pattern of undergraduate and graduate education.

With respect to research, it would imply that much more attention should be specifically directed toward how and where the systems methods can be successfully implemented. Existing emphasis on pure technique should, in engineering, be reduced. Rather, it would seem more fruitful to concentrate on identifying, by documentation, classes of problems to which the methods apply.

The proposed shift toward a concern for application, or shift for more emphasis on implementation within a department already oriented toward application does not mean that many advances in technique are not possible. Rather, the shift would acknowledge our collective failure so far to do enough in the way of showing where and how systems techniques are valid. It would recognize that, because the area has not been investigated extensively, efforts in this direction may have substantially greater results for the same amount of resources. The shift, in short, may be highly cost-effective.

As regards education, it would seem that we should develop curricula and courses which bring the student explicitly into contact with both the methods of systems analysis and design, the particular kinds of problems the person is likely to encounter, and the full socioeconomic environment in which these problems are likely to occur. This suggests the particular need for more multidisciplinary and more design subjects. Such projects are currently underway at MIT, as described below, and at some other universities, such as the University of California at Berkeley (3).

Developmental Sequence

It may be useful to speculate, also, on the process through which it may be practical to develop programs in systems analysis and design. There appear to be essentially two phases: the first involving what might be called the automation of traditional techniques; the second a major, and most difficult, revision of the design process itself.

For better or worse, the initial steps in the developmental process seem inevitably to center around the computer. In retrospect, it is not obvious that this step is necessarily wholesome, particularly since our hypotheses imply that computer-based techniques are only a smaller part of the entire effort. Yet, starting with the machine may have real tactical advantages.

Although the machine may not loom large in future systems programs, it often does, indeed, appear to provide a convenient basis for a start. For one thing, it provides a visible focus of what will develop into a fairly conceptual effort. An installation of this sort may, in fact, be necessary to define a systems group at its beginning. Such was certainly the case with our own Civil Engineering Systems Laboratory (CESL) at MIT, and seems to be the pattern in many other places. The computer also provides immediate opportunities for substantial improvements in many aspects of analysis and design. The routine computations of many different kinds of design can be automated. The design itself can thereby often be significantly improved, if only because the possibility of doing many more analyses almost ensures that better designs will be found. The first work at CESL, for instance, led to the development of programs for surveying (COGO, coordinate geometry), for earth calculations (DTM, digital terrain model), and for structural analysis (STRESS). As evidenced by these examples, this kind of work can be highly successful.

A related step in this sequence would lie in the development of a capability in analytic techniques that have been developed as a result of the computer, or have been made much more accessible and useful because of this tool. These are, in essence, what are generally known as the techniques of systems analysis. They are both deterministic, like linear and dynamic programming, or probabilistic, as with decision analysis, stochastic simulation, and statistical modeling.

Further development of useful computer tools is naturally associated with this phase. The existence of the computer allows one not only to program available knowledge, but also to apply new techniques to the traditional design process. Particular companies or laboratories may find it useful to develop their own programs directed toward their own problems. The Hydrologic Engineering Center of the U.S. Corps of Engineers has, for example, generated a number of simulation models for river basins. Other groups, universities in particular, may find it expedient to use more general programs. It was for this purpose that CESL developed the ICES (integrated civil engineering systems) set of programs. These include STRUDL (structural design language), TRANSET (transportation network analysis), and ROADS (highway design). As a commentary, it would appear that their first stage has been achieved reasonably widely. ICES programs, for example, have been implemented by now at something approaching 2000 installations.

The second phase in the development of systems analysis programs in engineering appears to be substantially different from the first. We would no longer be primarily concerned with how to automate procedures and how to apply computer-based techniques to the usual design process. We would be concerned with how to redefine the design process to make maximum use of the new capabilities. This task appears to be substantially more difficult than the first. The way to accomplish its objectives are not nearly as clear as they are for the automation of standard procedures. Indications are that this effort will require substantially more time, imagination, and insight.

The ability to manipulate huge quantities of data rapidly, to use powerful analytic techniques on large-scale problems, does open up new vistas of what kinds of issues can be addressed and what engineering might be. Fundamental changes in the analysis, design, and evaluation process are possible. Because of the enormity of the task, however, the changes are not likely to occur without substantial concentrations of effort.

This is the phase that would really integrate systems methods into the design process or, perhaps more accurately, refashion the design process so that it can profit fully from the available opportunities. This phase is directed toward answering the questions and hypotheses posed in the previous sections. It is a phase which appears to have just recently begun. Because of its extensive implications, it will probably last for many years. It is now the central concern of the MIT Civil Engineering Systems Laboratory.

MIT Experience

The MIT Civil Engineering Systems Laboratory provides a focus for work in the development and application of systems analysis and design. This concern spans from the definition of the problem, through the analysis, the design, the evaluation, and construction, to the operation of large-scale projects. Because of its explicit interest in applications, these activities are closely meshed with other portions of civil engineering and engineering more generally.

The academic program in civil engineering systems develops the student's capabilities in three complementary areas:

1. The theories and the methods of systems analysis and design.
2. The application of these tools to particular problems.
3. The definition of the social and the economic forces inherent in the environment in which these systems will be implemented.

In the belief that a full understanding of the strengths and the limitations of tools and theory comes only from learning how to mesh them with the structure of specific problems, each student is expected to focus his or her research and curriculum toward a particular area of application. A wide range of application areas is available, especially in transportation and water resources, in accordance with our view that these fields are most suited, at the moment, for systems applications.

Each student is expected to develop a basic understanding of the social, environmental, and economic forces which create the demand and the loads on civil engineering systems. The program often finds that students in systems study subjects outside the department in such fields as city planning, economics, management, operations research, political science, sociology, or urban studies, depending upon their individual interests.

The program originally developed at the graduate level but, since 1970, has also been integrated into the undergraduate curriculum in the department. Following introductory subjects in mechanics and civil engineering, students are expected to gain competence in a core of subjects in computer analysis covering probability, economics, systems analysis, and computer use. They are then expected to choose an area of concentration so that they will come to appreciate the essentials of the field and, thereby, to blend this with analytic knowledge to obtain problem-solving capabilities.

The faculty associated with the MIT Civil Engineering Systems Laboratory is attempting to develop a common understanding of how the many systems analysis techniques, made practical by the computer, should be fitted into the design process. Specific areas of emphasis include stochastic systems and statistical inference for the development of systems models; optimization techniques; and evaluation procedures, including multidimensional benefit–cost analyses and decision theoretic approaches.

Studies are now underway in which the systems approach is applied, in particular, to various transportation and water resource systems. These methods are, for example, currently being used for the design of airport systems for several major metropolitan areas. Members of the department have been engaged in a design project covering an entire river basin in Argentina and have conducted an analysis of New York City's billion dollar additions to its primary water supply system. These are typical of the large-scale projects to which these techniques and new methods have been and will continue to be applied.

It might be added that this kind of research work has also led to the preparation of a number of texts that attempt to present relevant elements of the systems approach. Texts on probability and statistics in civil engineering (2) and on systems analysis for engineers (4) have already been published. Following directly from ongoing research projects, this anthology of case studies of the application of systems analysis to planning and design has also been prepared in an explicit attempt to address the questions about the future of systems analysis raised here.

ACKNOWLEDGMENTS

The careful reviews and advice of our colleagues, Robert Stark and David Marks, are deeply appreciated. The support of the MIT Civil Engineering Systems Laboratory is also gratefully acknowledged.

REFERENCES

1. AU, T., AND STELSON, T. E., 1969. *Introduction to Systems Engineering, Deterministic Models*, Reading, Mass.: Addison-Wesley Publishing Company, Inc.
2. BENJAMIN, J. R., AND CORNELL, C. A., 1970. *Probability, Statistics and Decisions for Civil Engineers*, New York: McGraw-Hill Book Company.
3. DAVIS, H., private communication.
4. DE NEUFVILLE, R., AND STAFFORD, J. H., 1971. *Systems Analysis for Engineers and Managers*, New York: McGraw-Hill Book Company.
5. ELDIN, H. K., 1970. Education in Systems Engineering, *Engineering Education*, Vol. 60, No. 8, April, pp. 836–837.

6. GROSS, P. F., 1970. The Undergraduate Curriculum in Operational Research, *CORS Journal*, Vol. 8, No. 1, March, pp. 44–56.

7. JOHNSON, W. K., 1971. Current Use of Systems Analysis in Water Resource Planning, paper presented at the ASCE's 19th Hydraulic Division Conference, Iowa City, Iowa.

8. KAVANAGH, T., 1971. Systems Approach, *Engineering Issues*, Oct., pp. 75–82.

9. KLEINDORFER, P. R., 1972. Operations Research and Systems Analysis in the Public Sector, *Bell Journal*, Vol. 3, No. 1. Spring.

10. TABAK, D., 1972. Curriculum in Systems Engineering, *International Journal of Systems Science*, Vol. 3, No. 2, July, pp. 177–185.

11. VIDALE, R., 1970. Systems Engineering as an Academic Discipline, *Engineering Education*, Vol. 60, No. 8, April, pp. 832–835.

12. WAGNER, H., 1971. The ABC's of OR, *Operations Research*, Vol. 19, No. 6, Oct. pp. 75–82.

3

Implications of Alternative Institutional Arrangements for Implementation of Analysis*

MICHAEL RADNOR, ALBERT H. RUBENSTEIN, AND DAVID A. TANSIK

In this chapter we hypothesize that the environment in which analysis, and innovative activity in general, is carried out is an important determinant of the mode and effectiveness of its implementation. Specifically, we have considered how differences in the operationality of goals, such as may be found between government and business organizations, will influence the strategic behavior of the managers of innovative groups. Second, we have examined how the nature of the analyst's relationship with his client and the degree of his support from management may influence the likelihood of implementation.

Difficulty with implementation, or even, in many cases, the complete lack of it, is the visible consequence of many of the organizational problems of integrating systems analysis into the planning process. Although this problem has been frequently recognized as being of concern to both practitioners and organizational researchers, there is a divergence between this concern and the paucity of relevant literature. Thus Batchelor (1), in his 1959 and 1962 annotated bibliographies of operations research literature, did not include implementation in his subject index; Huysmans (8), in a 1968 dissertation on the implementation of operations research, found little in the literature on this subject; and, in a 750-item annotated bibliography on the

*Extensively adapted and condensed, for the purposes of this text, from Implementation in Operations Research and R&D in Government and Business Organization, published in *Operations Research*, Vol. 18, No. 6, Nov.-Dec. 1970.

management and organization of operations research and the management sciences prepared by Radnor and Mylan (10), only 15 items were specifically devoted to the topic, with most of these being of a speculative nature.

POSSIBLE EFFECTS OF ENVIRONMENTAL DIFFERENCES

The type of institutional environment in which the analysis takes place appears to have a significant effect on the likelihood of implementation. One of the potentially significant differences between the organizational environments is in the comparative operationality of their goals and outputs. If an organization's goals and outputs are less explicit than those of another group the influence of operations research or management science (OR/MS) projects on them may be much more difficult to demonstrate. In such cases, the nature of the relationship between the analyst and his client may be critical. To accomplish implementation, the operations researcher would then seemingly have to convince the client of the benefits of the activity on the basis of reputation, trust, past experience, or other intangible factors.

March and Simon state the hypothesis explicitly (9). Defining goals as operational "when a means of testing actions is perceived to relate a particular goal or criterion with possible courses of action . . .," they contend that:

> When a number of persons are participating in a decision-making process, and these individuals have the same operational goals, differences in opinion about the course of action will be resolved by predominantly analytic processes, i.e., by the analysis of the expected consequences of courses of action for realization of the shared goals. When either of the postulated conditions is absent from the situation (when goals are not shared, or when the shared goals are not operational and the operational subgoals are not shared), the decision will be reached by predominantly bargaining processes.

This idea is illustrated on the left-hand side of Figure 3.1.

Significant differences in the operationality of goals may exist between business and government organizations. These differences are not, however, absolute. Some government organizations, such as the Post Office, may have goals that are more operational than those of some business organizations, such as regulated utilities. Nevertheless, we could in general say that government organizations have outputs and goals which are less defined than those of business organizations. We could then expect implementation problems and procedures of a different sort between government and business.

In a business environment, the profit motive is dominant. Based upon survey data, George England contends that profit maximization is a primary goal of American business managers. Social welfare, which is a major goal of government, is at the bottom of the business manager's hierarchy of goals

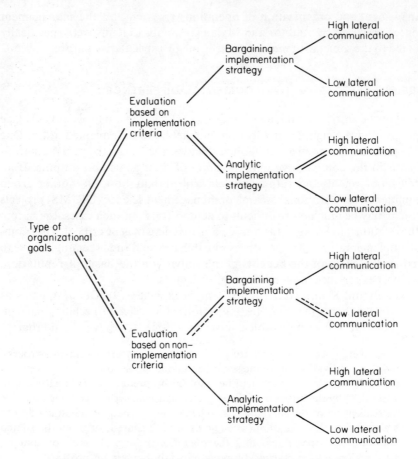

Figure 3.1 A model relating organizational goals and implementation strategies. The double lines indicate hypothesized results in organizations with operational goal structures; the dotted lines indicate hypothesized results in organizations with nonoperational goal structures.

(6). Operational goals for business organizations could thus be formulated on the basis of contributions to profits.

Government bureaus, however, do not for the most part operate in an economic environment. Although they must buy resources, they do not have markets for their outputs or products. This lack hinders them to the extent that they cannot use prices and consumer response as an objective guide for determining "proper" courses of behavior. As Anthony Downs has pointed out, the major yardsticks (income and profits) for decision making used by private firms are far less available to men who run bureaus (5). March and Simon similarly suggested that the evaluation of the performance of bureaucratic personnel therefore tends to rest ". . . heavily upon the opinions of each man expressed by his . . . supervisors" (9).

To the extent that evaluations of an analyst's performance are based on his contributions to outputs, he is motivated to cultivate client relations that would result in visible implementations of projects. As the analyst and his clients interact and learn more about each other, we might expect that they would begin to acquire shared beliefs and develop an atmosphere of mutual understanding (4, 7). Based on this more favorable climate, the analyst may be able to anticipate problems the client may have and to provide help in these areas. This further promotes implementation of the OR/MS results. Also, if industrial firms are profit-oriented, and if this fact leads to the evaluation of an OR/MS activity according to whether its recommendations are implemented, then industrial OR/MS activities may be expected to engage in more of the lateral communications that promote implementation. On the other hand, if implementation is not used to a great degree as a criterion for evaluation, as appears to be the case for government organizations, it is likely that there will be less lateral communication. This second hypothesis is expressed in the right-hand side of Figure 3. 1.

Complete confirmation of the usefulness of the full model shown in Figure 3.1 must await further research; such a program is in progress. We may, however, examine several of the individual propositions that it implies. These concern several factors:

1. The nature of the relationship between the client and the analyst.

2. The level and type of top-management support for the research activity.

3. The type of organizational and external environment in which the research activity is pursued.

EVIDENCE OF SURVEY DATA

We surveyed some 60 business and 40 federal civilian government organizations and conducted hundreds of interviews. We collected data that enabled us to examine, even if somewhat tentatively, the nature of the model in Figure 3.1. The results for each factor are reported below.

Client Relations and Implementation

Support for the hypothesis that the relationship between the client and the analyst influences the implementation of OR/MS projects is provided by Table 3.1. In both business and government organizations, activities reporting high implementation problems tend to be associated with reports of bad relations with clients. On the other hand, good client relations tend to be associated with low levels of implementation problems. These observations reinforce Bennis's remark that "Implementation is the problem, and the relationship between researcher and user is its pivotal element" (2).

Table 3.1 Influence of Client Relations on Implementation Problems in Government and Business Organizations

Type of Organization	Level of Client Relations	Organizations Having Implementation Problems		
		High	Low	Total
Government	*Bad*	4	1	5
	Medium	2	7	9
	Good	0	14	14
	Total	6	22	28
Business	*Bad*	15	1	16
	Medium	9	16	25
	Good	1	10	11
	Total	25	27	52

Although no inference is made as to the causal connection between these variables, it does seem reasonable to assume that the maintenance of a satisfactory relation by an OR/MS activity with clients is a critical factor in effecting implementation of results.

Top-Management Support and Implementation

In an earlier paper, we showed that the level of top-management support is a key factor in determining whether an OR/MS activity will be successful in an organization (11). In fact, the fortunes of specific OR/MS activities tend to follow rather closely the organizational fortunes of their specific top-management sponsors. In reference more specifically to problems of implementation, the level of top-management support of the OR/MS activity similarly seems important. Both the business and government data indicate that such support may encourage the implementation of the results of OR/MS projects. This is brought out by Table 3.2. Endorsements of OR/MS by top managers, or even pressures they bring to bear on clients in the organization, may then influence the implementation of OR/MS projects.

Combined Effects of Top-Management Support and Client Relations

In addition to their individual relations with implementation problems, client relations and top-management support themselves tend to be associated in our data, as evidenced by Table 3.3. Again the causal connections between

**Table 3.2 Influence of Top-Management Support on Implementation Problems
in Government and Business Organizations**

Type of Organization	Level of Top-Management Support	Organizations Having Implementation Problems		
		High	Low	Total
Government	Low	3	2	5
	Medium	1	5	6
	High	2	15	17
	Total	6	22	28
Business	Low	5	2	7
	Medium	12	12	24
	High	8	13	21
	Total	25	27	52

**Table 3.3 Client Relations and Top-Management Support Interrelations in
Government and Business Organizations**

Type of Organization	Level of Top-Management Support	Level of Client Relations			
		Bad	Medium	Good	Total
Government	Low	4	1	0	5
	Medium	0	2	4	6
	High	1	6	10	17
	Total	5	9	14	28
Business	Low	4	1	2	7
	Medium	8	13	3	24
	High	4	11	6	21
	Total	16	25	11	52

these variables are not determined here. In Table 3.4, we show the combined influence of top-management support and client relations on problems of implementation. In only 7% of the government cases and 18% of the business cases was a high level of implementation problems associated with either a high or medium level of client relations or top-management support. Thus these two variables seem to be critical areas of concern in analyzing problems of implementation for OR/MS projects.

Note, however, in Table 3.4 that there are 10 cases (8 businesses, 2 government) in which high levels of implementation problems were reported in conjunction with high levels of top-management support. In only one case (business), however, was there a report of high implementation problems and good client relations. This observation lends some tentative support to the speculation by Huysmans that, although both client relations and top-manage-

Table 3.4 Combined Influence of Top-Management Support and Client Relations
on Implementation Problems in Government and Business Organizations

Type of Organization	Level of (Top-Management Support)/(Client Relations)	Organizations Having Implementation Problems		
		High	Low	Total
Government	High/good	0	10	10
	High/medium	1	5	6
	High/bad	1	0	1
	Medium/good	0	4	4
	Medium/medium	1	1	2
	Medium/bad	0	0	0
	Low/good	0	0	0
	Low/medium	0	1	1
	Low/bad	3	1	4
	Total	6	22	28
Business	High/good	0	6	6
	High/medium	4	7	11
	High/bad	4	0	4
	Medium/good	1	2	3
	Medium/medium	4	9	13
	Medium/bad	7	1	8
	Low/good	0	2	2
	Low/medium	1	0	1
	Low/bad	4	0	4
	Total	25	27	52

ment support are important, perhaps the maintenance of good client relations is relatively more important in alleviating implementation problems (8). Several case studies in the literature of "failures to implement" illustrate this contention (3).

Effect on Evaluations, Implementation, Strategy, and Communication Behavior of Environmental Differences

We proposed that differences in goal operationality, such as may be found between business and government organizations, can influence the behavior of members of the OR/MS activity as well as the behavior of OR/MS clients. Figure 3.2 summarizes the data collected to test the client-researcher interaction model shown in Figure 3.1. These data support the hypotheses that higher levels of goal operationality lead to evaluations of the OR/MS activity based upon results of implemented projects, to instances of "analytic" implementation strategies, and to high levels of lateral communications between the analyst and his client. Conversely, nonoperational "goals" lead to

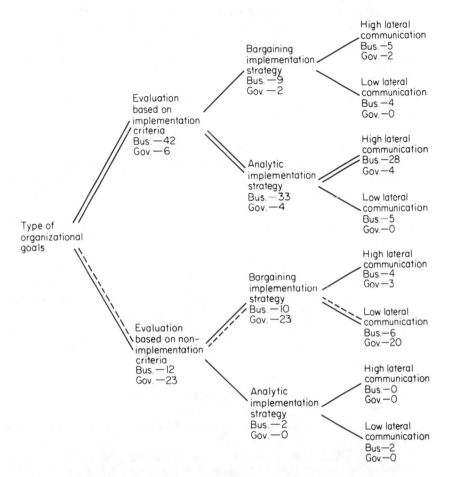

Figure 3.2 The data collected confirm the model relating organizational goals and implementation strategies. Instances in which each possible combination was observed in government and business are as indicated.

"nonimplementation" evaluations, bargaining implementation strategies, and lower levels of OR/MS client communications.

GUIDELINES FOR IMPLEMENTATION

In operations research, and innovative activity in general, barriers to implementation of results are critical to the success of projects and the program as a whole. No general methods have been found, so far, for assuring successful implementation. A number of methods have been used, some of which can be designed into the systems analysis process. Their usefulness is supported by the evidence of our model and the confirming data, but, naturally, they do not all work under all conditions.

Methods that have been helpful in bringing about successful implementation include the following:

1. Assuring that there is a clear and recognized need for the results at the time the project is undertaken.

2. Involvement of the ultimate user of the results early in the process, and maintaining communication throughout the project.

3. Focus of the direction or strategy for the project on an individual or a small group that can review progress and make decisions about changes in direction or level of effort.

4. Having top-management support and enthusiasm.

5. Allowing or encouraging researchers to follow projects into application and make careers there, if they so desire.

Many other specific devices or implementation strategies have been used or advocated. The wide diversity of methods, however, and the lack of consistent cause-effect relations between methods used and project success (some of the above methods have been used in projects that failed) suggest that we need to do much more research before we will have a reasonable theory, let alone a handbook on how to achieve implementation.

REFERENCES

1. BATCHELOR, J., II, 1962. *OR, An Annotated Bibliography*, St. Louis: St. Louis University Press, Vol. 2 (Vol. 1, 1959).
2. BENNIS, W., 1965. Commentary on "The Researcher and the Manager: A Dialetic of Implementation," *Management Science*, Vol. 12, Oct., pp. B13–B16.
3. CHURCHMAN, C. W., 1960. Case Histories Five Years After—A Symposium, *Operations Research*, Vol. 8, pp. 254–259.
4. CHURCHMAN, C. W., AND SCHAINBLATT, A. H., 1965. The Researcher and the Manager: A Dialectic of Implementation, *Management Science*, Vol. 11, Feb., pp. B69–87.
5. DOWNS, A., 1967. *Inside Bureaucracy*, Boston: Little, Brown and Company.
6. ENGLAND, G. W., 1967. Organizational Goals and Expected Behavior of American Managers, *Academy of Management Journal*, Vol. 10, June, pp. 107–117.
7. HOMANS, G. C., 1950. *The Human Group*, New York: Harcourt Brace and World, Inc.
8. HUYSMANS, J. H. B. H., 1968. The Implementation of Operations Research: A Study of Some Aspects Through Man–Machine Simulation, Internal Working Paper 78, Space Sciences Laboratory, Social Sciences Project, University of California, Berkeley, Calif., Jan.

9. MARCH, J. G., AND SIMON, H. A., 1966. *Organizations*, New York: John Wiley & Sons, Inc.
10. RADNOR, M., AND MYLAN, D. (in preparation). The Adoption and Diffusion of Operations Research/Management Science Activities: A Worldwide Annotated Bibliography.
11. RADNOR, M., RUBENSTEIN, A. H., AND BEAN, A. S., 1968. Integration and Utilization of Management Science Activities in Organizations, *Operational Research Quarterly*, Vol. 19, pp. 117–141.

4

Systems Analysis of Large-Scale Public Facilities: New York City's Water Supply Network as a Case Study*

RICHARD DE NEUFVILLE

This chapter explores the strengths and limitations of systems analysis by means of a case study: that of designing a large-scale civil engineering facility, the billion-dollar additions to New York City's water supply system. The systems analysis indicated that economies of up to 50% were possible over the anticipated cost of the original proposal generated by the traditional engineering design process. These opportunities became apparent through the use of mathematical models of the system which could be used, in a computer, to explore hundreds of possible configurations. The computer also made it possible to calculate several measures of effectiveness for each trial design and to evaluate the tradeoffs between each of these criteria. The planners could then specifically choose what kind of design they wanted and thus achieve significant increases in design effectiveness.

Significant institutional constraints did, however, preclude the use of some forms of economically desirable systems. To many it always appears perverse that public agencies are so often immune to analytic rationality. It might more sensibly be argued, however, that analysts simply fail to appreciate the more complex rationality of interest groups and social organizations (12). Until they do, it would seem wise to recognize that environmental restrictions, whether social, economic, or physical, are pervasive and characteristic of systems problems. Their overall effect is to limit the amount of

*Adapted from the *Journal of Systems Engineering* (England), Vol. 2, No. 1, Summer 1971.

savings that can be realized. In this case the overall economies achieved were up to $100 million (although accurate levels are difficult to measure) or two-thirds of what might have been technically possible.

PROBLEM FORMULATION

On Being Drawn into the Fire

The first lesson to be drawn from the New York City experience concerns the manner in which the problem was posed to the analysts. It came to us after extensive preliminary designs had been prepared by a traditional design agency; after their proposal had been initially rebuffed fairly unceremoniously by the authorities—in short, after a fair amount of bad blood. Positions had polarized around particular concepts leaving little room for discussion about the merits of alternatives. These are clearly not the most advantageous working conditions for rational analysis. But it would seem that such situations will be typical as long as systems analysts are in short supply and their use a rarity.

A little introspection soon indicates why it is unlikely that systems analysis will generally not, in the foreseeable future, be an integral part of the design or planning process from the start. Most groups do not now employ systems analysts. They could not even if they wanted to; there are not enough around. Consequently, most organizations would have to make a special effort to use systems analysis as a regular part of planning or design. Using analysts is also different from simply having them on the staff. Tame in-house analysts or computer experts, docile enough not to ask threatening questions, are catchy institutional adornments which are fun to trot out before stockholders or officials. Between times they can be conveniently tucked away in computer centers or long-range corporate planning units and have essentially no effect whatsoever on what actually happens. Most organizations will not make the special effort to use systems analysis unless they sense that there is some problem as a motivation to do so. As one's problems are generally perceived by comparing oneself to one's peers (the "I-don't-care-if-I'm-mediocre-as-long-as-I'm-above-average" syndrome), the need to bring powerful analysis to bear on a problem will generally not be felt until major difficulties, technical or political, are encountered in the conventional design process. By then the fat is in the fire. It is tempting to argue that this will continue to be the case as long as there are few qualified professional analysts.

This reasoning certainly applied to the use of systems analysis on New York City's water supply network. A municipal design agency, the Board of Water Supply, was allowed to proceed with its design and to spend several hundreds of thousands of dollars preparing specific plans. They developed a proposal for a five-stage addition to the existing water supply system which would cost about $1 billion. This price tag brought about an immediate reaction from the budgetary authorities. They claimed that the initial design, predicated upon a 25% population growth of New York City, was contrary to the evident stagnation of the central city. It was, therefore, suggested that the plan was extravagant and that no additions were needed. The Board of Water Supply claimed that it was reasonable and conservative to allow for growth over the next 40 years. The battle was joined. The systems analysts were then brought in.

The evidence indicates that the situation is quite similar elsewhere and in other matters. In England, for example, planning for the Third London Airport had to reach a substantial deadlock over a particular design (the Stanstead site in this instance) before the authorities commissioned an extensive research effort into the problem. As with the MIT analysis for New York City, the Roskill Commission on the Third London Airport brought together analysts for a one-time analysis in the middle of a controversy. The effort could be neither as dispassionate nor as organized as desirable.

The contention here is that systems analysts will often be confronted with messy, controversial situations: that they will practice in this arena as often, if not more often, as they will be able to work in carefully organized developmental programs. In short, a neatly ordered design process, of the kind we so often talk or write about, may be the exception rather than the rule.

The Problem as Presented

The original perception of the design problem was created by the traditional, institutionalized processes, as might be expected. In this case they were represented by the Board of Water Supply, the municipal agency responsible for supplying potable water to New York City.

The Board postulated that the City required substantial additions to its capacity for distributing water throughout the City (11). The need for the new facilities appeared to be based on two basic assumptions:

1. That the population of the city itself would increase steadily over the next 40 years.

2. That the existing facilities either would or did require extensive rehabilitation.

The postulated need for additional capacity then led the Board of Water Supply to evolve a plan to provide the required facilities. The Board was and is responsible by law to do so. Anticipating the next section, it is worth noting that this legislative delegation of authority unwittingly implied a significant design decision. Indeed, the Board was responsible for supplying potable water, not for distributing it. In practice, this meant that the Board was in charge of the "primary" water supply network (aqueducts approximately 10 ft in diameter or larger) and that another municipal agency built and maintained the secondary and tertiary water supply network (water mains 6 ft in diameter and the smaller conduits). Responsibility for the water system was legally divided. Therefore if the Board of Water Supply was to design additional capacity, it was essentially constrained to design additions to the primary system, whether or not these represented the most desirable investment to make in the system!

The preliminary design developed by the Board of Water Supply consisted of about 50 miles of tunnels approximately 28 ft in diameter. It was estimated in 1967 to cost about $1 billion and was slated to be built in five stages. The first stage was intended to comprise the largest tunnels and was estimated to cost $323 million. The whole affair was to be one of the biggest, if not the biggest, single civil works project ever.

Apart from the size of the project, the development of the design proceeded much like most other designs. There were some fairly sweeping fundamental assumptions and institutional factors which shaped the overall concept of the solution. The parallel with the evaluation of the plans for the Third London Airport is reasonably close. In that case the British Airports Authority assumed that airline traffic would grow steadily and should be encouraged; from this it followed that additional capacity would have to be found and further that, since the Airports Authority was involved, this capacity would be provided by means of an additional airport rather than any of the possible alternatives. The similarity between these two cases, and indeed the generality of the points to be made by examining the New York City example, is enhanced by comparison of the detailed assumptions made in the design.

The preliminary design prepared by the Board of Water Supply was, in effect, based on several specifications:

1. Only one overall geometrical configuration appeared to have been extensively considered as a possible location.
2. The design horizon was 40 years; that is, the facilities were supposed to meet the conditions anticipated for the year 2010.
3. The water would be forced through the network by gravity alone.

These simple specifications, just as those made about other plans, may sound innocuous enough. But they may also be quite arbitrary insofar as they are not based on any assessment of their implications for the operation or the cost of the system to be designed. Most of them needed to be changed. As shown later, this chapter argues that similarly arbitrary design or planning decisions need to be examined carefully and, in general, also adjusted.

This was the essence of the problem presented to the author and his colleagues at MIT. They were asked to do a systems analysis of the additions to the water supply system. They were expected to determine the most relevant criteria for evaluation, to develop appropriate models of the system so that alternatives could be explored, and to recommend an optimal design and implementation strategy.

Institutional Constraints

Design is, as politics, an art of the possible. Good design must therefore recognize what limitations are placed upon it. Some of the restrictions on the design of public facilities are always physical; these are generally searched out carefully by the designers, who are accustomed to deal in material terms. Many of the constraints are also, however, economic or social.

The first limitation on the systems analysis described here was in the nature of its power: it was essentially only negative. The systems analysis process as usually constituted serves many advisory functions: it can ask probing questions, provide means to explore alternatives, and suggest optimal solutions to a problem. But in terms of actual power, the analysts can usually only hold up bad decisions; they rarely can actually make people accept a plan they prefer. This, certainly, was the situation in the United States Department of Defense, where systems analysis was carried out on a grand scale with extensive executive backing (7).

In the New York City case, the systems analysts were brought in to work for the budgetary or review authorities. Through their clients, they could delay the approval of the project as advocated by the Board of Water Supply. But neither the analysts nor the budgetary process could substitute a different scheme without the consent of the Board of Water Supply. The Board had an effective monopoly on the detailed engineering data required in order to design and construct the project. The Board of Water Supply also had a veto power over what gets accomplished, as do most of the traditional design agencies.

Faced with this array of forces the systems analyst can at most hope to achieve a compromise. At a later date, when systems analysis may be more widespread, one might hope for better. Meanwhile, it is perhaps best to use the analysis process to determine which aspects are most critical to achieving improvements in the design, and to use this knowledge to advantage in the negotiations that will lead to the ultimate compromise. This was what was done in this instance, as is described below.

The second pervasive limitation on the application of systematic analysis which it is relevant to mention here arises in connection with the manner in which projects are funded. In the United States and, one suspects, elsewhere, it is common for there to be institutional mechanisms which bias investments in favor of capital-intensive projects. Certain agencies may, for example, be exempt from taxation and thereby can afford to substitute capital-intensive designs for more economical projects with relatively high operating costs. The individual Port Authorities responsible for most American airports constitute such a tax shelter: this accounts in part for the fact that American airports use extensive terminal buildings, whereas European airports (London, Paris, Lisbon, etc.) often use vehicles to carry passengers to and from aircraft.

In New York, the uneconomical bias toward capital-intensive projects exists in a different, but equally insidious, form. The city can raise money from three different sources in a system that inherently promotes monumentalism. These sources are as follows:

1. Current revenues (mostly taxes), which must pay for all operating and maintenance expenses of the city. Most of any government's current revenues are entailed to specified commitments (salaries, social security, public welfare, etc.) (2). Furthermore, politicians are loath to raise taxes, so there is relatively little money to be had from this source and much competition for it. Designs that imply heavy operating expenses therefore have little chance of approval.

2. Funds raised by bonds issued within the debt limit. These may pay for capital expenses, but their total amount is commonly limited by law as a device to impose financial responsibility. The competition for this money exists but is much less severe than that for current revenues.

3. Funds raised by bonds for capital purposes which are exempt from the bond limit. For special kinds of capital investments and, importantly, for projects sponsored by the Board of Water Supply, the city can issue bonds freely. The money raised in this manner is quite costless, politically. The tendency is to use it whenever possible, to build monumental civil works even if they are not economical.

THE SYSTEMS ANALYSIS APPROACH

Five-Step Iterative Process

The approach outlined here is essentially that described by the author elsewhere (4). For convenience it is described as consisting of five steps. The particular categorization of the activities is clearly arbitrary, however, and others are equally plausible.

The steps suggested are as follows:

1. Definition of the objectives or, equivalently, of the criteria of evaluation.

2. Specification of measures of effectiveness, the quantitative indices by which one can evaluate the degree to which objectives have been attained.

3. Generation of alternatives, both by overall types or class and exhaustively, especially if computer models are used, within each class.

4. Evaluation of the alternatives according to their achievement in terms of the measures of effectiveness.

5. Selection, as the result of a careful weighting of the several achievements of each alternative, of the best solution.

Overall, the analysis process was, and is, visualized as a search for optimal designs. Since clumsy or even false starts are inevitable, the process was also visualized as a learning experience in which the analysis team gains greater understanding of the system being examined, of its points of leverage, and of how optimal plans for it can best be formulated. In short, the analysis effort is seen as an iterative process in which one may cycle through the analysis several times.

In the New York City case, for example, the preliminary analyses helped us identify critical issues, such as reliability, which required their own measures of effectiveness. They also helped identify new classes of alternatives which could be considered to advantage. The discussion that follows could not hope, and does not try, to document this iterative process. For simplicity, the steps are discussed in sequence, although each of them did not develop fully before the subsequent one began.

Definition of Objectives

The Board of Water Supply indicated to us that they used, as most design agencies today, a single primary standard to evaluate the performance of their design. In particular, they specified that they had to deliver water at a pressure of 40 psi at the curb, which is equivalent to 100 ft of water. As can be imagined, such a standard is a sparse description of the quality of water delivered.

Most realistic descriptions of the outputs of a system must be in terms of many attributes. Municipal water, for example, serves many purposes. A fraction of it, about a quarter, is used by individuals at home. Most of it is used industrially, as in the cooling of steam in electric power plants or in breweries: and the remainder is sprayed on fires or used to wash the streets. The users of the water are concerned, in varying degrees, about many aspects of its quality. For drinking purposes it should be sanitary and have a reasonable taste and color; industrial users may be concerned about the total amount available and its chemical or corrosive properties; the fire department will naturally be most anxious about the reliability of its supply; the municipal council wil be concerned with the proper distribution of the services provided; and so on. A complete analysis should consider each of the important dimensions.

One of the first tasks of the systems analysis of the proposed Third City Tunnel was, therefore, to lay out explicitly what the several objectives of this large project could or should be. Objectively, some 5 to 10 major goals might reasonably be ascribed to the proposed facility. But the analysis was not able to consider as many for two reasons, which are probably typical of such an enterprise.

First, the terms of reference of our study were written quite restrictively, apparently as a result of political pressure generated by anxiety over what might be proposed. We were prohibited, for example, to question whether or not there would ever be enough water to supply the demands the new tunnels were supposed to fulfill. Although legal compacts with other states cast doubt on whether the additional water could be obtained (6), we were explicitly instructed not to concern ourselves with the availability of water. A parallel situation was observed in England in the case of the Third London Airport. Although there was no national strategy for the location and development of airports, the inquiry and the study team set up to review the question of the Third London Airport (the Roskill Commission) was instructed to assume that a new airport was, in fact, needed. A systems analysis born out of controversy, as that one and ours were, will almost certainly be overly restricted in scope and precluded from examining some of the fundamental objectives.

Second, the number of objectives we considered explicitly was restricted by our own limited capabilities and resources. Appropriate, quantifiable models were neither available nor possible to develop for several aspects of the water supply system. The available time and manpower likewise prevented us from doing all that appeared desirable. Our position in this respect was clearly typical.

The objectives that we finally considered were both qualitative and distributional. Specifically, the analysis focused on:

1. Overall performance of system.
2. Distribution of the service over the municipal area.
3. Reliability of the service.
4. Total cost of construction and operation.

Measures of Effectiveness

The degree to which objectives were attained by alternative designs was estimated using objective indices. In general such measures of effectiveness should be chosen carefully, as their very nature may easily bias the outcome of the planning process. If the cheapness of a transportation system is measured by the cost per ride instead of the cost per mile, for example, short, compact systems will be preferred over ones with longer routes and greater economies of scale. The choice of the measures of effectiveness is therefore not trivial.

In the New York City case, the effective range of choice of the measures of effectiveness was limited by what we could extract conveniently from our mathematical models of the system. The measures actually used were therefore somewhat crude. Nevertheless, they made it possible to explore the most important aspects of the design problem.

As a first-order estimate, performance was evaluated in terms of the pressure at which water could be supplied. Since static pressure measures the energy contained in the water, it can be transformed into velocity and rate of supply and is, therefore, a reasonable proxy for the quality, in a mechanical way, of the water. Overall performance was taken as an average of the pressure, p_i, at several key points of the water supply network, weighted by the quantity, q_i, of water desired at each of these i nodes. Specifically, the index was:

$$\text{overall performance} = \frac{\sum p_i q_i}{\sum q_i}$$

and was given in units of pressure.

A distributional measure of effectiveness should always accompany any index of overall performance. A situation that is satisfactory on the average may be attained at the expense of a number of extremely poor conditions (as illustrated by the proverbial man with one foot in boiling water and the other in ice who should, on the average, be comfortable). The measure used in this analysis was simply that of the performance at the extreme end of the supply network, where performance would inevitably be lowest.

The reliability was estimated in terms of the performance that could be expected if a major failure of the system, such as the closure of one of the three main tunnels, occurred. Rather than attempt to guess at the probability of failure, a usual measure of reliability but one that would be extremely subjective in this instance, it was felt best to focus directly on the quality of the failure mode. The measure of reliability used was, in fact, an index of whether the performance of the water supply network would degrade catastrophically or gracefully.

The total cost of the alternative designs investigated was expressed in terms of the present values, taken to a convenient base year, of the discounted expenditures. This is one of the two standard measures, the other being the annualized costs, which can be obtained by a simple mathematical manipulation. But traditional design agencies or planning groups, which must justify present expenditures, are naturally reluctant to discount future benefits. Thus many of them, and the New York City Board of Water Supply in particular, do not always discount future cash flows. This failure biases planning toward overinvestment and monumentalism. Analysts need to be sure that the time streams of costs and benefits are always properly discounted.

The validity of the principle of discounting future costs and benefits is certain. The technical argument revolves around what annual rate should be used. Theoreticians are agreed that the appropriate rate is the opportunity cost of the capital invested, but have so far not been able to provide accurate estimates of this quantity. The discount rate used thus varies significantly from place to place and from agency to agency. In general the discount rate used for public investment in the United Kingdom and Europe, about 10%, is about twice that used in the United States. The lower American figure appears to represent weaker central planning, which is unable to make its logical arguments overcome the reluctance of the construction groups (5), rather than a true difference in the opportunity cost of capital.

The discount rate used to determine total cost for the New York City analysis was 5% per year. While it seems that the true opportunity cost for New York City is closer to 10 or even 15% at the present time, the 5% rate was, in a perverse way, not too inaccurate given the particular way in which water supply projects could be funded. The use of the 5% discount rate in the systems analysis in any case represented an improvement over the alternative discount of zero.

Alternatives

Although systems analysis is, in many respects, not much more than the organized, possibly computer-based application of common sense, good systems analyses are hard to achieve. The difficulties lie not so much in the organization of the process, which is really quite simple, but in the choices that must be made. The analyst is constantly faced with choices between different forms of analysis. Some of these will be tedious and unproductive, whereas others may be very efficient. To be effective, the analyst must make good use of the limited resources at his disposal. The way a problem is formulated can be critical.

The importance of problem formulation is perhaps nowhere more evident than in the analysis of alternatives. This is where most of the resources are consumed. Here is where insightful choices can have the greatest significance.

The search for an optimal design for a system needs to look broadly at many different classes of alternatives and, within each class, iteratively at the different possibilities. The questions are: Which are the most important classes or kinds of alternatives? How can one identify the most promising individual designs?

The choice of which classes of alternatives to investigate, although important, maybe even crucial, cannot be decided by mathematical rule. A good choice will be one that provides the principal opportunities to improve the design. It may be the product of purely accidental qualities: insight, experience, or luck. The improvement of the choice requires that one search aggressively for attractive classes of alternatives, allow time for them to be found, and not foreclose the opportunity to examine them.

Considerable thought was devoted to the determination of the appropriate classes of alternatives for the design of any additions to the water supply system. Ultimately we came to three which may have extensive generality:

1. The physical configuration of the facilities themselves; in this case the route of the tunnels—their length, width, and type of construction.

2. The phasing over time of the provision of the capacity, being an exploration of which design horizon is the most economical.

3. The substitution of operating capabilities for constructed facilities, which would here be the improvement of performance (defined in terms of pressure) by use of pumps rather than by more tunnels.

The last two categories covered alternatives that had not initially been considered explicitly. The recognition of this omission in the systems analysis process led to substantial improvements in design, as shown below.

Sensitivity analyses were used to search for ways to improve alternatives within any given class. With regard to the configuration of the tunnel, for example, we differentiated the expression for energy loss in terms of the tunnel characteristics to determine the elasticity of energy loss with respect to diameter, initial head, and other design parameters. This gave us insight into the best ways to reduce pressure drops and, consequently, to increase performance.

Evaluation of the Alternatives

The primary vehicle for the evaluation of the alternatives in terms of the several objectives was a computer model. Specifically we had the capability, available as part of the ICES (Integrated Civil Engineering System) developed by the MIT Civil Engineering Systems Laboratory, to represent networks and to calculate flows and pressures at desired points (8). This model enabled us to investigate, in a few minutes, alternative designs which would take trained engineers several days to evaluate manually. Since the model was, furthermore, time-shared on an IBM 360/67, it was easy to follow up an effective improvement quickly. The systems analysis was therefore able to investigate literally hundreds of configurations, whereas the traditional planning process has been limited to relatively few.

The evaluation was conducted in terms of a cost-effectiveness analysis (3). The success of each alternative in terms of meeting the objectives was plotted to define a region of feasible design. The outermost limit of possibilities, which was the locus of the best performance that could be achieved for every level of cost, represented the cost-effectiveness function. (Technically, the cost-effectiveness function is a simplified version of the more general production function, the latter being the locus of the most effective possible designs given in terms of the physical resources used rather than the cost. See reference 4, chap. 2). Once the cost-effectiveness function was available it became possible to consider and choose the best design, the one that met all the objectives satisfactorily at a minimum cost.

The cost-effectiveness function in terms of overall performance is shown in Figure 4.1. As expected theoretically, it conforms to the empirical law of diminishing marginal returns: less improvement in design can be obtained for the same extra cost as the design becomes more costly. In such situations, which are pervasive, it is reasonable as a general rule to select a design or plan which is in the region where the money is well spent, such as the one indicated by point A. It certainly is not economically desirable to build a system in which money is spent relatively unproductively, as occurred for a preliminary design of the Third City Water Tunnel. As a first result, then, the systematic use of computer models had revealed that it might be possible to save 30% of the total cost. This large saving could be attained by reducing the size of the planned facilities without thereby significantly altering performance.

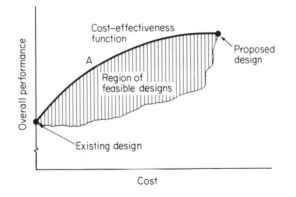

Figure 4.1 Cost-effectiveness function in terms of overall performance.

To validate this initial conclusion, a cost-effectiveness analysis was carried out for the second measure of effectiveness: the performance at the extremities of the system (Figure 4.2). This analysis indicated that it was still possible to obtain an effective design at substantially reduced cost. It also showed that there were two different types of configuration, I and II, and that it was important, regardless of the level of design, to choose the latter. Inspection of the alternatives showed that the group II designs were ones in which there were redundant tunnels. The existence of such facilities, like the proposed Third City Tunnel parallel to the previous two, evened out the loss of pressure and thus improved performance over the entire system.

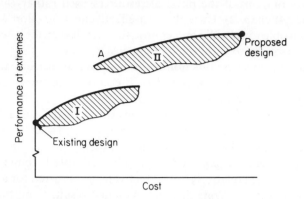

Figure 4.2 Distributional objective: cost-effectiveness function for
the performance at the extremities of the system.

The necessity for a Third City Tunnel of some size was finally confirmed
by examining the reliability of the system (Figure 4.3). Unless some form of
parallel tunnel were provided it appeared quite possible that the city would be
in danger of losing a major part of its supply as the existing tunnels continued to deteriorate. The Board of Water Supply was thus quite correct in
asserting the urgency of some form of tunnel. But, as indicated again by
region A, it was still quite possible to achieve substantial reductions in total
cost.

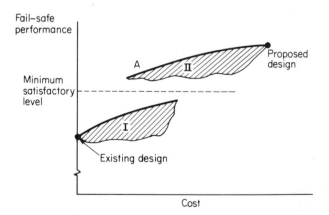

Figure 4.3 Reliability objective: cost-effectiveness function for fail-
safe performance.

With regard to the second class of alternatives, those referring to time
phasing of additions to capacity, it is possible to use a fairly standard approach.
The essential tradeoff is between possible economies of scale and the discount

rate. The presence of economies of scale impel a designer to plan larger facilities so as to increase the capacity obtained per dollar invested. The existence of
a discount rate, on the other hand, encourages him to defer expenses as much
and as far into the future as possible. If there were no discounting of future
costs, it would be wise to build as large a facility as might ever be needed. If
there were no economies of scale, it would be rational to defer construction
until the moment it was needed.

The optimal design horizon for different situations has been determined
by Manne (9). His solutions are given in terms of the discount rate, the rate
of growth of demand, and the economy of scale factor, α. The latter has to
be determined empirically by calibration of the formula:

$$\text{cost} = (\text{constant}) \times (\text{capacity size})^{\alpha}$$

For tunnels it appears that $\alpha \sim 0.6$. The most economical design horizon
can then be determined from Figure 4.4.

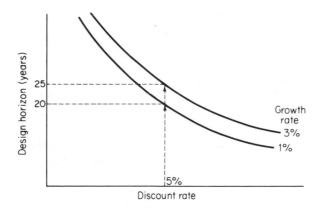

Figure 4.4 Nomograph for the determination of the best design
horizon (economy of scale factor, $\alpha = 0.6$).

The most economical period between additions to capacity turns out to
be between 20 and 25 years. This is roughly one-half of what is assumed in
the construction of many large-scale projects. By using this design period, it
is possible to capture substantial savings not possible if the longer period is
used, as was originally planned for the Third City Tunnel. These results were
used to confirm the desirability of substantially reducing the designed size
of proposed tunnels.

The statement that a 20-year design horizon is most desirable is not the
same as saying that the large, 40-year, capacity will not be needed. It is merely to say that it will be cheaper to build this capacity in two installments,
each planned for 20 years' worth of growth.

Construction in smaller stages is not only more economical, it is also more flexible. If the projected growth somehow fails to materialize, further substantial savings can be achieved by merely not building the capacity that is now known to be unnecessary. Even if the demand does occur, it may very well be of a different kind and in a different place than originally anticipated. (Indeed, the recently released results of the 1970 Census indicate that the population of New York City has remained static between 1960 and 1970, contrary to the previous assumption of rapid growth). The shorter design horizon permits the planner to adjust his program to such important developments, whereas a 40-year horizon does not.

Consideration of the final overarching class of alternatives, the use of operating improvements instead of additional capacity, indicated that further savings could be achieved. This is often the case. Experience indicates that engineers have a tendency to think of construction solutions to operational problems.

The basic operational problem of a water supply system is, as it is in many other systems, the existence of peak loads 20 to 30% above the average. The system must operate at particularly high levels for relatively short periods. In the New York City case, the system was designed to operate without degradation of quality at the peak hour on the peak day, which occurs with a frequency of about 0.0001. The capacity provided to meet this peak is thus wasted 99.98% or more of the time.

To meet peak demands by providing capacity that will be available all the time, rather than by using standby capabilities that will be rarely used, can be expensive. The operation of small-scale, relatively inefficient facilities over extended periods can also be costly. The most appropriate design represents a balance between operating costs and fixed investments (Figure 4.5) (1).

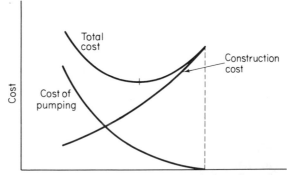

Designs with different tunnel sizes and amount of pumping

Figure 4.5 The most economical design for a water supply system is, in general, one that includes pumping to take care of peak loads.

For the New York City water supply system and the particular kind of peak loads it carried, a 30% reduction in total cost appeared to be possible by using pumping plants to provide the pressure (and performance) desired at peak hours. It should be noted that the role of pumping and similar operational measures is quite limited. Their costs rise quite rapidly when they begin to be used to meet continual, base-load demands.

IMPLEMENTATION

The plan ultimately selected for implementation was a compromise, as was to be anticipated. For the record, it is appropriate to describe the plan settled upon. In a larger sense, however, it is more pertinent to indicate why the compromise design took the particular form it did and how it came about. These elements constitute the guidelines for the future.

The final design for the first stage of the Third City Tunnel, the one now being implemented, was as follows:

1. It was located along the same plan and elevation as the preliminary design.

2. It was designed without pumping, just as the original design.

3. It was only three-fifths the size, as measured by the relevant cross-sectional area, being about 22 ft in diameter instead of the approximately 28 ft originally planned. As a result of this single change the cost of the first stage was reduced from an estimated $323 million to $223 million, an apparent savings of about $100 million (13).

It is revealing to examine the nature of the compromise. In many superficial aspects, the route and scheme of operation, there was no change at all. This would be important from the point of view of the Board of Water Supply; they had laid much of their prestige on the line in laying out their original concept and could hardly retreat from their position without severe loss of face. Being an engineering group, however, they could give much more easily on the cost aspects of the plan, which, one might say (although I would not), were not really technical or engineering matters. For the Board of Water Supply to reduce the size of the project was the easiest thing it could do. As it happened, it was also the step that the budgetary review authorities, with whom the Board of Water Supply were in conflict, most demanded. The compromise was arrived at, as might be expected, on the basis of mutual expediency. Optimality of the resultant design might not have been the prime mover of the compromise.

The significant reduction in the net cost was reason enough to justify the systems analysis, which cost only $50,000. But it would be a gross oversimplification to believe that the analysis was, or could be, primarily responsible for this outcome. No significant savings could have been achieved if the budgetary authorities had not been prepared to wage an administrative battle.

Conversely, greater savings could have been accomplished if the construction money had not been so easy to obtain (as explained previously) and thus if the Budget Bureau had been more determined.

The role of the analysis reported here, and perhaps the role of all similar systems analyses, was to clarify issues and indicate points of leverage in the preliminary plans. The analysis provided ammunition that might be used. The effectiveness of this ammunition depended, and depends in general, on many other factors besides the quality of the analytical work. Many of these factors would appear to be set in advance, before the analysis began. It is then inappropriate to judge the analysis itself on the size of the net results that occurred after the analysis was completed.

The analysis can, however, have significant longer-term effects, as did the New York City study. In the first instance, it may be expected that the agencies that were participants in the debate that the systems analysts was called in to investigate will develop more modern analytic capabilities. This was certainly the case in the Department of Defense during the McNamara period. It was also, apparently, true for the Board of Water Supply. Since the debate occurred, they are reported to have begun to use the concept of opportunity cost and to implement a few computer models of their network. Second, the analysis may draw attention to important problems that management needs to face. In this case the MIT effort emphasized the need to pay attention to the secondary pipe networks in the city; the volume of construction on these systems has since increased fivefold (10). Such secondary benefits may, ultimately, be equally as important as the immediate, more obvious primary results.

REFERENCES

1. BAYER, B., 1970. *Tradeoffs Between Tunnel Size and Pumping Capacity in Primary Water Distribution Systems*, C.E. Thesis, Department of Civil Engineering, MIT, Cambridge, Mass., Feb.
2. CRECINE, J. P., 1967. A Computer Simulation Model of Municipal Budgeting, *Management Science*, Vol. 13, No. 11, July, pp. 786–815.
3. DE NEUFVILLE, R., 1970. Cost-Effectiveness Analysis of Civil Engineering Systems: New York City's Primary Water Supply, *Operations Research*, Vol. 18, No. 5, Sept.-Oct., pp. 785–804.
4. DE NEUFVILLE, R., AND STAFFORD, J. H., 1971. *Systems Analysis for Engineers and Managers*, New York: McGraw-Hill Book Company.
5. FOX, I. K., AND HERFINDAHL, O. C., 1964. Attainment of Efficiency in Satisfying Demands for Water Resources, *American Economic Review*, Vol. 54, No. 3, May, p. 198.
6. HIRSHLIEFER, J., DE HAVEN, J. C., and MILLIMAN, J. W., 1960. *Water Supply: Economics, Technology, and Policy*, Chicago: Chicago University Press.

7. HITCH, C., 1965. *Decision-Making for Defense*, Berkeley, Calif.: University of California Press.
8. LIU, K. T., 1968. *Pipe Network Analysis in Integrated Civil Engineering Systems (ICES)*, Cambridge, Mass.: IBM Cambridge Scientific Center.
9. MANNE, A. S., 1967. *Investments for Capacity Expansion*, Cambridge, Mass.: MIT Press.
10. Pollution Fighter Marty Lang Does Battle in "Wicked Gotham," 1971. *Engineering News-Record*, Mar. 4, pp. 24–26.
11. *Report of the Board of Water Supply of the City of New York on the Third City Tunnel First Stage*, July 1966.
12. SIMON, H. A., 1959. *Administrative Behavior*, New York: The Macmillan Company.
13. Tunnel Goes for Record $222.6 Million, 1970. *Engineering News-Record*, Jan. 15, pp. 13–14.

PART II

Systems Modeling

5

Introduction

We selected the following cases in order to illustrate the principal issues in each of the three important elements of a system that an analyst may be called upon to model: production, supply, and demand functions.

Production functions represent the technically efficient means of using any combination of inputs to achieve outputs or results. For any situation, the production function is a special subset of all technically feasible solutions: the set of all combinations in which the resources are not wasted, that is, which are technically efficient. For example, if the most output that can be obtained from one man and one truck is 1000 ton-miles a day, that point is a part of the production function for transportation in terms of manpower and vehicles; other points, representing less output (perhaps because the driver sleeps on the job or the truck is poorly maintained), are technically feasible but not efficient and not on the production function.

Supply functions isolate the subset of the production function which is economically efficient for any particular set of relative costs of the inputs. In general, the supply function only describes a narrow portion of the production function. Steel skyscrapers, multistory concrete frame buildings, and quonset huts may each be designed in a way that is technically efficient, but only one is likely to be economically efficient for providing housing for a specified set of prices for land, labor, and materials.

Demand functions represent the schedule of the amount of output that will be consumed by the public. This amount will, naturally, vary according to the price and quality of the output, as well as to the price and quality of competitive products.

A proper understanding of each of the three elements is critical to the planning and design of public systems. Indeed, the loads on any such system, which by definition is intended to serve the public and be responsive to their desires, are determined by how well the public likes the kinds of services that the system provides. Conversely, the kind of design that should be implemented depends on the anticipated loads. In short, the proper configuration of a system requires that the planner or analyst simultaneously consider the interaction of the possible supply and demand and their dynamic equilibrium over time. We thus focus on the particular issues in modeling supply and demand.

To put this discussion into context, it should be pointed out that professional planners and designers have not, traditionally, considered the interaction of supply and demand in formulating a system. Their usual approach has been to assume that demands were fixed, independent of what was designed. (Frequently, these fairly arbitrary levels of desire have been called "needs"; in fact, these desires are no more needs, however acute, than my son's desires for more spending money). This simple approach requires less analysis and computation than the realistic consideration of both supply and demand jointly, and it is perhaps all that could be achieved in practice without the help of computers. But it is now becoming increasingly clear that, in order to obtain viable plans and designs for a system, demand and supply do, in fact, have to be considered simultaneously. Civil engineering, in particular, is now coming to this realization.

The approach in this section is explicitly problem-oriented. The focus is on showing how one can develop models that will be directly useful for planning and design. Specifically, we wish to indicate how one can build models of supply and demand both deductively and by inference, and to suggest how one can overcome attendant conceptual and technical difficulties in carrying out these tasks. The discussion does not extend to theoretically interesting typologies such as linear or dynamic systems. The methods for representing such systems can be useful tools in a variety of contexts. But we do not feel that a discussion of technique would really lead to interesting perspectives on what one might need to model to solve a problem, or on how one might validate such a model. In this section, as in the others, the concentration is on the choice of approaches available to solve a problem.

Two principal philosophical approaches to modeling are considered: the deductive and the inductive or inferential. The first approach builds up a model as a logical consequence of a set of initial facts. The procedure requires that we deduce the consequences of our premises or preliminary equations. The inductive approach, on the other hand, refers initially to a set of observations about how a system behaves or what it looks like, and attempts to reconstruct, or infer, the nature of the underlying mechanisms that cause the system to behave as it does. In practice, this second or inferential approach is largely statistical.

The three topics to be discussed are, then, the development of production and supply functions both by deduction and inference, and of demand functions by inference. (There does not appear to be any practical possibility, at present, of building demand models deductively). There are distinct methodological problems, and different limits to knowledge, connected with the development of each of the three types of models. These separate issues are each illustrated by a pair of case studies. In each pair one of the cases is relatively simple, and the other, being more advanced, suggests how far this kind of modeling can now be taken.

SUPPLY MODELS: DEDUCTIVE

The definition of the probable results of possible plans is an important aspect of any systems analysis. Specifically, it is frequently desirable to determine how to use available resources, X_i, to achieve certain objectives or outcomes. In particular, it is relevant to define the most that can be obtained from a specified set of resources. This maximum output, Z, represents a technically efficient use of the resources. The locus of all efficient points:

$$Z = g(X_1, \ldots, X_n) \qquad (5.1)$$

is known as the *production function*. This function may either be a continuous analytic function or a set of discrete functions or points which represent the entire range of possibilities. Knowledge of the production function can be extremely useful in systems planning and design.

The production function is an especially attractive model of a system because it contains all the possible system configurations that might be economically efficient, that is, all the supply functions. The *supply function* is, in effect, the locus of all points on the production function which achieve output at the least cost, for the costs that prevail in any particular situation. And for each particular design problem the analyst will be interested in the alternatives that are both technically and economically efficient, that is, those defined by the supply function. The systems planner will be interested in the production function because it enables him to obtain all the supply functions he might need for the different economic circumstances, and sets of relative prices for the resources, that he might encounter; the production function provides an economical means to obtain the supply functions which are ultimately desired.

As is demonstrated subsequently, models of the production function cannot be defined through statistical analysis of the operations of actual systems. Rather, production functions must be modeled by a deductive process. They must be built up through a detailed description of the underlying physical phenomena that generate outputs from the inputs. Fortunately, it is relatively straightforward, at least in principle, to construct models of the production process in this fashion. Indeed, because the production function deals with physical phenomena, it is frequently possible to determine its underlying detailed relationships through experimental procedures either in the laboratory or as applied to prototype installations. When these relationships are already understood and tabulated in detail, it is quite simple, if tedious, to construct the production function. Conversely, when it is not possible to determine these relationships, it is no longer possible to model the entire production function, and supply models must be formulated, one at a time, by statistical inference from particular situations.

Analytic Functions

In a number of situations it is possible to represent the entire production function by a single analytic function. This is most frequently possible when the system is governed by a continuous environment, such as a force field.

The case study in Chapter 6 of inland water transport illustrates this situation. In this instance the analysis proceeds directly from the results of laboratory experiments which define the basis for the production function. The case study itself represents a straightforward analysis of the characteristics of the production function under different circumstances. The virtue of the case, which is analytically simple, is that it describes the nature of the production function for this system for the first time. This case can be taken as a model of what should be possible to accomplish for any number of other systems.

Discrete Functions

In most situations it will not be possible to define a production function analytically; it will have to be built up through a detailed investigation of each of the possible designs. The development of the production function then requires much more effort and detailed knowledge than for the previous situation. The case study by Cootner and Löf illustrates the point. Here, they based their production function for thermal efficiency on the available extensive and detailed knowledge of thermodynamics, as tabulated in steam tables.

The case study in Chapter 7 of thermal efficiency is also interesting in that it shows how the supply function is to be obtained from the production function. By using a detailed description of the cost of the inputs and applying the optimality conditions, the authors pinpoint the locus of points on the production function which define the supply function.

ADDITIONAL READINGS

The fundamental theory of production functions is available in a number of texts on applied economics, such as

BAUMOL, W. J., 1965. *Economic Theory and Operations Analysis*, Englewood Cliffs, N.J.: Prentice-Hall, Inc.

This theory is presented in the context of planning and design in

DE NEUFVILLE, R., AND STAFFORD, J. H., 1971. *Systems Analysis for Engineers and Managers*, Chap. 2, New York: McGraw-Hill Book Company.
CHENERY, H. B., 1949. Engineering Production Functions, *Quarterly Journal of Economics*, Vol. 63, pp. 503–531.

SUPPLY MODELS: INDUCTIVE

Definite limitations exist on the amount of knowledge we can obtain by inference from observations on the way actual systems operate. Specifically, it is not possible to obtain a description of the entire production function. All that can be obtained by direct statistical inference is information about the supply function.

This limitation is caused by the fundamental rationality of the designers of existing systems. They can be presumed to attempt to make their systems as economical as possible. In so doing, they will, in effect, be trying to satisfy the optimality conditions, which themselves define the supply function. Consequently, observations on the operation of existing systems will, within a margin of error, be limited to observations on the supply function alone. So that is all that any subsequent statistical analysis can hope to reconstruct. Extrapolations beyond the supply function, although useful perhaps, would be conjecture.

Although one can only use statistical inference to specifically characterize a particular portion of the production function, it is nonetheless possible to obtain valuable insights into the nature of the production function. Most importantly, it is possible to determine whether the production process exhibits increasing or decreasing returns to scale, at least along the segment described by the supply function. Such information can be most valuable to the planner since the existence of economies of scale implies that the planner ought to engage in a deliberate strategy of early construction in anticipation of future growth so as to capture these benefits.

Case Studies

Murphy's analysis in Chapter 8 of productivity increases in the airline industry is a fairly standard application of statistical analysis to the estimation of the supply function. It provides a reasonable model of what can be easily done, with not too much more than appropriate data and a modicum of computer time. It is certainly an example that deserves to be followed extensively by planners and designers as they seek to describe similar systems.

The analysis by Nerlove in Chapter 9, on the other hand, is a much more advanced analysis of the possible variations in the shape of the supply function for a particular process. It is representative of the very best analysis of this type, and illustrates the kind of detailed results that can be derived from a carefully formulated analysis.

It should be pointed out that both case studies actually speak of cost functions, of the form:

$$\text{cost} = f(\text{output})^{1/R} \tag{5.2}$$

rather than supply functions, which are usually of the form:

$$\text{output} = g(\text{cost})^{R} \tag{5.3}$$

Reference to cost functions is traditional in the literature. As can be appreciated, however, the cost function is essentially an alternative representation of the supply function.

ADDITIONAL READINGS

SHEPHARD, R. W., 1963. *Cost and Production Functions*, Princeton, N.J.: Princeton University Press.

WALTERS, A. A., 1963. Production and Cost Functions; An Econometric Survey, *Econometrica*, Vol. 31, pp. 1–66.

An excellent presentation of advanced applications is also provided by

NERLOVE, M., 1965. *Estimation and Identification of Cobb–Douglas Production Functions*, Chicago, Ill.: Rand McNally & Company.

DEMAND MODELS: INDUCTIVE

In current practice, all demand models are developed from statistical analysis of data on actual choices made by potential users of a system, or by proxies for them. Consequently, it is only appropriate to consider inductive or statistical models. Eventually, perhaps, the development of deductive demand functions may be practical. Indeed, our current preliminary under-standing of how to estimate empirical multiattribute utility functions for individuals, as illustrated in Chapter 24, and related techniques of survey research on consumers suggest that it may become possible to construct a priori demand functions. But this would appear to be some time off.

Nonexperimental Research

An essential characteristic of the problem of estimating the public's response to a new system, that is, its demand function for the services pro-vided, is that we are involved in what is known as nonexperimental research. This does not mean that the analysis is not rigorous—quite the contrary; it means that the analyst does not have the possibility of performing con-trolled experiments on the general population. This makes the analysis procedure significantly different from that involved in trying to investigate the characteristics of a physical process such as represented by a production function.

The fundamental difficulty of nonexperimental research, and the single most important problem in the development of demand functions, is that we may have extreme difficulty in determining what kind or what functional form of a model we ought to estimate. Because we cannot generally control the environment in which we make our observations, as we can in the labo-ratory, we cannot be sure how the possible factors affect one another, or what causal mechanisms operate within the system. This weakness is important because it is easy to demonstrate that, for any given set of obser-vations, several different and often contradictory models can fit the same data. Consequently, in order to have any real confidence in the predictive power of a demand model, it is necessary to develop specific procedures that enable the analyst to gain confidence in and validate his particular model.

It should be remarked that the situation for demand models is, indeed, quite different than that for models of supply. In dealing with physical pro-cesses an extensive repertory of experimental results can be called up which indicate how elements of the system do combine, and what kind of func-tional models are most appropriate. Specifically, for example, there is ample evidence to suggest that the Cobb–Douglas models of the form:

$$Z = k \prod_i X_i^{\alpha_i} \tag{5.4}$$

are appropriate. These were used, in fact, in both the case studies of cost functions.

Appropriate forms of the demand models to be statistically estimated can be developed either from theory or a priori information. Generally both will be brought to bear and neither, singly or in combination, will be fully satisfactory. Consequently, much emphasis needs to be placed on the validation of the model.

As already indicated, mere correlation of a model with a set of observations is not a satisfactory demonstration of its validity. Although some inappropriate models can be sorted out because they fail to correlate with the data, several competitive models can generally fit the data closely. The real test of a model is, then, whether it does actually predict the response of the public to significantly new situations.

Case Studies

The first case study, that of the demand for airport access services in Chapter 10, gives an example of how one might go about developing a demand model from scratch. The emphasis is on two procedural aspects: first, the specification of reasonably controlled situations that will permit one to define fairly accurately the structure of the model to be estimated; and, second, the definition of a sequential analysis that permits the kind of validation of a demand model that is necessary. The value of this study is not in the analysis itself, which is rather simple, but in the example it suggests for the careful development of demand models for other situations.

Fisher's analysis in Chapter 11 of the demand for railroad passenger transportation bases a highly sophisticated analysis on a rare (but not unique) sequence of data. Because of the significant changes in the economy during the periods he investigates, he is able to validate his model by considering its predictive power in these new situations. Furthermore, he cleverly uses the first periods of his analysis to derive a priori information which he can then use to formulate a more complex model of demand for subsequent periods. As his success in doing this depended upon the existence of a strategic set of circumstances, his analysis cannot be cast as a universal paradigm for demand modeling. Yet it is a classic analysis worthy of emulation to the extent possible.

ADDITIONAL READINGS

The fundamental issues in nonexperimental research are presented, in the context of planning and design, in

DE NEUFVILLE, R., AND STAFFORD, J. H., 1971. *Systems Analysis for Engineers and Managers*, Chaps. 12, 13, New York: McGraw-Hill Book Company.

A more comprehensive exposition of the subject is given by

BLALOCK, H. M., Jr., 1961. *Causal Inferences in Non-experimental Research*, Chapel Hill, N.C.: University of North Carolina Press.

COCHRAN, W. G., 1965. The Planning of Observational Studies of Human Popula-
tions, *Journal of the Royal Statistical Society*, Ser. A, Vol. 110, pp. 234–255,

also in

ZELLNER, A. (ed.), 1968: *Readings in Economic Statistics and Econometrics*, Boston:
Little, Brown and Company.
WOLD, H. O. A., 1956. Causal Inference from Observational Data: A Review
of Ends and Means, *Journal of the Royal Statistical Society*, Ser. A, Vol. 119,
pp. 28–61.

A specific and most worthwhile illustration of the kind of ambiguities
that may occur when models with conflicting consequences fit the data equally
well is given by

FORBES, H. D., AND TUFTE, E. R., 1968. A Note of Caution in Causal Modeling,
American Political Science Review, Vol. 62, pp. 1258–1264.

For a clear exposition of the techniques of statistical inference appropriate
to demand modeling the reader is referred either to the fairly simple description of Kane, which requires little background, or to the more complete
presentation by the Wonnacotts.

KANE, E. J., 1968. *Economic Statistics and Econometrics*, New York: Harper and
Row, Publishers.
WONNACOTT, R. J., AND WONNACOTT, T. H., 1970. *Econometrics*, New York:
John Wiley and Sons, Inc.

FUTURE PROSPECTS

Many of the concepts and methods suggested by the case studies on
systems modeling are little known both in the practice and teaching of planning and design. Most usually, systems engineers and analysts have assumed,
implicitly if not explicitly, that the modeling process required not too much
more than a logical arrangement of specific facts, an assumption assisted,
occasionally, by the use of simple statistical analyses. Sometimes this approach
is sufficient. This is the case, for example, with many of the hydrologic representations of river basin systems, such as that used in Chapter 17 in the
case on the Delaware River. But very frequently the traditional approach
is not only insufficient but quite misleading. This would appear to be true
for a significant fraction of the systems models for urban transportation,
land use, housing, and other public services. We are suggesting that a different set of approaches should be brought into planning and design.
    Much of what we suggest as desirable for modeling by planners and
designers has already been developed. Indeed, a rich theoretical literature
on the definition of supply and demand models already exists in economics,

econometrics, and, as regards causal modeling, in sociology. An obvious question, then, is why this material has not already been extensively applied in systems planning if it is so available? The reason these approaches have not yet been applied is, we believe, that relatively few practical applications of the theory have been available until fairly recently. Indeed, although computers are not strictly required, in practice the availability of computers has significantly increased the use of the approaches recommended. As a consequence, a significant number of examples of applications have been described since the mid-1960s, and more are being developed increasingly rapidly.

It can now be expected that the experience that is becoming available in related fields will soon be transferred into the practice of systems planning and design. With due regard for the generational lags in the transfer of new developments from one field to another, we fully expect that the next decade will witness a significant expansion in the use of causal modeling and econometrics in systems analysis. The case studies have been selected to suggest some of the different issues and procedures that should be considered in this process.

6

Economics of Inland Water Transport*

JOHN F. HOFFMEISTER, III, and RICHARD DE NEUFVILLE

Ultimately, optimization of the design of any transportation system requires that the economics of supplying the system be considered jointly with the demand for the services the system provides. It is not possible to speak meaningfully of the best design without a clear understanding of the loads the design is intended to carry. Specifically with regard to inland water transport, the optimum configuration at any time depends on the demand for speed and frequency of service, and on whether or not there are significant backhaul loads. Likewise, the most desirable strategy for expanding capacity is influenced by the nature of the change in demand, and whether it is compatible with large tow shipments, for example.

In many practical situations, however, the planner is not able to antici pate demands accurately. He may, for instance, be called upon to suggest design configurations for a developing country, or to anticipate tow configurations for a new waterway or for some distant date. It is, then, most important for him to know what his supply possibilities are. With this knowledge he can at a minimum screen out dominated design strategies and focus on a narrow set of alternatives.

*Adapted from Optimizing the Supply of Inland Water Transport, *Journal of Transportation Engineering, ASCE*, Vol. 99, Aug. 1973, and from *Investment Strategies for Developing Areas: Models of Transport* by Richard de Neufville, John F. Hoffmeister, III, and David Shpilberg, MIT Report R72–48, 1972, and from *The Economics of Inland Waterway Transportation*, M.S. and Civil Engineer Thesis by John F. Hoffmeister, III, Department of Civil Engineering, MIT, Cambridge, Mass., 1972.

As it happens, the essential information about supply is embedded in the production functions. These are purely technical relationships which can be analyzed with comparative ease. First, one can examine them and determine whether or not they are characterized, for a particular transportation system, by the existence of diminishing marginal returns and returns to scale. Knowledge of these inherent features can be used to specify, independently of the particulars that may be relevant to given situations, what optimal designs and development strategies will be like. Second, by combining knowledge of the production functions with estimates of costs, it is possible to define expansion paths which indicate the least cost designs for different levels of capacity.

This study indicates how one can define optimal configurations for supplying inland water transport. As a first step, it shows how one can develop a production function, in this case by reference to direct physical measurements. Second, it explores the production function and observes that the supply of water transport is characterized by diminishing marginal returns and increasing returns to scale. The effect of particular environments which distort the production function, such as the existence of stream currents or of bends in the waterway, are also shown. Since the production function is a purely technological relationship, independent of costs, it and its implications are universally valid in any country.

DEVELOPMENT OF THE PRODUCTION FUNCTION

A production function gives a general statement of all outputs that can be obtained from all technically efficient combinations of resources or inputs. Its characteristics can be derived in two ways, either through detailed knowledge of the physical processes involved, or by some form of statistical analyses of data on actual operations. In this instance, we are able to define the physical relationships through detailed laboratory and field experiments (5, 6).

This analysis focuses on inland water transport, specifically on the use of relatively shallow draft towboats and barges. These craft typically carry significant volumes of cargo traffic in the United States and Europe. They constitute a mode of transport which generally is very inexpensive to operate. Further, they do not necessarily require the heavy capital investments in the development of rights-of-way, such as those associated with highways (although heavy expenditures are clearly required if one does choose to change rivers substantially as one has done in the United States and Europe). It may, thus, be expected that this form of transport will continue to be important. We may even anticipate that it will contribute significantly to the development of large areas, such as the Orinoco and Amazon basins, just as it did in the United States.

The engineering production function described below relates the output of a barge tow, in terms of ton-miles per hour, to the design parameters, such as the configuration of the barges and the horsepower of the towboat, and to the prevailing environmental conditions, such as the shape and velocity of the stream. At this level of analysis, these features are described in aggregate terms: width and depth of stream, length of barge, and so on. Neither the natural variations in the environment nor the effect of different types of hull design are considered in detail. These factors would, of course, have to be taken into account for the design of a particular vessel for a particular river. But when we are concerned with the productivity of inland water transport as a system, when we are interested in finding out what kinds of configurations make sense, and in screening out dominated solutions so that we can proceed to detailed design efficiently, the kind of aggregate model we describe is appropriate.

A detailed definition of the production function for inland water transport can be readily obtained because the dominant components of this activity consist of physical processes which are relatively easy to model and understand. Indeed, the main effort required to transport cargo by water consists of the propulsive energy needed to move the vessels through the water, and this is a process that is well documented. Further, because water is a continuous medium, it is possible to describe this process, and the total production function, in terms of continuous equations which are easy to manipulate as desired.

Other activities for which we might want to design public works, and other transportation activities in particular, do not lend themselves so easily to the definition of a production function. In air transport for example, although the aircraft also move through a continuous medium in which their performance can be described mathematically, the direct costs of operating the aircraft only account for half the resources consumed in operating an airline; the rest occur on the ground in a variety of activities. The costs of highway transport, on the other hand, depend so intimately on the particular topology and geology of a route that they are essentially impossible to define a priori in a production function.

The fundamental physical situation for water transport is that, for any barge–towboat flotilla to proceed at constant speed in a given waterway, the effective push generated by the engines, EP, must equal the resistance, R, of the flotilla:

$$EP = R \qquad (6.1)$$

The effective push depends both on the horsepower of the engines and the speed of the vessel through the water. The resistance depends both upon the characteristics of the vessel, such as its speed, breadth, length, and draft, and upon the cross section of the waterway. The narrower and tighter the river, the more difficult it is to push a boat through it. Both the nature of the push and the drag can and have been defined by towtank tests.

The production function derives, in detail, from Eq. (6.1). The equilibrium tow speed of a barge flotilla relative to the water, S^*, has been derived and refined by Howe (8–10). It is, approximately:

$$S^* = -1.14HP$$
$$+ \frac{\{1.3HP^2 - 4\beta[(-1)^{\delta+1}F - 31.8HP + 0.0039HP^2 - 0.38(HP)D]\}^{1/2}}{2\beta}$$

$$(6.2)$$

where HP is the brake horsepower of the towboat and D is the depth of the channel. The slope-drag force, F, due to the slope of the river, is:

$$F = 0.00086(S_w^2)(D)^{-4/3}[(52 + 0.44H)HLB + 24{,}300 + 350HP - 0.021HP^2]$$

$$(6.3)$$

where H, L, and B are the draft, length, and breadth of the vessel. The coefficient β is defined as:

$$\beta = 0.0729e^{1.46/(D-H)}H^{0.6+(50/W-B)}L^{0.38}B^{1.19} + 172 \qquad (6.4)$$

where D and W are the depth and width of the waterway. Finally, the speed of the flotilla relative to the ground, S, can be determined as:

$$S = S^* + (-1)^\delta S_W \qquad (6.5)$$

The δ is a dummy variable, indicating the direction of travel of the flotilla ($\delta = 0$ for downstream, $\delta = 1$ for upstream). The output of the barge flotilla is, finally, nothing more than its speed times its capacity. The output is then defined in terms of the configuration of the barge and of the prevailing environment of the stream.

The inputs to the production function of water transport are basically of two types: the capital costs, mainly associated with the construction of the vessels and dependent upon their size, and the operating costs, mainly attributable to the energy required to push the vessel and closely dependent upon the power of the engines. We can, in fact, define two design parameters that correspond to these categories: the deck area, describing the size of the barge, and the horsepower, describing its power. This dichotomy of inputs is particularly interesting because it greatly facilitates the graphical description of the nature of the production function. For specific problems, as indicated below, one might choose to include more components, such as barge width and depth. However, those refinements do not affect the basic conclusions of this study. Therefore, the total output, Z, can be defined as:

$$Z = h(\text{deck area, horsepower}) \qquad (6.6)$$

For easy computation, a short FORTRAN program was written for an IBM 1130 of 16K byte storage capacity. This program utilizes a structure of consecutive DO loops to evaluate the equations describing the production function. It calculates output, in ton-miles per hour, as a function of the user-specified inputs to the production function, specifically in terms of the two generalized inputs, deck area and horsepower. If it were desired to explore specific problems in detail, the program could readily be expanded to consider the barge width and depth, as well as other features.

EXPLORATION OF THE PRODUCTION FUNCTION

This section explores the underlying properties of the production function for water transport. The existence of diminishing marginal returns, returns to scale, the effects of stream currents, and the effects of constraints such as a maximum limit on the length of the barge flotilla are examined. The effect of a stream current is to distort the isoquants of the production function in favor of increased horsepower. Returns to scale are shown to vary depending upon the nature of the waterway. In general the effect of the diminishing marginal returns, and of the decreasing returns to scale associated with the constraints, is to yield a convex feasible region for output, for which it is possible to find specific optimal designs.

Base Case: Stream Current = 0 and No Constraints

As a first step in the examination of the production function we considered situations where the stream current equals zero (still water) and there were no constraints on the configuration of the barges. This might occur, for example, on a lake. The effect of other stream conditions are subsequently examined by comparisons to this base case.

Diminishing marginal returns were found to exist throughout the production function. It will be recalled that the marginal return or product of a single specific input X_i, MP_i, is:

$$MP_i = \frac{\partial Z}{\partial X_i} \qquad (6.7)$$

As shown in Figure 6.1, the slope of the production function with respect to either input is strictly concave. In this case the underlying technical reason for the diminishing marginal returns of output with respect to horsepower

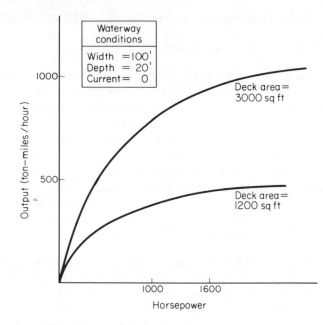

Figure 6.1 Typical diminishing marginal returns of water transport production function.

is that the rate of increase of equilibrium speed in still water with respect to an increment in horsepower is continually decreasing. As the speed increases, the water flow to the propeller screws of the towboat is restricted, so maximum advantage cannot be drawn from the extra horsepower. Also, the resistance of the barge flotilla increases because more water is drawn from under the flotilla, causing the barges to "squat." Although it is to be expected that a production function will conform to the empirical law of diminishing marginal returns, this is not necessarily so, and should be verified.

The question of whether a production function has increasing returns to scale is important because their existence has significant implications for design of facilities for which a growth in demand is expected. If there are increasing returns to scale, which means that one can get more output per unit input using larger processes, it is then often desirable to overdesign initially so as to capture these future economies (4, 11). It must be remembered that returns to scale refer only to the relationship between changes in the physical quantity of output and changes in the physical quantity of all inputs simultaneously and in the same proportions. If doubling or halving all inputs always results in exactly doubling or halving efficient output, the production function is said to possess constant returns to scale. If doubling all inputs more than doubles output, it has increasing returns to scale; if it less than doubles output, it has decreasing returns to scale.

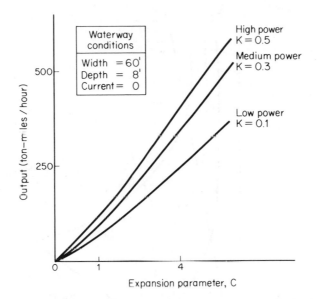

Figure 6.2 Increasing returns to scale for water transport (design is $C[K$ (horsepower), barge deck area]).

To test for the presence of returns to scale, three prototype designs were considered for high-, medium-, and low-powered vessels. The ratio of horsepower to deck area, K, was held constant for each case. Typical results appear in Figure 6.2. As can be seen, the production function in this case exhibits slight returns to scale, especially for the lower-powered craft. This should not be unexpected; it is, indeed, the phenomenon that justifies increasingly larger ocean tankers.

The production function itself can be represented by means of its isoquants or contours of equal output. A typical illustration is shown in Figure 6.3. At any point, the slope of the isoquant is equal to the marginal rate of substitution of the inputs, MRS; which is a negative quantity as input being subtracted by the other is added to keep output constant.

$$MRS = -\frac{MP_i}{MP_j} \tag{6.8}$$

This is equal to the rate at which inputs must substitute for each other to maintain a given level of output.

The expansion paths are the loci of optimal design for a specific set of economic circumstances and, specifically, of costs for the several inputs. Each point on the expansion path is defined by the optimality conditions—that the ratio of marginal product to marginal cost for all inputs should be

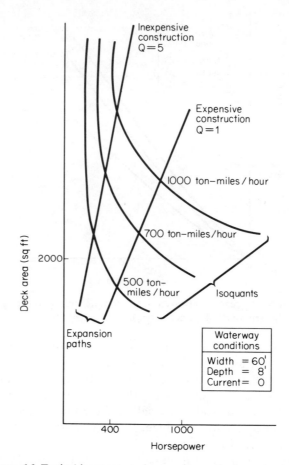

Figure 6.3 Typical isoquants and expansion paths for water transport in still water.

equal. Assuming that the unit costs are linear, as they approximately appear to be over the range of craft used for inland water transport (1–3), the optimality conditions are that:

$$MRS = -Q \tag{6.9}$$

where Q is the ratio of the cost per unit of horsepower to the cost per unit of barge deck area. Typical expansion paths for different cost ratios, such as might appear in different countries or regions, are shown in Figure 6.3.

For this case, of water transport over still water where there are no constraints on the waterway, the expansion paths are approximately straight lines through the origin of the graph. This means that the ratio of the inputs in the optimal design combination does not change for different scales of design. Technically, the production function is, then, said to be linear and homogeneous. Under these circumstances, it makes sense to speak of optimal design ratios. For typical circumstances, where $Q = 0.2$, the optimal design for still water such as a lake is thus about 4 hp per displacement ton. Where engines are especially expensive compared to the costs of construction of the barges, as they might be in less-developed areas distant from manufacturing plants, so that for $Q = 0.1$ or less, the optimal design ratio could easily be as low as 2 hp/ton.

Effect of Stream Current

Stream currents proportionately increase or decrease the velocity of the flotilla relative to the ground, depending on whether one is going upstream or downstream [Eq.(6.5)]. What is more significant is that the average velocity for a round trip, $\bar{S}$, decreases nonlinearly with increasing stream velocity, S_W. By simple calculation we can obtain:

$$\bar{S} = S^* - \frac{S_W{}^2}{S^*} = S^*\left[1 - \left(\frac{S_W}{S^*}\right)^2\right] \qquad (6.10)$$

Typical results are shown in Figure 6.4. This implies that the horsepower needed to maintain a constant average velocity, and thus output, for a given

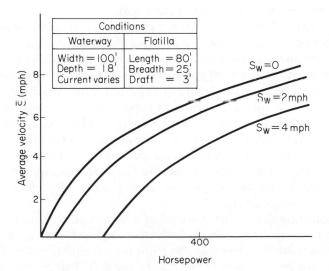

Figure 6.4 The effect of a current is to lower average round trip speed and output nonlinearly.

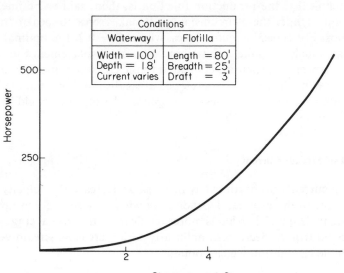

Conditions	
Waterway	Flotilla
Width = 100' Depth = 18' Current varies	Length = 80' Breadth = 25' Draft = 3'

Figure 6.5 The horsepower required to maintain a constant output for water transport increases exponentially with the velocity of the stream current.

barge must increase nonlinearly and more than proportionately than the increase in stream current velocity (Figure 6.5).

The primary effect of the existence of a stream current is a distortion of the production function. Besides lowering the total product that can be obtained from any combination of inputs, it has the effect of causing the marginal products of some inputs to become negative. It may now easily be technically inefficient to add more of any input. These results can be seen in Figure 6.6, which shows total product curves, that is, cross-sectional cuts through the production function, for both the cases where the stream current does and does not exist. The distortion of the production function is especially evident where the total product curves for flotillas of different deck areas cross, for the case of a stream current. For the conditions shown in Figure 6.6, this occurs at around 600 hp for $S_w = 6$ mph. This effect does not occur when the waterway is still.

The distortion due to the existence of a stream current transforms the production function from a linear, homogeneous function to a linear, non-homogeneous function. The expansion paths for this situation are still linear, but they do not pass through the origin (Figure 6.7). This means that, even given that the relative prices of the inputs are fixed and linear in quantity, the ratio of the inputs for the optimal design will change as we change the

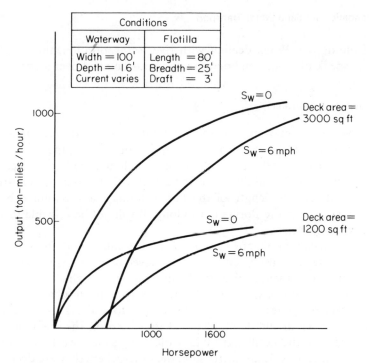

Figure 6.6 The production function is distorted and the total product curves lowered by a stream current.

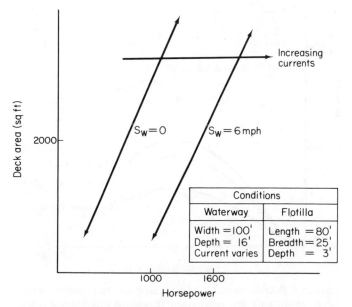

Figure 6.7 The expansion path and the optimal design for water transport is shifted by stream currents.

scale of the design. When dealing with river transportation, thus, we may no longer speak of an optimal design configuration valid over a range.

Waterway Constraints

The waterway may impose several limitations on the design of the barge flotilla. For instance, in order to allow for passing, the design must not exceed half the width of the channel. Nor should the barge exceed the depth, of course. More subtly, but often a factor that is the first constraint to be encountered, the length of the barge flotilla may be limited by the bends of the river: if the flotilla is too long, it will not be possible to navigate safely along the channel.

When a maximum length or breadth of a flotilla is specified, increases in barge deck area can ultimately only be achieved by increases in one dimension. The output is, then, subject to the diminishing marginal returns which have been shown to exist. Consequently, the design of the flotilla will, beyond a critical size, be subject to decreasing returns to scale. This critical size depends on the width and depth of the channel and on the maximum permissible length of the flotilla. It divides the design into two halves, one in which the returns to scale are increasing, the other where they are decreasing

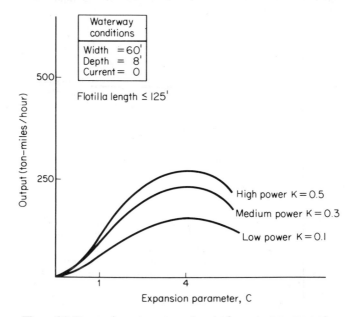

Figure 6.8 Decreasing returns to scale exist for water transport when barge dimensions are constrained (design is $C[K$ (horsepower), barge deck area]).

(Figure 6.8). This result will always hold, for as the breadth of the flotilla approaches the width of the channel, or as its draft approaches the depth of the channel, resistance increases without bound. Further, as the horse-power increases relative to a particular cross section of the channel, the effec-tive push decreases because the flow of water to the propeller screws of the towboat is restricted. Also, the resistance of the flotilla increases because an extreme drawing of water from under the barges causes them to "squat" deeper into the waterway. On smaller waterways, the onset of decreasing returns would probably be a very real operating constraint.

A constraint on the maximum length of a flotilla leads to the existence of dominated designs (Figure 6.9). This is a natural consequence of the de-creasing returns to scale that can exist for this situation. The production function now encloses a convex feasible region over the significant range of larger designs. Therefore, as a result of this constraint there are now design configurations which maximize absolute output from a flotilla. These did not exist for the previous situations; a bigger design could always carry more.

The constraints also distort the production function and, therefore, alter the expansion paths. This time, however, the expansion path is rotated clockwise instead of, as where a stream current exists, being translated later-

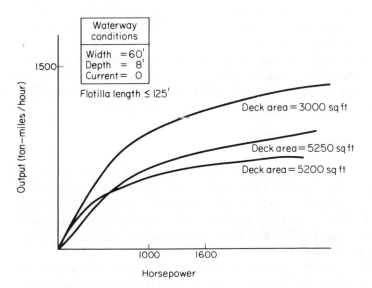

Figure 6.9 Dominated designs exist where the dimensions of the flotillas are limited.

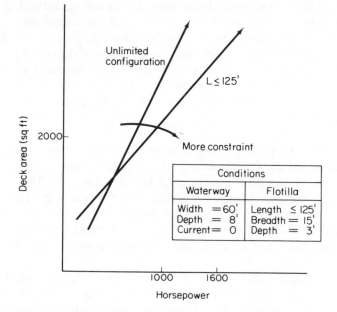

Figure 6.10 The expansion ratio and the optimal design for water transport is rotated by length limitations.

ally relative to those that exist for unconstrained designs for still water (Figure 6.10). That is, the existence of constraints promotes designs that have higher ratios of horsepower per square foot of deck area and which are, therefore, higher powered.

Again in this situation, the production function is no longer linear and homogeneous. Although remaining linear to all appearances, it is now nonhomogeneous. As before, this means that the ratio of inputs in the optimal design varies according to the scale of the design.

River Transport

For river transport, the designer must usually reckon both with currents and with limitations on the length of the flotilla. The production function appropriate to this environment is, essentially, produced by the combination of the effects described previously. There will be decreasing returns to scale for the larger designs and there will be designs that can maximize output for specified conditions.

The expansion path for river transport will be both rotated and translated with respect to the expansion paths that are appropriate to the unconstrained situations for still water. A typical result is shown in Figure 6.11. For this particular case, the expansion path does appear to pass through the

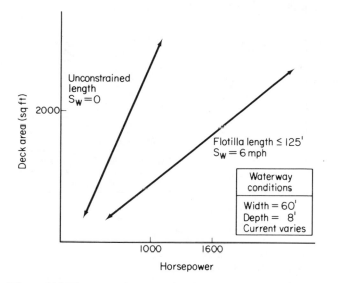

Figure 6.11 The expansion path for river transport will be lower than for still water, and may be homogeneous.

origin so that the production function would be homogeneous. This is, indeed, possible for river transport since, in this regard, the effects of stream current and of the limitations on the dimensions of the flotilla tend to cancel each other out. In any event, however, the expansion path is lower: the optimal design for river transport requires considerably more horsepower than a lake flotilla of equivalent size. This is expected.

Tradeoffs

Some tradeoffs between different levels of service, such as speed and cost, can be generated explicitly using the production function and previous results. For rivers and other situations where there is a maximum output which can be produced, it is possible to calculate, for specified costs, the unit cost of output for different designs. In particular, this can be compared to different levels of horsepower, which, in turn, correspond to different speeds. This is shown in Figure 6.12, from which we can observe the asymmetry in the cost function: it is costly to be underpowered, but the extra expense of having more power than that necessary to minimize unit costs of transport is not great. The cost of extra horsepower is all the less when one recognizes that speed may be desirable for its own sake, so that it is even worthwhile, from the point of view of maximizing total consumer satisfaction, to be overpowered.

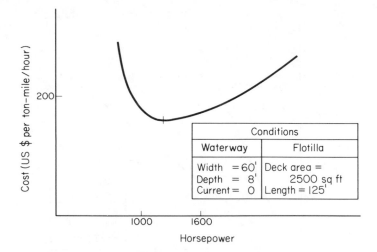

Figure 6.12 Typical tradeoffs between least cost of transport and horsepower and thus speed.

VALIDATION OF MODEL

As already suggested, the validity of the production function as an accurate description of the physical relationship between inputs and outputs has already been demonstrated elsewhere. As described in detail by Howe (9), the validity of this model has been shown both through towtank tests and by comparison of actual operations on the Ohio, Mississippi, and Illinois Rivers.

We now ask to what degree the conclusions about optimal design of the barges conform to the designs that are actually used successfully. Is the model effective in identifying useful regions of design? Can we use this production function to define optimal designs, to screen out definitely poor designs, for regions where we have no experience? Can we, in short, use the model to provide reasonable guidelines for transport programs in less-developed regions?

The evidence suggests that we can. First, reanalyses of the operations on the Ohio, Mississippi, and Illinois Rivers indicate that the designs that the models would have recommended for these waterways were, indeed, close to the configurations that operators actually found to be most economical and used (7). Admittedly, the margins of error are quite large because there exists considerable uncertainty both as to the costs of the inputs some 20 years ago and the actual configuration of the rivers. Nonetheless, the evidence does support the model.

Further evidence of the usefulness of the model in defining optimal designs was obtained from field work in Colombia. Specifically, we were able to compare the operations of two companies on the Meta River of the Amazonic basin. The first, Navenal, began serious operations in this area around 1950 and used fairly large, slow vessels. The second company, Expreso Ganadero, came into service in the late 1960s, using much smaller, faster flotillas. Both companies are based at Puerto Lopez, a shipping center at the headwaters of the Meta River, and both operate over the same waterways. Relevant data on the companies appear in Table 6.1.

Table 6.1 Design Ratios for Navenal and Expreso Ganadero Vessels

Company	Vessels	Date of Vessels	Ratio hp/ton
Navenal	El Patiño	ca. 1950	$2\frac{1}{2}$
	Meta		4
Expreso	A	ca. 1970	$5\frac{1}{2}$
Ganadero	B		$7\frac{1}{2}$

The development of the Meta basin by 1973 has made it both easier and cheaper to acquire marine engines at the riverhead, since there is now a good road to Puerto Lopez, and has also made it much more expensive to operate barges as wages have risen. We may thus infer that the optimal designs for river transport for the area became much more capital intensive between 1950, when Navenal configured its fleet, and the late 1960s, when Expreso Ganadero started operations. We can expect that the best designs would now have higher ratios of horsepower to size, just as Expreso Ganadero does. And in effect we found that Expreso Ganadero is much more efficient. It has, to all intents, now eliminated Navenal as a competitor on the Meta, just as our model would suggest. This is tangible evidence that the model can generate valid guidelines for optimal transport on inland waterways.

CONCLUSIONS

The analysis of the production function for a transportation service leads to specific guidelines as to how one can best configure the supply of that service. This knowledge, together with an understanding of the nature of the demand for a particular situation, then enables one to choose the best single design for the circumstances. This analysis of the production function for inland water transport leads to the following specific conclusions:

1. Because of the existence of diminishing marginal returns in the production process, one can essentially be guaranteed that a single optimal configuration exists. In practical terms, this also implies that it will be much easier to find the optimal designs.

2. The existence of increasing returns to scale for conditions of zero stream current (lakes, oceans) implies that for these environments it is desirable to build as large as possible and, when demand is expanding, to overbuild beyond immediate needs so as to take advantage of the increasing returns to scale.

3. For rivers where strong currents exist, increasing returns to scale do not exist. Further, the physical constraints of the waterway may place definite limitations on the optimal size of the barges and tows.

4. Fixed guidelines for the configuration of barge tows, in terms of hp/ton ratios, can be developed for lakes and other environments where currents are negligible. The ratios naturally depend on relative prices, and must be expected to be different between countries. These optimal ratios will, however, be essentially independent of the size of the tow.

5. For rivers, however, it will not be possible to develop optimal hp/ton ratios independent of size. The optimal amount of power with respect to tonnage depends strongly upon the physical dimensions of the tow.

6. Finally, it appears that total costs are relatively insensitive to additional amounts of horsepower beyond that required for the most economical design. This means that it may often be desirable to design relatively high-powered vessels, which would have operational advantages such as speed.

ACKNOWLEDGMENTS

The financial support of the U.S. Department of Transportation, Office of International Programs, and of the MIT Civil Engineering Systems Laboratory are thankfully acknowledged. Also, we are particularly grateful to Grace Finne for her encouragement and advice, and to our friends in Colombia, especially Jorge Acevedo, who was Director of Transportation at Planeacion at the time of this study, and Luis Mira, who helped us examine local operations in detail.

REFERENCES

1. APRON AND DUQUE, Ltda., and Frederic R. Harris Engineering Corp., 1968. *Mejoras a la Navegación y los Puertos del Rio Magdalena* (Improvements to the Navigation and Ports of the Magdalena River), report prepared for ADENAVI, the Colombian National Association of Shippers.
2. BARRERA, M., 1971. Private interview with John Hoffmeister, at the Union Industriale in Barranquilla, the largest barge construction company in Colombia, June.

3. CATERPILLAR CO., 1971. Technical literature regarding specifications and costs of marine engines supplied by R. Zimmerman, Peoria, Ill., Sept.

4. DE NEUFVILLE, R., AND STAFFORD, J. H., 1971. *Systems Analysis for Engineers and Managers*, New York: McGraw-Hill Book Company.

5. DE NEUFVILLE, R., AYALA, U., ACEVEDO, J., AND MIRA, L., 1972. *Role of Air Transport in Sparsely Developed Areas*, Cambridge, Mass.: MIT Report R72-49, Department of Civil Engineering, Sept.

6. DE NEUFVILLE, R., HOFFMEISTER J. F., III, and SHPILBERG, D., 1972. *Investment Strategies for Developing Areas: Models of Transport*, Cambridge, Mass.: MIT Report R72-48, Department of Civil Engineering, Sept.

7. HOFFMEISTER, J. F., III, 1972. *The Economics of Inland Waterway Transportation*, M.S. and C.E. Thesis, Department of Civil Engineering, MIT, Cambridge, Mass.

8. HOWE, C. W., 1965. Methods for Equipment Selection and Benefit Evaluation in Inland Waterway Transportation, *Water Resources Research*, Vol. 1, pp. 25-39.

9. HOWE, C. W., 1967. Mathematical Models of Barge Tow Performance, *Journal of Waterways and Harbors, ASCE*, Vol. 93, No. WW4, Nov., pp. 153-166.

10. HOWE, C. W., et al., 1964. *Inland Waterway Transportation*, Washington, D.C.: Resources for the Future (distributed by The Johns Hopkins Press, Baltimore) pp. 23-47.

11. MANNE, A. S., 1967. *Investments for Capacity Expansion*, Cambridge, Mass.: MIT Press.

7

Supply Curve for Thermal Efficiency*

Paul H. Cootner and George O. G. Löf

The construction of facilities for managing the quantity and quality of the nation's water resources continues to command large and growing investments. Yet those responsible for the planning and management of regional water resource systems are often inadequately equipped to estimate and forecast the benefits and costs associated with various existing and potential water uses. That this situation exists is no reflection on the ability of water resource planners, but rather reflects a comparative lack of research on these questions.

The result of this state of affairs has been that projections have usually been made in terms of "requirements" derived from the application of current ratios or coefficients, possibly adjusted to projected levels of future output. These "requirements" do not reflect in any systematic way the influence of economic factors such as the costs of inputs and the costs associated with waste disposal. Since in many cases waste loads and water uses can be very sensitive to costs, and thus are not requirements in any legitimate sense, it is important that methods be developed for forecasting these responses.

This analysis shows, by example, that it is possible to develop powerful and robust methods for predicting the potential nature of future developments. This can be done by carefully constructing models of the underlying

*Extensively adapted from Chapters 1 and 3 and Appendix B of *Water Demand for Steam Electric Generation*, Resources for the Future, Washington, D.C., 1965 (distributed by The Johns Hopkins Press, Baltimore).

technical processes as represented by the production function, and of the economic cost functions. Combinations of these two models into supply functions provide a simple approach to defining the nature of optimal designs and their economic potential. Exploration of these models, finally, provides a systematic perspective on the nature of future loads on a system.

THE PROBLEM

It is important, at the outset, to recognize some of the dimensions of water use. First, industry withdraws huge amounts of water from its sources. In the United States as a whole, about twice as much water is withdrawn by manufacturing industries as by municipalities. And over two-thirds of the industrial use is accounted for by steam power plants, which are the focus of this analysis. All but a negligible percentage of the water used in thermal electric plants is for cooling. On the order of half of the nation's use of water is, thus, used for the waste disposal of excess heat.

The amount of water used for cooling in the production of steam electric power depends upon the amount of heat that is thrown off or wasted by the generation of power. This is determined by two factors: the technical nature and parameters of the production process, and the economies of construction and operation of steam power plants. Both aspects need to be investigated in detail to understand the future demands for water for thermal wastes.

The key notion of thermal efficiency needs to be defined in this context: it is the percent of energy available in a fuel that is converted into electric power. It is the complement of the percent of energy in a fuel that is wasted and discharged, as heat, to some cooling medium, such as water. By definition, the higher the thermal efficiency of a power plant, the less water will be used as a coolant.

But one should not conclude that higher thermal efficiencies are necessarily desirable or less wasteful. Plants with high thermal efficiency may, indeed, not be economically efficient; what they save on fuel may be significantly counterbalanced by much higher costs for the construction of a more sophisticated plant. The level of thermal efficiency chosen for new plants is customarily the result of careful balancing of the costs and gains involved in the choice. As a result, the thermal efficiencies used in practice are essentially never as high as what might be technically possible.

That it does not always pay to save that extra amount of fuel and maximize thermal efficiency seems hardly debatable: the real question is, how high should we push thermal efficiency? The answer to this is provided, in great part, by the supply curve for thermal efficiency, which indicates just how much one has to pay to obtain higher thermal efficiency. The costs defined by this function can be compared with the savings incurred by higher efficiency, equivalent to its demand curve, so that a desirable level of design can be estimated.

The supply curve for thermal efficiency is of inestimable value in determining the likelihood of continued rapid advance in the level of thermal efficiency that has prevailed in the past, and thus in forecasting regional demands for cooling. To define this supply curve, the production function of the generation of steam power was examined in detail. Once the costs of attaining different kinds of design were allowed for, it was possible to define economically efficient plants to attain any desired level of thermal efficiency. We could then define the supply function for thermal efficiency.

THEORY

As suggested above, the optimal designs are based upon the consideration of two models: the technical process itself, which is represented by the production function, and the economic model of the costs of construction and operation. The nature of these arguments is examined in turn.

Concept of a Production Function

The production function is simply defined as a relationship embodying all technological knowledge regarding the production of the commodity involved. It describes all the "most efficient" technology known to the producer, where efficiency is defined without recourse to any economic information about prices. It is a composite of many different methods of production. For example, the production functions for electric power will, in general, be defined in part by steam, hydroelectric, and nuclear generation processes; each of these will represent technically efficient ways of utilizing inputs over a given range, but only one of them will be economically efficient for a given set of prices.

The achievement of higher thermal efficiency, H, depends upon temperature, T, and pressure, P. These three quantities are, indeed, related through well-known thermodynamic considerations (6). Symbolically, the form of the production function for thermal efficiency is

$$H = h(T, P) \qquad (7.1)$$

where the levels of temperature and pressure may be considered the inputs into the process.

It is convenient to portray the production relationships by means of isoquants, in this case of isoefficiency curves. For each level of efficiency, all the combinations of temperature and pressure were connected by a smooth curve. Reflecting the empirical law of diminishing marginal returns—that successive increments of an input are continuously less effective—it is to be expected that the curves will be convex to the origin.

Cost Model

The factors of production inherent in the formulation of Eq. (7.1) are necessarily quite abstract. There is certainly no market for the purchase of "pressure" or "temperature." We can, however, construct a schedule indicating the alternative levels of temperature and pressure which can be obtained with any given amount of capital, C. We can represent this as:

$$C = c(T, P) \qquad (7.2)$$

This cost function can be represented by isoquants just as the production function. These isocost curves, usually referred to as "transformation" curves, represent the various combinations of factors that will require the expenditure of a given amount of capital. Unlike the isoquants for efficiency, however, the transformation curves are usually expected to be concave to the origin, reflecting the increasing difficulty and cost of raising the level of one input relative to another.

Locus of Optimal Designs

Our objective is the construction of a supply schedule for thermal efficiency of the form:

$$H = s(C) \qquad (7.3)$$

representing the rising cost of increasing thermal efficiency in the most economical manner. The locus of optimal design for which this supply curve is determined follows from the maximization conditions: that the marginal productivity of each input per unit of cost be equal. Dealing with just the two inputs of temperature of pressure, these conditions can be derived from Eqs. (7.1) and (7.2):

$$\frac{\partial h/\partial P}{\partial c/\partial P} = \frac{\partial h/\partial T}{\partial c/\partial T} \qquad (7.4)$$

Equivalently,

$$\frac{\partial h/\partial P}{\partial h/\partial T} = \frac{\partial c/\partial P}{\partial c/\partial T} \qquad (7.5)$$

This is the same as saying that, for the optimal design, the isoefficiency and isocost transformation curves ought to have the same slope:

$$\left(\frac{\partial P}{\partial T}\right)_h = \left(\frac{\partial P}{\partial T}\right)_c \tag{7.6}$$

and, thus, be tangent (5, 9, 16).

This formulation, in terms of capital, sidesteps the matter of operating revenues and costs, always a tricky issue in calculating a supply function. There are almost certainly decisions to be made in such an analysis about whether certain items should perhaps be added to costs, shifting the supply curve upward, or rather deducted from revenues, shifting the demand function downward. How we distinguish between increases in costs and decreases in revenues will affect the costs of increasing thermal efficiency, but they will have a corresponding effect on the demand (benefit) curve for thermal efficiency with which we will compare our supply curve. Since any consistent choice of procedure in these cases does not effect the final result, we are free to choose the most convenient approach.

Implications for Changing Costs and Technology

We can deduce from Eq. (7.4) that as the relative costs of securing any level of inputs change, so that $\partial c/\partial P$ and $\partial c/\partial T$ change relative to each other, the levels of $\partial h/\partial P$ and $\partial h/\partial T$ must also change to maintain optimality. In short, price changes entail a shift of the locus of optimal design along the production function. This change in the point at which production takes place does not imply a change in the production function, even if a new method of production is used. As long as all the possibilities at our disposal, all represented in Eq. (7.1), remain the same, the production function does not change, even though the particular technique does.

This does not mean that all changes in the technique of production are due solely to economic factors. Indeed, an invention might take place that does alter the production function. But the distinctive feature of technological change is not the shift in process, as from steam to nuclear generation of power, but the change in known techniques for producing an output. Change in the use of techniques or methods of production over time is not sufficient evidence to prove that technological progress has been made.

The distinctions between the movements along, and shifts of, the production function are not idle ones, since they have vastly different implications for any project concerned with forecasting the future path of technology. In the one case the determining elements of the study must be economic, the evaluation of the effect of changing input prices upon the method of production, and in the other it must be technological, the likelihood of certain improvements in technique. As in most cases in the real world, the electric industry's growth has been typified by both types of changes, but the ability to distinguish between these two elements should improve our ability to analyze the patterns of change.

To the extent that changing techniques of production merely reflect changing economic incentives, the necessity of formally predicting the rate of invention and innovation is eliminated. In such circumstances, the projected movements of the principal economic factors will be the principal determinants of economic change. If, on the other hand, the rate of change of technological knowledge is expected to be rapid, it may be possible to neglect the effect of changes in the economic parameters without introducing any important error. It is obvious that the real world is not as polar as all this, and that we are likely to find any number of possible intermediate positions more tenable than either of these extremes. Under these circumstances, the problem reduces to measuring the kinds of technical change that might be expected both from moving along the curve, on the one hand, and shifting the production function on the other. Although it is very difficult to predict the course of technological change, we can get an idea of the probable importance of innovation by sensitivity analyses investigating the nature of the most likely innovations and their effects on production techniques. Similarly, we can see what the effect of certain changes in the relative prices of factor and product inputs would be.

ANALYSIS

Production Function

Even under the most ideal circumstances, attempts to estimate the shape of a production function, particularly if it is to be made in detail, are fraught with difficulty. Some kinds of problems arise because the number of important inputs is usually very large, or because in most industrial processes the number of outputs is also substantial. Others arise because the technical secrecy imposed by economic competition inhibits the availability of data, particularly that portion which describes the frontier of knowledge in a field. In some segments of economic activity, the difficulties are aggravated by localized or rapidly changing techniques of production.

The situation in the electric generating industry is more favorable for estimating purposes than in most industries. Two inputs, capital and fuel, account for over 90% of total annual generating costs. The level of these costs is, in turn, determined by only a handful of factors. The industry, being a regulated public utility, is considerably less secretive than most about its technology, although its status as a public utility makes it more sensitive about certain other kinds of information. Furthermore, although there are regional differences in generating techniques used and, in a few companies, technical procedures which distinctively reflect the influence of a single dominant executive, the industry is remarkably standardized.

Despite the relative availability of information, the complexities of investigating the technological procedures are such as to prevent a precise mathematical formulation. The industry data are not usually organized in a form useful to the analyst; moreover, little information is provided on the costs of equipment that is over the horizon of present use. As a result, some of the material that follows is semi-intuitive in nature. If we are to draw rational conclusions from this study, the questions raised, however, must be answered. In so doing the authors draw on the best information available to them.

The first step in the definition of the production function was a graphical portrayal of the engineering production relationships. Since the level of thermal efficiency of a steam plant is an intricate function of a large number of engineering parameters, some simplification of the problem was necessary. The simplifying approach used was that commonly followed by industry engineers. The level of the less flexible parameters was taken as given, thus permitting the study to focus on the main factors affecting the level of efficiency, the temperature and pressure of the steam in the plant.

By use of temperature and pressure as the central factor inputs in the "production" of thermal efficiency, the production relationships were portrayed by a family of isoefficiency curves, each representing a different level of efficiency (Figures 7.1 and 7.2). These curves, calculated from engineering data and published estimates of industry engineers (1–3, 12–14), represent, to the best of our knowledge, the first formulation of the data in the form customarily used by economists, but the data themselves are the ones generally used in the industry. The curves should be considered as composites or rough averages of current plant performance, because there is plant-to-plant variation by virtue of such differences as types of fuel, cooling water temperatures, and the extent of heat economizing measures. For plants of more advanced design, especially those using double reheating of steam and boiler operation at supercritical conditions, additional curves must be added to the graphic portrayal, but the principle of the approach is not affected.

As an aside, it should be pointed out that although the curves in Figures 7.1 and 7.2 do not appear to be convex to the origin, they are. The distortion arises from the use of semilogarithmic scales, which were required by the relative magnitude of the variations of temperature and pressure.

Cost Model

To represent the data about the costs involved in attaining different levels of temperature and pressure, we developed a set of curves similar to

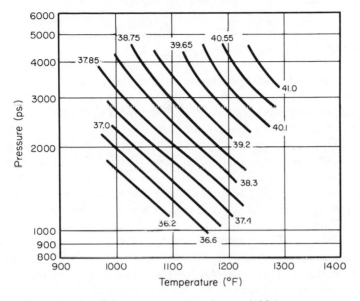

Figure 7.1 Isoefficiency curves; one reheat to initial temperature.

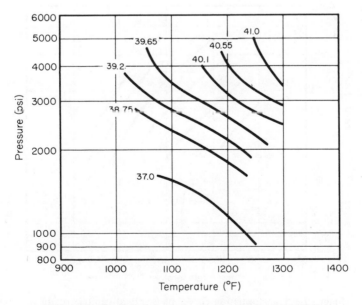

Figure 7.2 Isoefficiency curves; double reheat to 1,050° F.

those representing the engineering relationships (8, 12, 15, 17). Examples of these transformation curves are shown in Figure 7.3.

SUPPLY CURVE

The technical and cost considerations are united in Figure 7.4. The dashed line, which connects all the points of tangency between the isoefficiency and transformation curves, defines the most economical and optimal designs to achieve various levels of thermal efficiency.

The supply curve itself is obtained by plotting the marginal costs of achieving each unit rise in thermal efficiency (Figure 7.5). It is to be noted that the resultant supply curve is not only upward sloping, as might be expected if there are diminishing returns to scale, but rises sharply in the vicinity of a thermal efficiency of 40%. The cost of reaching a thermal efficiency of 42%, for example, is 10 times higher than the cost of reaching a thermal efficiency of 39%. Since a thermal efficiency of 40% is roughly that of the best new plants being installed, the sharp increase in costs in this area becomes important in determining prospects for improvements in thermal efficiency and, consequently, for reductions in thermal pollution.

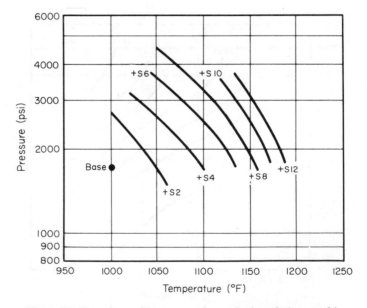

Figure 7.3 Transformation curves of marginal capital cost of increasing temperatures and pressure (measured from a base of 1,800 psi, 1,000°F).

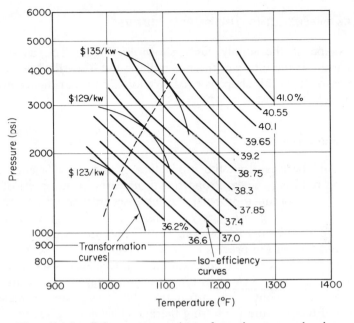

Figure 7.4 Isoefficiency curves and transformation curves, showing points of maximum efficiency.

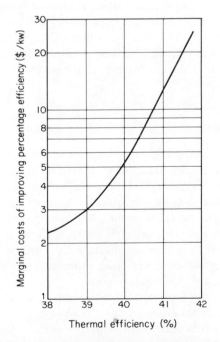

Figure 7.5 The supply curve of thermal efficiency in 1957.

FACTORS INFLUENCING THE SUPPLY CURVE

As the shape of the supply curve shown in Figure 7.5 has significant consequences for the future of thermal pollution, it is worthwhile examining the degree to which this shape is fixed. We look in turn at the technical and economic relationships.

Technical Factors

It should be clear that the production of thermal efficiency exhibits diminishing returns to scale. Proportional increases in the levels of inputs, P and T, do not produce proportional increases in thermal efficiency. For instance as can be seen in Figure 7.1 increasing P and T by a quarter from 2000 psi and 1000°F to 2500 psi and 1250°F only increases the theoretical thermal efficiency of the process from 36.6 to 40% or by less than a tenth. And, indeed, the maximum theoretical thermal efficiency is less than 100%, whereas there are no known upper limits to temperature and pressure.

There is also a continuing shift in the relative usefulness of increases in pressure and temperature for raising thermal efficiency. To see this we may refer to Figure 7.1. Let us focus on any base, say the design point where $P = 1800$ psi and $T = 1000$°F. It is then easy to see that increases in temperature are much more conducive to raising thermal efficiency than equivalent changes in pressure. A 20% increase in pressure from the base raises the thermal efficiency from only 36.2 to 37%, while a 20% increase in temperature raises thermal efficiency to 38.3%. The nature of the production function thus encourages one to design for higher temperatures in order to achieve higher thermal efficiency.

Economic Factors

The cost of achieving higher pressures and temperatures also becomes more and more expensive. Because of the technical characteristics of the metals used in the design of high-pressure boilers, the costs of providing higher temperatures rise especially quickly. As can be seen from Figure 7.3, marginal temperature increases in the 1200°F range, which are necessary if the thermal efficiency of steam electric plants is to rise above 40%, are substantially more costly than increases in the 1050°F range. In the case of pressure increases, the rising costs are much more modest—only slightly greater than would be indicated by constant costs.

The reason for the high cost of temperature, which is by no means at the frontier of the capabilities of current equipment, lies in the universal structural problems inherent in the combination of high temperatures and pressures. Because ordinary alloy steels tend to deform ("creep") under these conditions, the metals that prove satisfactory for power-generating equipment at temperatures below 1200°F are not adequate for the higher temperatures. To satisfy the structural requirements of higher temperatures, special high-temperature alloys are needed which require substantial additions of expensive metals to steel for the purpose of resisting creep and corrosion (1, 4, 7, 11, 18). These metals, such as nickel, may comprise 25 to 50% of the high-temperature alloys, costing up to 10 times as much as the ordinary steels. The alloys required for high temperatures will therefore be much more costly than the metals used below the high-temperature barrier. This cost disadvantage is further amplified by the greater difficulty in making the alloys: their very strength makes them hard to fabricate.

The increased costs of providing higher temperatures in particular is illustrated by the shape of the transformation curves, which are pulled in along the temperature axis (Figure 7.6). The net result is that the tendency to increase thermal efficiency by going to higher temperatures, which appears to be technically sound, is effectively blocked by the economics of the situation.

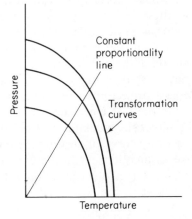

Figure 7.6 Transformation curves showing more rapid rise in cost of increasing temperature than of pressure.

Implications

Both the technical and economic factors combine to increase the slope of the supply curve of thermal efficiency at about 40%. The properties of the underlying production function and the economies of alloy production all point to a sharply rising supply curve. Short of a major technological change in the future, these forces will become increasingly strong and effectively limit the thermal efficiency of steam electric plants to around 40%.

SENSITIVITY ANALYSIS: POSSIBLE EFFECTS
OF TECHNOLOGICAL CHANGE

Although the implications of the current supply curve for thermal efficiency do not bode well for reductions in heat pollution, its shape is not necessarily fixed. We therefore explore the possible results of improvements in the design of existing power plants, and of the introduction of newer methods of production: gas turbines and atomic power.

Conventional Steam Power

There is, of course, some possibility for reducing costs in the future. As the high-temperature alloys become more widely used, the costs of making and fabricating them will doubtless be reduced. Based on an elimination of most of the extra compounding and fabricating costs and a reduction of 25% in the costs of the alloying elements, we have estimated a second supply curve for conditions we would expect to prevail in the year 1980 (curve B in Figure 7.7). Although these conditions cannot be predicted with any great accuracy, they do have a basis in the expected future costs of production and the changed conditions of industrial organization, now coming to fruition, induced by recent expansion in the nickel industry.

Since technology has a way of changing in unforeseeable ways, it cannot be expected that these cost projections will prove to be faultless. In order that our supply curve for thermal efficiency will not be invalidated completely by deviations in one of the assumptions from future reality, we have added to Figure 7.7 two other curves indicating the costs of improving efficiency under alternative assumptions. If in the coming years scientific advances and diligent research serve to speed up the rate of cost reduction, a lower cost curve, such as curve C, can be utilized. It there is no change at all, curve A can be used. It will be noted, however, that in all the alternatives, the sharp increase in costs around current best levels remains. We are still considering that the cost of attaining a thermal efficiency of 42% is an order of magnitude higher than that of realizing only 36 or 37%.

Moreover, we find that improvements in thermal efficiency, at least since about 1955 or so, have been more due to economic factors such as the cost of fuel, than to basically new technology. Indeed, the technology already exists that would permit us to achieve higher thermal efficiencies. The theory and the metals with which to construct more efficient plants than those in existence at any time have been available over most of the twentieth century. The reason plants with higher thermal efficiency were not built in the past, and are not being built now, is not that technology did not exist, but that they have been economically inefficient.

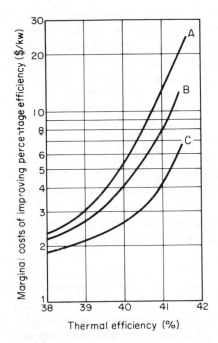

Figure 7.7 The supply curve of thermal efficiency in 1980 (1957 dollars). Curve *A* assumes no change in the supply of thermal efficiency. Curve *B* implies a 3 % annual improvement in real costs of supplying advanced plants. Curve *C* optimistically assumes both an improvement in technology plus a secular decline in factor prices and projects, an annual reduction in the real cost of 5 % for "frontier" plants and of 1 % for more standard plants.

This is not to say that there has been no real technological change in the electric utility industry. To mention only a few advances, improved fabricating techniques have been found, knowledge of metallurgy has improved, and the reliability of boilers has been increased. Nevertheless, our central point remains that changes in the costs of fuel, metals, and labor, and increases in the quantity of power needed have been even more important in determining optimum thermal efficiency. This suggests that the element of unpredictability in technological change is probably less important in determining future costs of thermal efficiency supply than might be indicated by superficial study of the industry.

Gas Turbine Power

The gas turbine, like the steam turbine, is a device for transforming thermal energy into electricity. Unlike the steam turbine, however, it is powered by combustion gases directly, instead of first transferring the combustion heat to steam. The primary advantage of the gas turbine in producing commercial electricity is its lower capital cost where only a few megawatts are required. It is therefore useful in very small power plants, or in large plants requiring operation only a small portion of the total time. It has become important in handling peak loads of utility systems, especially in those regions where the capital costs of alternative peaking power is highest.

Offsetting these advantages are two disadvantages, one mechanical and the other thermal. From a mechanical standpoint, the problem is turbine blade erosion. Since the powering force is the heated combustion products, this mixture is likely to contain any noncombustible impurities that were originally present in the fuel. Although this is not an important problem if the fuel used is natural gas or, to a lesser extent, residual oil, the mineral impurities in coal have prevented the use of gas turbines in coal-fired stations. Oxidation and other high-temperature damage to blades and other metal components also may occur and adversely affect the life of the equipment.

The thermal problem arises from the fact that combustion gases, unlike steam, do not condense to a liquid when cooled. As a result, the pressure at the turbine exit must be above atmospheric pressure, and the exhaust temperature of the gas turbine is generally higher than it would be in the case of a steam turbine. Accordingly, for any given inlet temperature the thermal efficiency of the gas turbine will be lower. Offsetting this drawback, however, is the considerably higher heat input temperature possible in the gas turbine and the increase in thermal efficiency due to this factor. Further, the gas turbine and steam turbine can be used in conjunction: the combustion gases give up part of their energy in the gas turbine before being used to generate steam for a conventional power plant. The thermal efficiency for such a plant is better than that of a conventional plant because of a greater total useful temperature range in these two heat engines.

The implications of the potential adoption of the gas turbine are especially important for water use. Since there is no water cooling of the gas turbine exhaust, any electricity produced by the use of gas turbines will not involve water. In effect, the adoption of this innovation would substitute air cooling for water cooling in the electric power industry. But, perhaps unfortunately, it is not likely that the gas turbine will significantly affect the production of electricity in the near future. Based on the engineering and economic data we have examined, it appears that whatever gas turbine capacity is installed on large systems will be used to supply peaking power and will, as a result, be operated at a low load factor. It does not seem, therefore, that the gas turbine will lead to significant reductions in thermal pollution.

Atomic Power

The potentialities for atomic power development pose less of a problem for water projections. There are two reasons for this. First, the "conventional" atomic power station, which utilizes a steam turbine to transform heat energy into electricity, introduces no radical changes into the nature of steam electric water use, and second, the projected costs of atomic power over the next two decades will prevent it from becoming a major factor in water use on economic considerations alone.

Despite substantial differences in design among the various existing and projected atomic power stations, they all utilize the same basic technique for the production of electricity. Heat from the atomic reactor is transferred to steam at high temperatures and pressures and the steam is, in turn, used to drive the turbine–generator. Conceptually, the sole differences between atomic and organic fuel stations are that the furnace is replaced by the atomic reactor and the transfer of heat to steam usually takes place with an intermediate stage of heat exchange—hence the terms gas-cooled or water-cooled reactors.

Once the heat is transferred to steam the two processes are the same. The steam expands through the turbine, and economic considerations require that it be condensed. Given a reactor and an organic fuel cycle of equivalent thermal efficiency, the cooling demand of each will be identical. There are no major differences in the nature of water use.

But while atomic and organic fuel cycles of equal efficiency would have the same cooling demands, it is unlikely that atomic power plants will be operated at thermal efficiencies as high as those prevailing in modern fuel-fired plants because of problems associated with high pressure and temperature inside a nuclear reactor. For an atomic cycle to be economically preferable to the optimum cycle for conventional fuel in a given location, it must offer lower fuel costs. This is because a nuclear reactor is more complex and, with its costly materials and radiation protection requirements, is fundamentally more expensive than the conventional boiler and furnace, while the costs of the generators are identical and of the turbines not widely different. Thus, if the capital costs of the atomic plant must be greater than those for the conventional plant, it can be more economic only if its fuel costs are lower. Lower fuel costs have, however, been shown to discourage increases in thermal efficiency (6, 10). Under these conditions, it is to be expected that atomic plants would have lower thermal efficiency and hence cause greater thermal pollution than conventional plants.

CONCLUSIONS

Based upon a detailed modeling of the production function for the generation of electric power, and of the prevailing and expected costs, it has been possible to develop a supply curve for thermal efficiency. The underlying technical and economic factors imply that this supply curve rises very sharply in the vicinity of a thermal efficiency of about 40%, which is about the level of design of current plants. This shape indicates that the costs of thermal efficiency above 40% will be about an order of magnitude higher than current costs. This conclusion still holds even for optimistic assumptions about technologic improvements.

The exploration of the simple technical production function and economic cost models leads to important conclusions about the future of thermal pollution. First, we may see that, despite available technology, thermal efficiency will not be raised appreciably because it would be economically inefficient to do so. Second, we can consequently expect that the total amount of thermal pollution will increase essentially as fast as the consumption of electric power. Finally, since it is impractical to increase thermal efficiency, it appears necessary to deal with thermal pollution by other means.

ACKNOWLEDGMENTS

Special appreciation is expressed to Allen V. Kneese and to Irving K. Fox, who headed the water resources program at Resources for the Future during the course of this study, for their suggestions, encouragement, and contributions. The authors are also indebted to the many persons in the electric power industry and elsewhere, too numerous to mention in detail here, who helped in this analysis.

REFERENCES

1. BABCOCK & WILCOX COMPANY, 1955. *Steam: Its Generation and Use*, New York.
2. BARTELS, J., 1954. Supercritical Pressure Steam Power Cycles, *Proceedings of the American Power Conference*, Vol. 16.
3. BARTLETT, R. L., 1958. *Steam Turbine Performance and Economics*, New York: McGraw-Hill Book Company.
4. BETHLEHEM STEEL COMPANY, undated. Quick Facts About Alloy Steels, Bethlehem, Pa.
5. CARLSON, S., 1956. *Study on the Pure Theory of Production*, New York: Augustus M. Kelly.
6. COOTNER, P. H., AND LÖF, G. O. G., 1965. *Water Demand for Steam Electric Generation*, App. A, Washington, D.C.: Resources for the Future (distributed by The Johns Hopkins Press, Baltimore).
7. CURRAN, R. M., AND RANKIN, A. W., 1956. Application of High Temperature Metals to Modern Large Steam Turbines, *Proceedings of the American Power Conference*, Vol. 18.
8. DAVIES, R. W., AND CREEL, G. C., 1963. Economics of the Selection of 2500-psig Double Reheat for a 300-MW Unit, *Proceedings of the American Power Conference*, Vol. 25.
9. DE NEUFVILLE, R., AND STAFFORD, J. H., 1971. *Systems Analysis for Engineers and Managers*, New York: McGraw-Hill Book Company.
10. *Electrical World*, July 20, 1959.
11. FAIRCHILD, F. P., 1956. Eight Years of Experience with Austenitic-Steel Piping Materials at Elevated Steam Conditions, paper No. 56-A-181 presented at the Annual Meeting of the American Society of Mechanical Engineers.

12. FLEISCHMAN, J. J., GIBELING, A. H., MERGY, A., AND MILLER, E. R., 1956. A Comparative Study of a Large Steam Turbine Application for Supercritical and Conventional Pressures, *Proceedings of the American Power Conference*, Vol. 18.

13. FRANCK, C. C., Sr., 1957. Superpressure Steam Turbines, *Proceedings of the American Power Conference*, Vol. 19.

14. KEENAN, J. H., AND KEYES, F. G., 1936. *Thermodynamic Properties of Steam*, New York: John Wiley & Sons, Inc.

15. PETERSON, H. J., 1963. The Economics of 2400 psig Versus 3500 psig for Large Capacity Units, *Proceedings of the American Power Conference*, Vol. 25.

16. SAMUELSON, P. 1948. *Economics: An Introductory Analysis*, New York: McGraw-Hill Book Company.

17. SEELYE, H. P., AND BROWN, W. W., 1954. The Economy of Large Generating Units, *Proceedings of the American Power Conference*, Vol. 16.

18. WILSON, R. M., FOLEY, R., AND FOLEY, F. B., 1956. Metallurgical Requirements of Metals for Steam Service Above 1000°F, *Proceedings of the American Power Conference*, Vol. 18.

8

Sources of Productivity Increases in the U.S. Passenger Airline Industry*

NEIL B. MURPHY

The purpose of this chapter is to estimate cost functions for the United States passenger airline industry in order to determine the sources of recent productivity increases. This productivity has indeed been rising rapidly: as measured by the output per employee, for example, its average rate of increase from 1961 to 1965 was nearly 11 % compared to 3.2 % annual increase for the entire economy (4). A particular effort was made to determine if these increases in productivity were due to increasing returns to scale as the airlines became larger, as this would have important implications for optimal size of the airlines and, consequently, for a national policy on airline mergers. A special effort was also made to determine the effect on costs of new technology, in particular of the introduction of jet aircraft.

THE MODEL

The study of increasing returns or economies of scale and of the effect of technical change concerns the essential nature of the production process. If we could, we would really like to be able to model the production function, which describes all the possible levels of output that can be achieved by a

*Extensively adapted from Sources of Productivity Increases in the U.S. Passenger Airline Industry, *Transportation Science*, Vol. 3, No. 3, August, 1969, pp. 323–328, and the comments by J. P. Mayberry and the author's reply, *Transportation Science*, Vol. 6, No. 1, Feb., 1972, pp. 94–97.

technically efficient use of the inputs. If we could describe the production function completely, we could readily describe its specific characteristics and, for example, whether or where it has increasing returns to scale. Unfortunately, direct observations of the variables entering the production function are rarely available.

An alternative method of studying the nature of the production function, and of economies of scale in particular, is to estimate the parameters of a cost function. If one assumes that firms attempt to minimize the cost of achieving any particular level of output, one can use observations on their performance and output to define a specific portion of the production function: the cost function. This is the locus of economically efficient designs for prevailing prices. Technically, it is a reduced form of the production function and uniquely related to it. As it happens, the cost function, being related to current economic conditions, describes the relevant portion of the production function and is sufficient to give us a reasonable estimate of the economies of scale that might exist.

To suggest how this works, assume that we have a particular production function, say a Cobb–Douglas production function of the form

$$Z = \alpha X_1^{\beta} X_2^{\alpha} X_3^{\partial} \tag{8.1}$$

where Z is output and X_1, X_2, and X_3 are the usual three factor inputs of labor, capital, and materials. Assuming that cost minimization holds, Eq. (8.1) can be transformed into a cost function of the form:

$$C = hZ^{1/r} P_1^{\beta/r} P_2^{\alpha/r} P_3^{\partial/r} \tag{8.2}$$

where P_1, P_2 and P_3 are the prices of the input factors. This cost function, being linear in its logarithms, will be easy to estimate by ordinary least squares. As it happens, the parameter r denotes whether economies of scale exist: if $r > 1$, there are increasing returns to scale. [See the discussions of this by Shephard (3) and Walters (6), as well as by Nerlove (2) and in Chapter 9 of this text.]

For this problem, a cost function was specified for the production of air travel, and the parameters are estimated. The dependent variable of the cost function for this analysis was taken as total direct operating costs (DOC). This quantity accounts, in general, for only about half of an airline's operating costs, the rest being consumed in the "indirect" operating costs, including reservations and sales efforts, advertising, and so on. The DOC may, however, be taken as a reasonable estimate of the actual costs of operating the fleet of aircraft, and we focus on that aspect of the airline.

The specification of output is somewhat difficult since there is no physical unit produced by an airline. In this chapter, output is defined as total available seat-miles flown, as opposed to some other measure, say the number of passenger miles carried. That is, it is assumed that efficiency in production should be evaluated independently of whether the service is consumed or not. However, not all seat-miles are homogeneous since the length of the trip on which the seat-miles are produced may also affect costs. That is, certain components of cost, such as ground handling, are related to the number of trips rather than seat-miles. With seat-miles held constant, an increase in length of each trip, its stage length, is accompanied by a decrease in the number of flights. Therefore, average stage length is included as an independent variable, modifying the output, and a negative coefficient is expected. Average fleet size of the particular aircraft was, likewise, also entered as an independent variable. It was thought that, with total available seat-miles held constant, this variable would account for the variations in capacity utilization due to differences in route structure and scheduling opportunities.

It is also necessary to specify the prices of the inputs as independent variables. In general, there are three factors of production: labor, capital, and materials. In this chapter it is assumed that unionization tends to standardize wage rates across airlines and that capital equipment prices are identical for all airlines. As indicated elsewhere, for example in Chapter 9, this implies that the factor prices for these variables can be implicitly included in the constant terms and, thus, excluded from explicit consideration in the cost model (2). To account for differences in wage and capital costs that might occur over time, as by inflation, and specifically between the two years for which we had cross-sectional data, a dummy variable representing the shift in time within the sample is included as an approximation of the shift in the wage rate and the price of capital. As prices for materials, can be assumed to vary significantly across the country, their price has to be accounted for explicitly. To do this, a variable was included representing the price of fuel, which is the most important material consumed by an airline.

Finally, the level of technology, or type of aircraft, influences cost. For this reason, a set of dummy variables is specified to test the impacts of shifts by an airline from one type of aircraft to another, holding constant the influence of other variables. Four types of aircraft were relevant to the period under study. These were, proceeding from the earliest to the most advanced: piston, turbo-prop, turbo-jet, and turbo-fan.

In summary, the cost function describing the direct operating costs of the U.S. passenger airline industry was taken to be primarily determined by the total output (available seat-miles), output modifers (average stage length and fleet size), a fuel cost, inflation or other temporal effects, and type of aircraft. Specifically:

$$C = aM^b L^c F^d P^f T_1{}^{g_1} T_2{}^{g_2} T_3{}^{g_3} Y^h \qquad (8.3)$$

where C = total direct operating costs

M = total available seat-miles

L = average stage length

F = fleet size

P = cost of fuel per gallon

and where the following dummy variables are equal to 1 or 10 so that, after the logarithmic transformation which will make the cost function linear, they will equal 0 or 1:

$$Y = \begin{cases} 10 \text{ for } 1966 \\ 1 \text{ for } 1965 \end{cases}$$

$$T_1 = \begin{cases} 10 \text{ for turbo-prop aircraft} \\ 1 \text{ for all others} \end{cases}$$

$$T_2 = \begin{cases} 10 \text{ for turbo-jet aircraft} \\ 1 \text{ for all others} \end{cases}$$

$$T_3 = \begin{cases} 10 \text{ for turbo-fan aircraft} \\ 1 \text{ for all others} \end{cases}$$

As there is no dummy variable for piston aircraft, coefficients of the dummy variables for aircraft are interpreted as meaning the effect on cost of shifting to that aircraft from a piston or propeller aircraft, holding constant the influence of all other variables. As the higher subscripts on the aircraft dummy variables indicate the newer aircraft, it is expected that each of the coefficients, g_i, will be negative, and that $|g_3| > |g_2| > |g_1|$.

For the particular production function we have used, the value of the coefficient of the output, b, determines whether economies of scale exist. This is one of the attractive characteristics of the Cobb–Douglas production function. This can be shown by differentiating Eq. (8.3). Since it is multiplicative, we obtain

$$dC = \left(\frac{b}{M}\right) C \, dM \tag{8.4}$$

so that

$$\frac{dC/C}{dM/M} = b \tag{8.5}$$

and the coefficient of output is the elasticity of cost with respect to output. If it is less than unity, economies of scale are present; if it exceeds unity, diseconomies of scale exist; and if it is equal to unity, we have constant returns to scale.

ANALYSIS AND RESULTS

Data

The source of data for this analysis was the *Aircraft Operating Cost and Performance Report* of the U.S. Civil Aeronautics Board, a recurrent publication which tabulates all the basic costs on domestic airline operations (5). Data were taken for two years, 1965 and 1966, and were pooled to form a single cross section. Observations were by aircraft type, by manufacturer, and by airline. Four aircraft classifications, 11 domestic trunk airlines, and 13 local service carriers yielded 175 observations, since not every airline used aircraft of each type.

Statistical Analysis

Taking the logarithms of both sides of the cost function Eq. (8.3) yields a linear, additive expression for which the exponents of the cost function are the coefficients. These coefficients can then be estimated by ordinary least-squares analysis, once the observations are also logarithmically transformed. The results are shown in Table 8.1.

Table 8.1 Results of Statistical Estimation of Parameters of the Cost Function, Based on 1965 and 1966 Data

Variable	Estimated Exponent	Standard Errors in Estimate	t Statistic
M	1.023	0.027	37.883
L	−0.047	0.066	−0.707
F	0.068	0.042	1.612
P	0.051	0.019	2.658
Y	0.031	0.017	1.808
T_1	−0.051	0.027	−1.885
T_2	−0.208	0.040	−5.190
T_3	−0.246	0.036	−6.790
	$R^2 = 0.944$	$F = 352.6$	

The t statistics associated with the estimates in Table 8.1 can be interpreted in two ways. In a classical sense, we can say that they indicate whether a parameter is significantly different from zero. From this point of view, the regression coefficient for output is highly significant and the coefficient or elasticity of the price variable is also statistically significant. The coefficients for the output modifying variables had the expected signs but could not, with 90% confidence, be said to be statistically different from zero.

In a Bayesian sense, however, we might more simply say that, while the estimates of the other variables were not too tight, such that little confidence could be placed on their accuracy, the expected range on the estimate of the output variable was tight.

The most important immediate observation from the results is that the estimate of the coefficient of output, M, is not appreciably different from unity. This parameter, being the elasticity of cost with respect to output, is the measure of possible returns to scale. Since it is approximately equal to unity, this suggests that there are no economies of scale in the airline industry when the influence of other variables is held constant. This finding indicates that a rapid growth in the demand for air travel, bringing about a corresponding growth in the size of the airlines, will not, by itself, bring about lower unit operating costs.

Effect of Dummy Variables

When dealing with purely linear equations, each dummy variable is usually interpreted as a shift in the intercept of the linear equation due to some discrete event that is thought to influence the variable. In this situation, however, the intercept of the estimated linear transformed equation is a multiplicative factor in the actual cost function, Eq. (8.3), which of course, has no intercept. For this reason, the impact of technical change, or of time, on cost cannot be determined from the equation directly. The dummy variables are, in effect, shifting the multiplicative factor.

The procedure for calculating the percentage change in cost due to a dummy variable and, in particular, due to a shift in technology is as follows:

1. Using some values for each of the continuous independent variables, use the estimated cost function to calculate cost. In this case the mean values of the logarithms were used. If the value of each of the dummy variables is unity, then this calculated cost is interpreted as the expected annual cost of running a propeller-driven plane for the sample geometric mean number of seat-miles, fuel cost, and so on. Call this C_1.

2. Replicate this procedure, but setting one of the dummy variables in Eq. (8.3) equal to 10. The result will be interpreted as the expected annual cost, given that the particular effect indicated by the selected dummy variable is operative, for the sample geometric mean values of the continuous independent variables. Call this C_2.

3. The percentage change in cost due to a shift in technology holding constant the influence of all other factors is $(C_1 - C_2)/C_1$.

As an example of this procedure, consider a simplified cost function:

$$C = aM^b T^k \qquad (8.6)$$

where T is some 1 or 10 dummy variable. The estimated effect of the phenomenon represented by the dummy variable is, then

$$\frac{C_1 - C_2}{C_1} = \frac{aX^b - aX^b 10^k}{aX^b} = 1 - 10^k \tag{8.7}$$

The effects of using different aircraft on the cost function were estimated by this procedure. The estimated percentage change in costs which result from the shifts from propeller to different types of jet aircraft are shown in Table 8.2. These results were all statistically significant with the expected

Table 8.2 Decreases in Cost Resulting from Shifts in Type of Aircraft, All Other Variables Held Constant

Shift		% Change in
From	To	Total Direct Costs
Prop	Turbo-prop	−11.1
Turbo-prop	Turbo-jet	−33.5
Turbo-jet	Turbo-fan	−8.4

negative sign. In addition, the rank order of the coefficients is identical with the technological ranking of the aircraft. That is, operating costs are lowered when a shift is made from propeller aircraft to turbo-prop, from turbo-prop to turbo-jet, and from turbo-jet to turbo-fan.

Finally, the coefficient on the time-shift variable is positive and statistically significant at the 10% level. The order of magnitude of the coefficient is reasonable in that it reflects a 3% increase in cost, which is rather in line with the annual rate of inflation prevailing in the United States at that time.

CONCLUSIONS

The results suggest that there are no economies of scale for the operating costs of passenger airlines in the United States. This finding appears to hold over a broad range of sizes of airlines, from among the largest in the world to the reasonably small local service carriers existing in 1965–1966. Consequently, as higher levels of output are accompanied by proportionately higher costs, there are no compelling reasons for airline mergers on economic efficiency grounds.

The shift to more productive aircraft does, however, lead to increases in productivity, as would be expected. Thus the observed decreases in unit costs during the 1960s can be attributed to the change in the mix of aircraft, which, indeed, changed rapidly from piston to turbo-jets and turbo-fans in this period. We must conclude that the opportunities for significant future cost reductions, given the present aircraft types, are limited because of the lack of substitution possibilities. This is consistent with the statements of airline officials that the decline in unit costs will "bottom out" because ". . . the substitution of jets for piston aircraft has largely been accomplished" (1). The introduction of the large "stretch-jets," such as the Boeing 747, seems to be the best opportunity for increased productivity. The effect of the introduction of these aircraft can be tested as the data become available.

ACKNOWLEDGMENTS

The author would like to acknowledge the assistance of Barbara Gisler and Elaine Kokiko in preparing this paper. Frederick W. Bell made a number of helpful comments that substantially improved the content of the paper.

REFERENCES

1. KERLEY, J. J., 1967. Financial Problems of the Airlines, *Financial Analysts Journal*, Vol. 23, pp. 73–84.
2. NERLOVE, M., 1965. *Estimation and Identification of Cobb–Douglas Production Functions*, Chicago, Ill.: Rand McNally & Company.
3. SHEPHARD, R. W., 1963. *Cost and Production Functions*, Princeton, N.J.: Princeton University Press.
4. U.S. Civil Aeronautics Board, 1966. *Trends of Productivity in the Airline Industry*, Washington, D.C.: U.S. Government Printing Office.
5. U.S. Civil Aeronautics Board, 1967. *Aircraft Operating Cost and Performance Report*, Washington, D.C.: U.S. Government Printing Office.
6. WALTERS, A. A., 1963. Production and Cost Functions: An Econometric Survey, *Econometrica*, Vol. 31, pp. 1–66.

9

Estimation of a Cost Function: Returns to Scale in Electricity Generation*

MARC NERLOVE

This chapter shows how characteristics of the production function can be inferred from statistics about actual operations. Whereas the previous study by Murphy considered only one possible form of the production and resultant cost functions, this analysis considers alternative possibilities in some detail. The case study illustrates how careful theoretical and practical arguments can be used to distinguish between models which are almost equivalent statistically, and indicates how the detailed structure of a production function can be modeled by statistical inference. The particular question addressed by this study is the extent of increasing returns to scale in the generation of electric power in the United States. (Editors' Preface)

THE PROBLEM

Returns to Scale

The question of whether there are increasing or decreasing returns to scale in an enterprise or plan, that is, whether larger units of production produce more per unit invested, is fundamental to design. It has an impor-

*Adapted from Chapter 6 of *Estimation and Identification of Cobb–Douglas Production Functions*, Rand McNally & Company, Chicago Ill., 1965. A prior version of this material is also given in C. Christ et al., *Measurement in Economics: Studies in Mathematical Economics and Econometrics in Memory of Yehuda Grunfeld*, Stanford University Press, Stanford, Calif., 1963. Errors in the regression analyses that were present in the original versions have been corrected by the author.

tant bearing on the optimal allocation of resources. As Chenery and Manne have pointed out, the extent of returns to scale is a determinant of investment policies in growing industries (1, 5). If there are increasing returns to scale and a growing demand, firms may find it profitable to add more capacity than they expect to use in the immediate future.

The existence of economies of scale also has an important bearing on the institutional arrangements necessary to secure an optimal allocation of resources. If, as many believe, there are increasing returns to scale over the relevant range of outputs produced by a public project, in particular by utility undertakings, larger quantities of outputs can be produced more cheaply, and marginal costs decrease as output rises. Following the criterion of marginal cost pricing, these companies must either receive subsidies or resort to price discrimination in order to cover costs at socially optimal outputs.

In studying the problem of returns to scale, the first question one must ask is: To what use are the results to be put? It is inevitable that the purpose of an analysis should affect its form. In particular, the reason for obtaining an estimate of returns to scale will affect the level of the analysis: industry, firm, or plant. For many questions of design, for example, the plant is the relevant entity. On the other hand, when questions of taxation are at issue, the industry may be the appropriate unit of analysis. But if we are concerned primarily with the general question of public regulation and with investment decisions and the like, it would seem that the economically relevant entity is the firm. Firms, not plants, are regulated, and it is at the level of the firm that investment decisions are made.

Previous empirical investigations that have a bearing on returns to scale in electricity supply are those of Johnston (2), Komiya (3), Lomax (4), and Nordin (8). All are concerned with returns to scale at the level of the plant, not the firm, and present evidence suggests that there are increasing or constant returns to scale in the production of electricity. It can be shown, however, that because of transmission losses and expenses of maintaining and operating an extensive transmission network, a firm may operate a number of plants at outputs in the range of increasing returns to scale, and yet be in the region of decreasing returns when considered as a unit (6). Although firms as a whole have been treated in this investigation, the problem of transmission and its effects on returns to scale has not been incorporated in the analysis, which relates only to the production of electricity. The results of this analysis are in agreement with those of previous investigators and suggest that the bulk of privately owned U.S. utilities operates in the region of increasing returns to scale, as is generally believed.

Characteristics of the Industry

The production of electric power has been carried out for the period covered by this analysis in three main ways: by internal combustion engines, which produced a negligible amount of power; by hydroelectric installations, which produced about one-third of all U.S. power; and by steam-electric installations, which accounted for the remaining two-thirds of U.S. power production. [Recent statistics show, in fact, that hydroelectric plants only supply about 5% of the U.S. consumption.—*Eds.*] Few firms rely solely on hydroelectric production because of the unreliability of supply. Furthermore, suitable sites for hydroelectric installations are rather limited and, except for those sites requiring an immense capital investment, almost fully exploited. Because of the great qualitative difference between steam and hydraulic production of electricity, this analysis is limited to steam generation.

The costs of steam-electric generation consist of energy costs and capacity costs. The former consist mainly of the costs of fuel, of which coal is the principal one. Energy costs tend to vary with total output and depend little on the distribution of demand through time. Capacity costs include interest, depreciation, maintenance, and most labor costs; these tend to vary, not with total output, but with the maximum anticipated demand for power (i.e., the peak load). Unfortunately, available data do not permit an adequate treatment of the peak-load dimension of output. To the extent, however, that a larger firm with a greater diversity of customers is likely to have a peak load which is a smaller percentage of output than a small firm, it follows that capacity costs per unit of output tend to be less for larger than for smaller firms. But this is a real economy of scale, and one reason for looking at firms rather than plants is precisely to take account of such phenomena.

The technological and institutional characteristics of the U.S. electric power industry that are important for our problem are:

1. Power cannot be economically stored in large quantities, and, with few exceptions, must be supplied on demand.

2. Revenues from the sale of power by private companies depend primarily on rates set by utility commissions and other regulatory bodies.

3. Much of the fuel used in power production is purchased under long-term contracts at set prices. The level of prices is determined in competition with other uses.

4. The industry is heavily unionized and wage rates are also set by contracts that extend over a period of time. Over long periods, wages appear to be determined competitively.

5. The capital market in which utilities seek funds for expansion is highly competitive, and the rates at which individual utilities can borrow funds are little affected by individual actions over a wide range. Construction costs vary geographically and also appear to be unaffected by an individual utility's actions.

Two conclusions may be drawn from these characteristics. First, it is plausible to regard the output of a firm and the prices it pays for factors of production as exogenous, despite the fact that the industry does not operate in perfectly competitive markets. Second, subject to some qualifications discussed elsewhere (6, 7), the problem of the individual firm in the industry would appear to be that of minimizing the total costs of production of a given output, subject to the production function and the prices it must pay for factors of production.

THE THEORY

As indicated, the characteristics of the electric power industry suggest that a plausible model of behavior is cost minimization in an environment where output and prices for the inputs may be taken as given. Such a situation is quite common to a very large number of industries, and they all share a common problem. Largely because the firms can each be assumed to try to minimize the costs of production, it is not possible to derive a production function by statistical analysis of the data solely on inputs and output. The observed relationship between output and cost is a confluent relationship that does not describe the production function at all; it describes the cost function.

The cost function can, however, be used to estimate important parameters of the production function. This is what we do in this study. The method for doing this is comparatively simple, as shown below.

Suppose that a firm has a production function of the Cobb–Douglas type:

$$y = a x_1^{\alpha_1} x_2^{\alpha_2} x_3^{\alpha_3} \tag{9.1}$$

where y is the output and the x_i are the inputs for the three factors: labor, capital, and materials. Suppose also that the costs of production, c, are given by:

$$c = \sum_i p_i x_i \tag{9.2}$$

where the p_i are the prices for the several inputs.

Minimization of costs, which is assumed, implies certain optimality conditions. Specifically, these are the familiar conditions that the ratios of marginal productivity to unit cost must be equal for each factor. For the Cobb–Douglas form of the production function, the marginal product for any input is equal to $(\alpha_i/x_i)y$, as can be seen by simple differentiation of Eq. (9.1). The optimality conditions can therefore be stated as:

$$\frac{\alpha_1}{p_1 x_1} = \frac{\alpha_2}{p_2 x_2} = \frac{\alpha_3}{p_3 x_3} \tag{9.3}$$

Solving Eqs.(9.3) and (9.1) for each of the x_i and substituting in the cost function, we obtain:

$$c = ky^{1/r} \, p_1{}^{\alpha_1/r} \, p_2{}^{\alpha_2/r} \, p_3{}^{\alpha_3/r} \qquad (9.4)$$

where

$$r = \sum \alpha_i$$
$$k = r(a\alpha_1{}^{\alpha_1} \, \alpha_2{}^{\alpha_2} \, \alpha_3{}^{\alpha_3})^{-1/r}$$

The parameter r measures the degree of returns to scale. If it is greater than unity, there are increasing returns to scale throughout the range over which Eq. (9.1) is supposed to hold; if it is less than unity, decreasing returns; and if it is equal to unity, constant returns. It is rather important to note that r can be obtained directly by statistical estimation of the cost function: it is simply the inverse of the coefficient of output.

The fundamental duality between cost and production functions, when we assume that the firm minimizes costs, assures us that the relation between the cost function, obtained empirically, and the underlying production function is unique (9). Under the cost-minimization assumption, they are simply two different but equivalent ways of looking at the same thing.

Note that the cost function must include factor prices if the correspondence between cost and production functions is to be unique. There is a difficulty in this, since the prices paid for inputs may change over time or differ in a cross section. In past studies, this problem has been handled by deflating cost figures by an index of factor prices, a procedure that typically leads to bias in the estimation of the cost curve unless correct weights, which depend on the unknown parameters of the production function, are used (2). However, given that price data are already available for the construction of an index, we can take the obvious step of including factor prices directly in the cost function. This is indeed what we shall do, although elementary economics textbooks usually treat cost functions as relations between cost and output alone, factor prices being parameters of the problem.

THE MODELS

What form of production function is appropriate for electric power? The Cobb–Douglas function of Eq. (9.1) is attractive for two reasons: First, it leads to a cost function that is linear in the logarithms of the variables:

$$C = K + \frac{1}{r}Y + \frac{\alpha_2}{r}P_1 + \frac{\alpha_2}{r}P_2 + \frac{\alpha_3}{r}P_3 \qquad (9.5)$$

where capital letters denote logarithms of the corresponding lowercase letters. Its linearity makes it especially easy to estimate. Second, a single estimate of returns to scale is possible.

The question is whether a Cobb–Douglas function accurately characterizes the condition of production in the electric power industry. A number of alternative models are possible and, indeed, plausible (3, 6, 7). But such approaches do not appear to afford sufficient advantages to offset the formidable computational problems associated with them. For the present purpose, it has seemed best to begin with an ordinary Cobb–Douglas function.

Relatively simple estimation procedures can be devised for evaluating the parameters of the assumed production function. The cost function, Eq. (9.4), incorporates all the required conditions except that the coefficients of the prices must add to unity. It is a simple matter to incorporate this restriction by dividing costs and two of the prices by the remaining price. It does not matter either economically or statistically which price we choose. When fuel price is used as the divisor, the result is:

$$C = P_3 = K + \frac{1}{r}Y + \frac{\alpha_1}{r}(P_1 - P_3) + \frac{\alpha_2}{r}(P_2 - P_3) \qquad (9.6)$$

which will be called model A.

Model A assumes that we have relevant data on the "price" of capital and that this price varies significantly from firm to firm. If neither is the case, we are in trouble. Most of the results presented here are based on model A, but the data used for this price of capital are actually inadequate, as indicated in the original text (7). If one supposes, however, that the price of capital is the same for all firms, which is not implausible, one can do without data on capital prices The assumption that capital "price," P_2, is the same for all firms implies that $(\alpha_2/r)P_2$ is a constant and thus that Eq. (9.5) can be rewritten as:

$$C = K' + \frac{1}{r}Y + \frac{\alpha_1}{r}P_1 + \frac{\alpha_3}{r}P_3 \qquad (9.7)$$

where

$$K' = K + (\alpha_2/r)P_2$$

since the exponents of the input levels in Eq. (9.1) are assumed to be the same for all firms. Equation (9.7) is called model B.

STATISTICAL RESULTS

Preliminary Analysis

Estimation of model A from cross-sectional data on firms requires that we obtain data on production costs, total physical output, and the prices of labor, capital, and fuel for each firm. For model B we do not need the price of capital since it is assumed to be the same for all firms. These data were obtained for a sample of 145 privately owned utilities in 1955, as discussed in detail elsewhere (6).

The results from the least-squares regressions on all the data are given in Table 9.1; the interpretation of these results in terms of the parameters of the production function is given in Table 9.2.

Table 9.1 Results from Regressions on Data for 145 Firms in 1955

Model	Regression Number	Coefficient (and Standard Error) of			
		Y	$P_1 - P_3$	$P_2 - P_3$	R^2
A	I	0.721	0.562	0.003	0.931
		($\pm$0.175)	($\pm$0.198)	($\pm$0.192)	
B	V	0.723	0.483	0.496	0.914
		($\pm$0.019)	($\pm$0.303)	($\pm$0.106)	

Table 9.2 Returns to Scale and Elasticities of Output with Respect to Various Inputs Derived from Results in Table 9.1

Model	Regression Number	Returns to Scale	Elasticity of Output with Respect to		
			Labor	Capital	Fuel
A	I	1.39	0.78	0.00	0.61
B	V	1.38	0.67	−0.03	0.69

The correlation for model A is high, $R^2 = 0.93$, which is somewhat unusual for such a large number of observations. Increasing returns to scale are indicated, and the elasticities of output with respect to labor and fuel have the right sign and are of plausible magnitude. However, the elasticity of output with respect to capital price has the wrong sign; fortunately, it is statistically insignificant.

The difficulties with capital in model A may be due in part to the difficulty of measuring both capital costs and the price of capital. The former were measured as depreciation charges plus the proportion of interest on long-term debt attributable to the production plant; the figure for capital price was compounded of the yield on the firm's long-term debt and an index of construction costs. Depreciation figures reflect past prices and purchases of capital equipment, whereas the price of capital as constructed for this study does not. It is perhaps not so surprising then that the price has little effect on costs.

Model B is designed to evade the difficulty with data on capital. It is apparent that its estimates of returns to scale and the elasticities of output with respect to labor and fuel are little different from those of model A.

A Statistical Difficulty: Nonlinearity

The first regressions embody a statistical difficulty which is not apparent from the examination of the coefficients and their standard errors. As part of these analyses, the residuals from the regressions were plotted against the logarithm of output. The result in both cases is schematically pictured in Figure 9.1. It is clear that neither regression relationship is truly linear in logarithms. To test this visual impression, the observations were arranged in order of ascending output, and Durbin–Watson statistics computed; the values of the statistics indicated highly significant positive serial correlation, confirming the visual evidence of nonlinearity.

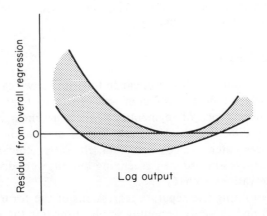

Figure 9.1 Residual pattern from the overall regression, 145 firms.

Although several conjectures may be advanced (7), a simple hypothesis does explain the observed result—that the degree of returns to scale is not independent of output but varies inversely with it. Figure 9.2 illustrates this explanation: the solid line gives the traditional form of the total cost function, which shows increasing returns at low outputs and decreasing returns at high outputs. If we try to fit a function for which returns to scale are independent of the level of output, for example models A and B, which are linear in logarithms, a curve such as the dashed one will be obtained. The shaded areas show the output ranges, high and low, for which total costs are underestimated.

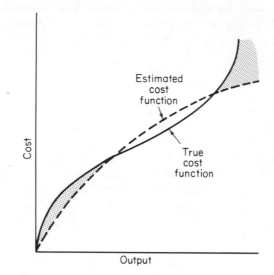

Figure 9.2 A "textbook" explanation for the residual pattern obtained from the overall regression.

More Refined Analyses

If the true cost function is not linear in logarithms, we can either fit an overall function which reflects this fact or we can attempt to approximate the actual function by a series of segments of functions linear in logarithms. Because of fitting difficulties and the problem of determining the form in which factor prices enter the cost function, the latter course was chosen initially. Firms, arrayed in order of ascending output, were divided into five groups of 29 observations each (6).

The results of fitting five separate regressions of the form indicated by models A and B and the corresponding implications for the parameters in the production function are presented in Tables 9.3 and 9.4.

The results of these regressions with respect to returns to scale are appealing. Except for a statistically insignificant reversal between groups C and D, returns to scale diminish steadily, falling from a high of better than 2.5 to a low of slightly less than 1, indicating increasing returns at a diminishing rate for all except the largest firms in the sample. However, in the case of regression III, the elasticity of output with respect to capital behaves very erratically from group to group and has the wrong sign in groups A and E. In regressions VI both the elasticity of output with respect to labor and with respect to capital behave erratically, the latter having the wrong sign in group E.

Table 9.3 Results from Regressions Based on Models A and B for 145 Firms in 1955 When Stratified into Five Categories of Output

Model	Regression Number	Coefficient (and Standard Error) of			
		Y	$P_1 - P_3$	$P_2 - P_3$	R^2
A	III A	0.398	0.641	0.093	0.512
		($\pm$0.079)	($\pm$0.691)	($\pm$0.699)	
	III B	0.668	0.105	0.364	0.635
		($\pm$0.116)	($\pm$0.275)	($\pm$0.277)	
	III C	0.931	0.408	0.249	0.571
		($\pm$0.198)	($\pm$0.199)	($\pm$0.189)	
	III D	0.915	0.472	0.133	0.871
		($\pm$0.108)	($\pm$0.174)	($\pm$0.157)	
	III E	1.045	0.604	0.295	0.920
		($\pm$0.065)	($\pm$0.197)	($\pm$0.175)	
B	VI A	0.361	0.212	0.655	0.438
		($\pm$0.086)	($\pm$1.259)	($\pm$0.350)	
	VI B	0.661	0.401	0.490	0.672
		($\pm$0.106)	($\pm$0.333)	($\pm$0.134)	
	VI C	0.985	0.014	0.330	0.647
		($\pm$0.180)	($\pm$0.261)	($\pm$0.138)	
	VI D	0.927	0.327	0.426	0.884
		($\pm$0.106)	($\pm$0.228)	($\pm$0.064)	
	VI E	1.035	0.704	0.643	0.934
		($\pm$0.067)	($\pm$0.272)	($\pm$0.132)	

Table 9.4 Returns to Scale and Elasticities of Output with Respect to Various Inputs Derived from Results Presented in Table 9.3

Model	Regression Number	Return to Scale	Elasticity of Output with Respect to		
			Labor	Capital	Fuel
A	III A	2.52	1.61	−0.02	0.93
	III B	1.50	0.16	0.53	0.81
	III C	1.08	0.44	0.27	0.37
	III D	1.09	0.52	0.15	0.42
	III E	0.96	0.58	−0.29	0.67
B	VI A	2.77	0.59	0.37	1.81
	VI B	1.51	0.61	0.15	0.74
	VI C	1.02	0.01	0.67	0.34
	VI D	1.08	0.35	0.27	0.46
	VI E	0.97	0.68	−0.33	0.62

Breaking the sample into five groups does significantly reduce the residual variance. Because of the erratic behavior, however, of the coefficients of independent variables other than output, it appears that we may have gone too far. Regressions III and VI are based on the assumption that all coefficients differ from group to group. This is equivalent to presuming that the scale of an operation affects not only returns to scale but also marginal rates of substitution.

As a compromise, it is possible to assume what is known as neutral variations in returns to scale; that is, the returns to scale vary with output but the marginal rates of substitution do not. This is equivalent to assuming that the coefficients for the various prices in the individual group regressions are the same for all groups while allowing the constant terms and the coefficients of output to differ (7). The hypothesis of neutral variations in returns to scale is tested in this way only in the context of model A.

The results of these regressions, number IV, and the implications for the production function are presented in Tables 9.5 and 9.6. An analysis of covariance comparing this regression with the previous results for model A, regressions III, gives an F ratio of 1.576. With 8 and 125 degrees of freedom, a ratio this high is significant at better than the 90% level; hence we cannot confidently reject regressions III on statistical grounds. Examining the results derived from regressions IV, however, we find that the degree of returns to scale steadily declines with output until, for the group consisting of firms with the largest output, we find some evidence of diminishing returns to scale. Furthermore, the elasticities of output with respect to the various input levels are all of the correct sign and of reasonable magnitude, although it is

Table 9.5 Results from Regressions on Model A when Neutral Variations in Returns to Scale Are Assumed

Regression Number	Coefficient (and Standard Error) of			
	Y	$P_1 - P_3$	$P_2 - P_3$	R^2
IV A	0.394 ($\pm$0.055)			
IV B	0.651 ($\pm$0.189)			
IV C	0.877 ($\pm$0.376)	0.435 ($\pm$0.207)	0.100 ($\pm$0.196)	0.950
IV D	0.908 ($\pm$0.354)			
IV E	0.062 ($\pm$0.169)			

Table 9.6 Returns to Scale and Elasticities of Output with Respect to Various Inputs Derived from Results Presented in Table 9.5

Regression Number	Returns to Scale	Elasticity of Output with Respect to		
		Labor	Capital	Fuel
IV A	2.52	1.10	0.25	1.17
IV B	1.53	0.65	0.15	0.73
IV C	1.14	0.50	0.11	0.53
IV D	1.10	0.48	0.11	0.51
IV E	0.94	0.41	0.09	0.44

still felt that the elasticity with respect to capital appears implausibly low. Thus, on economic grounds, one might tentatively accept the hypothesis of neutral variations in returns to scale.

A Revised Cost Function

If one accepts the hypothesis of neutral variations in returns to scale, a somewhat more refined analysis is possible since we may then treat the degree of returns to scale as a continuous function of output. That is, instead of grouping the firms as we did previously, we estimate a cost level function of the form:

$$C = K + \frac{1}{r(Y)}Y + \frac{\alpha_1}{r}P_1 + \frac{\alpha_2}{r}P_2 + \frac{\alpha_3}{r}P_3 \qquad (9.8)$$

where $r(Y)$, the degree of returns to scale, is a function of the output level. Since neutral variations in returns to scale are assumed, the coefficients of the prices are unaffected.

A preliminary graphical analysis indicated that returns to scale as a continuous function of output might be approximated by a function of the form:

$$r(Y) = \frac{1}{\beta_0 + \beta_1 \log y} \qquad (9.9)$$

Thus, instead of the regressions of the form suggested by Eq. (9.6), model A, we fit model C:

$$C - P_3 = K + \beta_0 Y + \beta_1 Y^2 + \frac{\alpha_1}{r}(P_1 - P_3) + \frac{\alpha_2}{r}(P_2 - P_3) \qquad (9.10)$$

and, instead of Eq. (9.7), model B, we fit model D:

$$C = K' + \beta_0 Y + \beta_1 Y^2 + \frac{\alpha_1}{r} P_1 + + \frac{\alpha_3}{r} P_3 \qquad (9.11)$$

The results for the regressions on models C and D are given in Tables 9.7 and 9.8. Note that returns to scale and the other parameters have been computed at five output levels only, so that the results in Table 9.8 may be readily compared with those in Tables 9.4 and 9.6. These five levels in Table 9.8 were taken as the median output for each group.

Table 9.7 Results from Regressions Based on Models C and D for 145 Firms in 1955; Continuous Neutral Variations in Returns to Scale

Model	Y	Y^2	$P_1 - P_3$	$P_2 - P_3$	P_1	P_3	R^2
			Coefficient (and Standard Error) of				
C	0.151 ($\pm$0.062)	0.117 ($\pm$0.012)	0.498 ($\pm$0.161)	0.062 ($\pm$0.151)	—	—	0.958
D	0.137 ($\pm$0.064)	0.118 ($\pm$0.013)	—	—	0.279 ($\pm$0.224)	0.255 ($\pm$0.054)	0.952

Table 9.8 Returns to Scale and Elasticities of Output with Respect to Various Inputs Derived from Results Presented in Table 9.7

Model	Group	Returns to Scale	Labor	Capital	Fuel
			Elasticity of Output with Respect to		
C	A	2.92	1.45	0.18	1.29
	B	2.24	1.12	0.14	0.98
	C	1.97	0.98	0.12	0.87
	D	1.84	0.92	0.11	0.81
	E	1.69	0.84	0.10	0.75
D	A	3.03	0.85	1.41	0.77
	B	2.30	0.64	1.07	0.59
	C	2.01	0.56	0.94	0.51
	D	1.88	0.52	0.88	0.48
	E	1.72	0.48	0.80	0.44

Perhaps the most striking result of the assumption of continuously and neutrally variable returns to scale of the form suggested in Eq. (9.9) is the substantial increase in our estimate of the degree of returns to scale for firms in the three largest groups. Whereas before we found nearly constant returns to scale, it now appears that they are increasing. In addition, all the coefficients in both analyses are of the right sign, and the results based on model D yield results of plausible magnitude for the elasticity of output with respect to capital as compared with the elasticities with respect to labor and fuel.

Analyses of covariance, comparing model C and regression I and model D and regression V, yield F ratios of 1.631 and 9.457, respectively; both are highly significant. A comparison of model C with regression III yields an F ratio of 1.032, which, although not significant, does suggest that neutral variations in returns to scale of the form used are indistinguishable from nonneutral. Hence the hypothesis of neutral variations in returns to scale may be accepted both on economic grounds and on grounds of simplicity.

CONCLUSIONS

The study presented illustrates the use of cost functions for estimating characteristics of the production function. Based upon both theory and the evidence of the data, it would appear that the appropriate model for a firm in the electric industry is, indeed, the statistical cost function, which includes factor prices and is uniquely related to the underlying production function.

Further, it would appear that, at the firm level, it is appropriate to assume a production function such as a generalized Cobb–Douglas, which allows substitution among factors of production. When a statistical cost function based on this kind of production function is fitted to cross-section data on individual firms, there is evidence of such substitution possibilities.

Finally, the major substantive conclusions of this chapter are (1) that there is evidence of a marked degree of increasing returns to scale at the firm level in U.S. steam-electricity generation, but that the degree of returns to scale varies inversely with output and is considerably less, especially for large firms, than that previously estimated for individual plants; and (2) that it seems that variations in returns to scale in electricity generation may well be neutral in character, that is, that although the scale of operation affects the degree of returns to scale, it may not affect the marginal rates of substitution between different factors of production for given factor ratios.

REFERENCES

1. CHENERY, H. B., 1952. Overcapacity and the Acceleration Principle, *Econometrica*, Vol. 20, pp. 1–28.
2. JOHNSTON, J., 1960. *Statistical Cost Analysis*, New York: McGraw-Hill Book Company, pp. 44–73.
3. KOMIYA, R., 1962. Technological Progress and the Production Function in the United States Steam Power Industry, *Review of Economics and Statistics*, Vol. 44, May, pp. 156–166.

4. LOMAX, K. S., 1952. Cost Curves for Electricity Generation, *Economica*, Vol. 19, pp. 193–197.
5. MANNE, A. S., 1967. *Investments for Capacity Expansion*, Cambridge, Mass.: MIT Press.
6. NERLOVE, M., 1963. *Measurement in Economics: Studies in Mathematical Economics and Econometrics in Memory of Yehuda Grunfeld*, C. Christ et al., eds., Stanford, Calif.: Stanford University Press.
7. NERLOVE, M., 1965. *Estimation and Identification of Cobb–Douglas Production Functions*, Chicago, Ill.: Rand McNally & Company.
8. NORDIN, J. A., 1947. Note on a Light Plant's Cost Curves, *Econometrica*, Vol. 15, July, pp. 231–235.
9. SHEPHARD, R. W., 1953. *Cost and Production Functions*, Princeton, N.J.: Princeton University Press.

10

The Demand for Airport Access Services[*]

RICHARD DE NEUFVILLE

This study reports on a long-term effort to define the characteristics and patterns of behavior of airport access travelers and, most particularly, their potential response to changes in the nature and quality of airport access service. The immediate objective was to develop and validate models of the demand for airport access trips so that airport planners could effectively predict the volume of traffic on new services and thus design these services efficiently. This study also illustrates practical ways to validate the predictive power of models of traveler demand and, by example, indicates some of their difficulties.

THE PROBLEM

The effective design of public projects, and of transportation systems in particular, requires insightful understanding of the nature of both the technological possibilities and of the preferences of the potential users for the services to be provided. The technical possibilities can be represented by supply functions, which describe the cost of achieving different volumes of service in terms of the nature of the system and the quality of service it may offer. The desires of the users can be represented by demand functions, which describe the amount of service they will purchase according to its price and quality. The loads on a public system, and consequently its design

*Reprinted from *Traffic Quarterly*, Vol. 27, No. 4, Oct. 1973.

and its profitability, are thus an intersection between the prevailing supply and demand.

Supply functions are relatively easy for engineers to define. Many are available for different modes of transport, and for airport access systems in particular (1). Demand functions are, however, much more difficult to define. The problem is not that of being able to construct something that looks like a demand function, such as

$$\text{volume of trips} = f(\text{price, travel time, income}, \ldots) \qquad (10.1)$$

This, in fact, can be done very easily using any of a variety of statistical techniques. Nor is there much difficulty in finding some function to fit with observations on traffic, such as those obtained from an airport access survey. Indeed, an essential difficulty in obtaining a correct demand function lies precisely in the fact that it is quite easy to develop several relationships which look like demand functions, which fit the data equally well, but which are, however, quite contradictory in their implications about the effect on the volume of trips, of changes in price or speed, or other measure of the quality of service.

The development of correct models of demand is, in great part, the determination of which models do actually predict how potential users will react to different systems. The process for doing this may be seen as consisting of four steps. First, it is useful to obtain, by survey or other method, an understanding of the public's current patterns of use with respect to a system. The model must, at a minimum, describe this behavior. Second, it is desirable to stratify the data into homogeneous controlled groups so as to reduce the possible sources of variation and thus of ambiguity in the results. Third, we can estimate the relative strength of the explanatory variables of a demand model which appears reasonable on theoretical grounds. Finally, one can test the validity of the model by seeing to what degree it does predict the changes in demand that do actually occur when elements of the system are altered. This can be expediently done by before and after studies.

This is the process described below. The focus is on the development of a demand model for airport access, and the immediate results are relevant to the planning and design of airports. More generally, the study suggests how demand models could be developed and validated for other public systems, such as urban transportation.

EXISTING AIRPORT ACCESS PATTERNS

Overall surveys of the patterns of travel of persons going to and from airports have been conducted at many sites. These are briefly described below.

As they fail to define the detailed movements of passengers to the airports, four further in-depth surveys were conducted at Kennedy and La Guardia airports in New York City. These provided the basis for the specific demand models considered.

Previous Studies

Several important airport access studies have been conducted. The most notable ones are those carried out by the Port of New York Authority (now the Port Authority of New York and New Jersey), the Baltimore–Washington survey, and the Cleveland survey. Other important studies have been undertaken at Philadelphia, Los Angeles, Toronto, London, Salt Lake City, and at numerous other locations. Detailed comparisons of these studies have been given by Koller and Skinner (5).

The airport access studies have, almost exclusively, gathered data by means of questionnaires handed out to passengers or other persons. Because of the difficulty of getting these people, often in a rush, to stop for long, the survey instruments are typically quite short. They thus provide fairly superficial information about who is traveling, where, and how. But this can be interesting. For example, we now know (1) that only about one-third of airport access trips are made by travelers, the rest being employees, visitors, or deliverymen; (2) that just about one-half of the travelers go to or come from the central business district of a city, and the rest of the travelers and essentially all the visitors and deliverymen, that is, about five-sixths of the airport trips, have trip ends distributed around the suburbs; and (3) consequently that most airport trips will inevitably be associated with some mode of transport, such as the automobile, which can distribute trips around the metropolitan area. This is most useful to recognize, but it does not provide us with detailed knowledge about access patterns.

Additional data about airport trips have, of course, been generated by urban transportation studies. But these cannot be relied upon. Sample surveys have been carried out on all travelers who enter or leave most metropolitan areas in the United States but these cordon counts sample too few of the airport trips to be meaningful. And the home interviews contact only residents, who are known to go to and from the airport quite differently from nonresidents.

MIT Surveys

To obtain in-depth information about how airport access trips were made, the MIT Civil Engineering Systems Laboratory conducted four airport access surveys around New York City in 1970 and 1971. The procedures were devised to obtain specific information about exactly how particular travelers moved to the airport, how much they paid, how long it took them, why they did not use alternative modes of transport, and so on.

The surveys were conducted in close cooperation with two operators of airport access transportation: Wilder, providing limousine service from Kennedy and La Guardia to points throughout Westchester County; and Carey, which operates buses between both airports and downtown Manhattan. We exploited the facts that the trip between the airport and the first stop was quite long (and boring), and that travelers were going over a fixed route, to obtain very detailed responses to over 30 different questions tailored according to the trip, person, and time. In practice, personable college students would meet each passenger, request their cooperation, assist them, and collect the responses at the end of the trip. This individual approach led to very high response rates, over 95%, and to quite complete answers. Consequently, this effort was able to obtain uniquely precise descriptions of the airport access patterns of the travelers.

The kind of information requested by our questionnaires was similar to, but far more detailed than, that in other airport surveys. To tie in with other surveys, the usual questions about age, sex, income, occupation, and destination were asked. In addition, respondents were requested to describe precisely what mode of travel they took on each leg of their airport access trip; how much it cost them; their exact address (so that the cost of alternatives could be computed); how long they had spent on the trip and when they were scheduled to be at their destination; why they did not or could not choose other alternative modes; and so on. With these data, it was possible to reconstruct their choices and preferences quite precisely.

By surveying both Wilder and Carey, it was possible to compare central-business-district- and suburban-oriented travelers. Three surveys were conducted on Wilder as part of a sequence of before and after studies of the effect of changes in fare and travel time.

Results of MIT Surveys

In addition to the information needed to construct the total cost, in money and time, of each traveler's trip and his alternatives, the surveys provided extensive data on their patterns and preferences. Complete descriptions of these are given by Koller and Skinner (3,5) and Yaney (11). In particular, however, three sets of observations are relevant.

First, the observations indicate that passengers on their way to the airport are not, as a group, prepared to spend much money to save travel time on this trip. Certainly, there are numerous occasions when one is late for a departure and extremely sensitive to travel time. But, in general, passengers arrive well in advance of a flight and are unlikely to be willing to spend much for a faster trip to the airport. As shown by Figure 10.1, approximately 50%

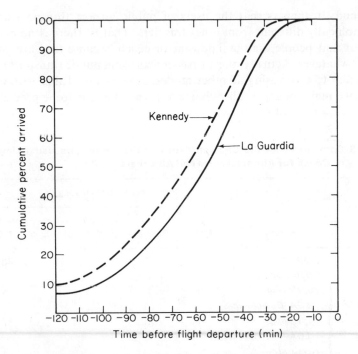

Figure 10.1 Cumulative arrival time distribution of Wilder passengers by airport—January, 1971.

of the passengers arrive at the airport 50 minutes ahead of flight time. This varies, of course, depending upon the nature of the flight and the experience of the passengers. People arrive earlier for long-distance and international flights; air commuters arrive much closer to schedule. But, as several other observations confirm (5, 11), people do arrive well ahead of departure time and are quite willing to pay the penalty of waiting. Additional evidence about people's reluctance to pay for speed can be obtained by looking at how people arrive at the airport access mode that will deliver them to the airport. As shown in Table 10.1, this secondary mode is often quite slow and cheap. This finding is contrary to the general assumption that airport travelers must have a high value for time (and would, consequently, always use taxis) (7); a different model is thus required.

Table 10.1 Secondary Modes of All Carey–La Guardia Passengers (%)

Carey Terminal	Private Auto	Taxi	Bus	Subway	Other (Walk)
East Side	2	49	18	3	28
Grand Central	0	28	8	17	47

Second, it appears that the users of public airport transport are not demographically different from other travelers. That is, there is no evidence to suggest that people use the limousine or coach because they are poor or infirm or whatever. Actually, the air passengers using public transport appear quite similar to those using all other modes, as shown by Table 10.2. Consequently, we may not expect that choice of access mode is to be predicted by income, age, or sex of the travelers.

Table 10.2 Comparison of the Characteristics of the Passengers Using Limousine or Coach for Airport Access and All Others

Airport Access Mode	Survey	Passenger Characteristics (%)			
		Male	On Business	25–55 Years Old	Income over $20,000
All	*1968 PONYA, domestic*	71	57	65	40
Limousine or coach	*1971 Carey, LGA–Downtown*	71	76	66	55
	1971 Wilder, LGA–Westchester	71	61	54	57
	1971 Wilder, LGA–Connecticut	72	54	67	64

Third, a significant faction of the users of public transport for airport access are, to a degree, captive. Although over 80% of the limousine riders from Westchester and Connecticut owned cars, and thus had them "available" in some sense (in that they could either be left off by car, park a car, or both), most of them really could not use them for the trips. They could either not leave their cars at the airport, perhaps because their wife needed it, or did not have someone who was free to drop them off at the airport, or both (Table 10.3). By focusing on the concept of auto availability rather than

Table 10.3 Classification by Auto Availability of Westchester Resident Limousine Passengers Going to the Airports (Jan. 1971)

Auto-Dropoff Available (%)	Auto-Park Available (%)		
	Yes	No	Total
Yes	35	19	54
No	29	17	46
Total	64	36	100

on the more usual one of auto ownership, one can explain the strong asymmetry in the use of public transport to and from the suburbs. Indeed, there are far more passengers, mostly local residents, on the limousine on trips from the airport than on trips to the airport. By actual count, there are approximately as many nonresident passengers in either direction of travel, and about three times as many residents used the limousine from the airport as to the airport. Consequently, as shown in Table 10.4, the composition as well as the volume of traffic changes remarkably by direction.

Table 10.4 Wilder Passenger Composition by Residency by Direction

Direction	Residency	% of Total Jan. 1970	Jan. 1971
To airport	Westchester	40	44
	Non-Westchester	60	56
From airport	Westchester	72	71
	Non-Westchester	28	29
Total	Westchester	58	61
	Non-Westchester	42	39

Many local residents typically get driven to the airport but, for a variety of reasons such as the uncertainty in the time of their return, must find their way home by public transport. These residents thus tend to use the limousine service only one way, causing a significant imbalance in the loads. These travelers' responses confirm that this phenomenon is caused by the unavailability, at the time they arrive, of anyone to pick them up. Consequently, it would seem that auto availability, a factor that system designers can measure but rarely influence, might be a strong determinant of modal choice for the airport access trip.

MODEL FORMULATION

The insights obtained from the surveys were combined with a priori theory to develop demand models. These would be treated as hypotheses and subsequently tested to see to what extent they did accurately predict the effects of changes in the airport access system. The formulation problem itself had two stages. First, we examined the effect of auto availability on the use of public transport to the airport. Second, we considered the effect of price and travel time, both of which are design variables which can be influenced by the planner, on the use of different modes of airport access.

Effect of Auto Availability

Auto availability is not measured on a continuous scale. It is, rather, discrete and binary: either the auto is available or it is not. Auto availability is, furthermore, inherently a characteristic specific to a person. It therefore appears essential to model this effect on a disaggregate basis, considering its influence on individuals. Several approaches to doing this have been described by now (6), and they each generate similar results from any set of data (10). For convenience, discriminant analysis was selected.

Discriminant analysis calibrates an equation of the form

$$Y = a_0 + a_1 X_1 + \cdots + a_n X_n \qquad (10.2)$$

where the X_i are explanatory variables. The dependent variable, Y, is known as the discriminant. The quantity Y is transformed into the probability that an individual traveler will choose to use a specified mode of transport, $P(M)$, as follows:

$$P(M) = \frac{e^Y}{1 + e^Y} \qquad (10.3)$$

$P(M)$, as defined, is essentially a continuous S-shaped cumulative distribution function which varies from zero to unity. For $Y = 0$, $P(M) = 0.5$ and a person is equally likely to choose either of two modes. For $Y > 0$ one mode is chosen, and for $Y < 0$ the other. The a_i assumed to be are calculated so as to minimize the number of incorrect assignments of choice of mode to an individual, that is, the number of times Y is calculated to exceed zero when it should not, and vice versa. This is a standard procedure for statistical classification. Although it has only recently been introduced into transportation demand analyses, it has been extensively used in physical sciences, such as biology. When $P(M)$ is aggregated over all individuals in any area, it can be considered as an indication of the modal split for that zone.

On theoretical grounds it would seem desirable to use at least travel time, T, and cost, C, in addition to auto availability, A, as explanatory variables. However, time and cost are both linear functions of distance in this case. Formally, these variables are collinear, so it is impossible to separate the effect of one from the other (see Table 10.5, for example). Consequently, either of them must stand as a proxy for both. The final model calibrated was, thus

$$Y = -1.40 + 0.08T - 4.7A \qquad R = 0.61 \qquad (10.4)$$

and its implications for the effect of auto availability are shown graphically in Figure 10.2. For residents at any specified distance from the airport, the probability of their using limousine service or public transport more generally is significantly influenced by whether or not a car is available. As can be seen, the question of whether or not people drive to the airport is dominated by the availability of the car.

A few comments on the implied effect of travel time to the airport on modal split are appropriate in connection with Figure 10.2. First, as described in detail subsequently, the travel-time variable is collinear with travel cost: both variables move together and are thus approximately linear functions of each other. Consequently, Figure 10.2 indicates that an even larger fraction of the airport passengers use the limousine service as taxis became relatively more expensive. This is to be expected.

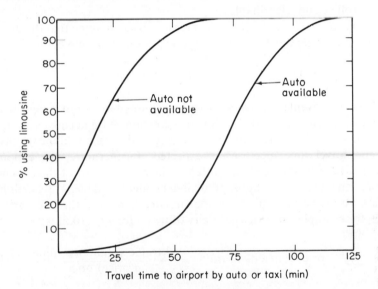

Figure 10.2 Limousine share for non-Westchester resident air passengers not residing in Westchester.

Effect of Time and Cost

The effect of changes in the time and cost of travel are especially interesting to the airport planner because these are measures of the level of service which he can control in order to improve problems in airport access. To estimate these effects accurately, it is necessary to control for the effect of auto availability. This was done by estimating a demand function for nonresident businessmen, who can be presumed not to have private cars available to them. The nonresident businessmen were readily identified from our detailed questionnaires of the users.

A correct statistical estimation of a demand function requires that we be assured that all the data used do, in fact, lie on the demand function. If they do not, the results can naturally be spurious. Technically, the demand function must be identifiable according to specific conditions (2). These are met, in this instance, because the travel time and cost of the limousine are fixed for any site, that is, the supply function is fixed, independent of the demand. In more general situations, such as for urban travel by car, we know that congestion occurs when the volume of traffic picks up, and thus that supply does vary. In such circumstances, statistical estimation of the demand function may be meaningless. The case considered here is, indeed, quite special from the point of view of determining a demand function accurately: it is controlled and identifiable.

From economic theory, it appears reasonable to assume that the demand is a jointly exponential function of the explanatory variables:

$$\text{volume of traffic} = A(\text{price})^a (\text{travel time})^b \qquad (10.5)$$

This form is convenient because once logarithms have been taken of both sides, it can easily be estimated by linear regression. With negative exponents, Eq. (10.5) shows that the number of trips will decrease nonlinearly as the costs and travel time increase, as expected. Furthermore, the exponents in Eq. (10.5) can be directly interpreted as the elasticity, η_x, of the dependent variable with respect to the appropriate independent variable, X. By definition, the elasticity is the ratio of the proportionate changes in the two variables, so that, for example, the price elasticity of travel for Eq. (10.5) is:

$$\eta_p = \text{price elasticity of traffic} = \frac{\partial(\text{traffic volume})/\text{volume}}{\partial(\text{price})/\text{price}} \qquad (10.6)$$

Simple substitution shows that $\eta_p = a$ for Eq. (10.5), and the elasticity of the volume of traffic with respect to travel time, $\eta_t = b$. The functional form of Eq. (10.5) is, thus, most convenient for estimating demand functions.

As the total volume of traffic that will go to an airport, V_{TO}, depends significantly on the state of the economy and other factors beyond the scope of the airport access system, it is convenient to focus on the market share of the limousine service, V_L/V_{TO}. As V_{TO} can be considered essentially independent of the costs of the access trips (people do not choose to fly from New York to Chicago on the basis of the taxi fare), we can determine it outside the model and use it and observations on the limousine traffic to obtain data on the market share.

The choice of the specific variables to be used to represent the concepts of price and travel time in Eq. (10.5) is not trivial. A common problem is that variables are collinear, that is, that one independent variable, say X_i, is a linear function of one or more of the other variables, X_j. Thus

$$X_i = \sum c_j x_j + c_0$$

with some reasonable degree of correlation. When this happens, it is impossible to distinguish between the effects of the variables in a statistical analysis because they are such close substitutes. Stated formally, the effect of collinearity is to make the standard errors on the coefficients of the affected variables very broad and thus make the estimates useless. If the variables are then significantly collinear, the analysis would be worthless.

An analysis of the degree of correlation between variables that could represent the costs, P_L and P_T, and travel times, T_L and T_T, of limousine and taxi, which are the alternative public modes for nonresidents, was performed. The results are shown in Table 10.5. In view of the high collinearity between most variables, it was decided to use the ratios of prices and travel times as the most appropriate variables.

Table 10.5 Collinearity Between Possible Explanatory Variables

Variables Compared	Coefficient of Correlation, R
P_L vs. $\begin{cases} P_T \\ T_T \\ T_L \end{cases}$	0.60 0.38 0.59
P_T vs. $\begin{cases} T_L \\ T_T \end{cases}$	0.92 0.81
T_L vs. T_T	0.62
P_L/P_T vs. T_L/T_T	0.38

The demand function actually estimated was thus of the form

$$\frac{V_L}{V_{TO}} = K\left(\frac{P_L}{P_T}\right)^a \left(\frac{T_L}{T_T}\right)^b \tag{10.7}$$

The results of the statistical estimation for the suburbs and downtown are given in Table 10.6, where a and b can be interpreted directly as the price and travel-time elasticities. All coefficients are significantly different from zero at a 95% confidence level.

The estimates easily meet the minimal theoretical requirements: the elasticities of travel by limousine with respect to its fare and travel-time are negative, as expected. Further, it would appear, as might be inferred from the way people arrive early at the airport, that these airport access passengers are much less sensitive to time than to cost. Finally, it also seems that persons traveling shorter distances, as to the business district, are less

Table 10.6 Estimates of Coefficients of the Demand

Area	Survey	Estimate of Coefficients			Coefficient of Correlation, R
		K	$\eta_p = a$	$\eta_t = b$	
Suburbs	Wilder 1970	0.15	−2.3	−1.3	0.59
	Wilder 1971	0.17	−1.5	−0.7	0.75
Downtown	Carey 1971	0.12	−1.0	−0.2	0.56

Table 10.7 Comparison of Price and Travel-Time Elasticities Estimated for Different Environments

Area	Analysis	Range of Elasticity	
		Price	Travel Time
Suburban airport access	Wilder	−1.5 to −2.3	−0.7 to −1.3
Intercity travel	Quandt and Baumol (8)	−2.3 to −3.2	−1.7 to −2.0
Intercity travel	Yu (12)	−2.2 to −3.2	−1.3 to −2.1
Downtown access	Carey	−1.0	−2.0
Washington airport access	Rassam et al. (9)	−0.4 to −1.0	0.0 to −0.1
Urban travel	Domencich et al. (4)	−0.1 to −0.4	−0.6 to −1.2
Urban travel	Warner (10)	−1.0	−1.26

sensitive to either factor. As shown by Table 10.7, the implications of this model are in reasonable agreement with other analyses. It should be noted, however, that the relative importance of the cost and travel-time variables are just the reverse of what is ordinarily experienced in urban transportation, which, naturally, is also a situation significantly different in many respects from airport access.

MODEL VALIDATION

The demand model, as formulated, fits reasonably well with the data and prior theory. But it is, as indicated earlier, quite easy to fit some model, even contradictory models, which satisfy these conditions. The real test of a demand model is whether it does provide the designer with accurate information about the consequences of his plans—whether, in fact, the model does predict.

To test our model, its predictions were compared to the results of specific major policy changes effected by Wilder in 1970 and 1971. As these changes occurred over the short run, while the rest of the system remained essentially fixed, it was possible to really measure the effect of changes in fare and travel time. In fact, Wilder was chosen as a focus for analysis because it was known in advance that these changes would occur in such a controlled environment, and that Wilder thus presented a strategic opportunity for the development of demand models.

Effect of Fare Changes

As a result of labor negotiations in the summer of 1970, Wilder decided it was forced to raise fares some six months after our initial 1970 survey. The research team undertook to advise Wilder as to its best strategy, predict its effect, and test its results in an identical survey in 1971. Because of the passengers' sensitivity to fare increases, especially when competitive fares are low, we recommended a graduated price change instead of the across-the-board differential originally proposed. Although this might appear to be an intuitively obvious move to economists, it was not to the management of Wilder. We were able to use the demand model to suggest that, if we were correct, gross revenues could be 30% higher than they would be under the alternative policy.

Fortunately, the model proved to be reasonably correct in this instance. After fare increases ranging from 8 to 19%, with a weighted average of 16%, the model predicted gross revenues and passengers to within $3\frac{1}{2}$%. At this stage, the model seems to be validated reasonably well as far as the effect of trip cost on demand.

Based on this success, Wilder's management decided to trust the demand model more fully. Specifically, they decided to reduce fares some 20 to 30%. If the absolute value of the elasticity of travel with respect to price is significantly greater than unity, $|\eta_p| > 1$, this strategy should enable them to raise revenues considerably. This conclusion follows directly from the definition of price elasticity, Eq. (10.6). Since the revenue equals volume of traffic times the fares paid, the change in revenue is

$$\partial(\text{revenue}) = \text{price} \times \partial(\text{volume}) \times \text{volume} \times \partial(\text{price}) \qquad (10.8)$$

so that the rate of change of revenue with respect to price will be

$$\partial(\text{revenue})/\partial(\text{price}) = \text{volume} \times (\eta_p \times 1) \qquad (10.9)$$

and total revenues will increase when prices are lowered for $\eta_p < -1$. As we believed the price elasticity for the limousine service to be about -2, we had actually suggested that Wilder lower fares after the labor negotiations for higher wages. But this strategy seemed so counterintuitive that Wilder had initially rejected the idea.

In addition to providing another opportunity to test the model by altering fares, Wilder also decided in 1971 to discard the use of 11-passenger limousines in favor of 40-passenger coaches. As an eventual consequence of this greater capacity, they decided to cut frequency in half, and to serve La Guardia and Kennedy every hour instead of every half-hour. An opportunity now existed to test the effect of increased travel time.

Effect of Travel-Time Changes

As anticipated, the changes in frequency in the Wilder service, which severely degraded its overall speed of service, also cut deeply into its volume of traffic. The predictive power of the model was satisfactory, although not as powerful as when considering the effect of price alone. This is possibly due to the magnitude of the changes involved.

Both the changes in fare and travel time were quite large. The fare decrease suggested a doubling of the travel volume, and the schedule change suggested halving it. As these changes work at cross purposes, one should expect that the net change in volume would be about zero. Actually, comparisons show that the actual volumes are about 20% lower than the predicted volumes, depending upon how one includes the effect of schedule change on travel time. If this error is attributed to each factor individually and taken as a percent of the large volume of traffic which the fare changes suggest we could attain, or which the schedule changes indicate we could lose, the error is down to about 5%, which is certainly within reason.

We may assume, consequently, that although the model did not accurately predict the joint effect of two significant changes, it could have quite well predicted each of their effects separately. Therefore, it seems appropriate to assume that the model still appears reasonable but to recognize that it is still extremely simple and subject to error.

CONCLUSIONS

An immediate result of this work is a first-order model of a demand function for airport access trips. Although the model does not have a sophisticated appearance, it appears to be robust. It is one of the very few urban transportation models which have been systematically tested for real predictive capability. As a consequence of these tests we may conclude that auto availability, as distinguished from simple auto ownership, is a prime determinant of how people get to an airport. Also, it appears that airport access travelers are quite highly cost-sensitive in their modal choice to the airport, apparently much more so than they are time-sensitive.

The analysis also suggests how one might formulate and validate demand models in practice. As the example suggests, the process of determining which model really can be useful to planners as a predictive tool can be time-consuming, messy, and inconclusive by parts. As such, it is little different from most experimental studies. Yet, like other experimental studies, it can lay the foundation for an effective description of the phenomena of interest.

ACKNOWLEDGMENTS

The work reported here has been sustained over several years by friends in the industry, at the Port Authority of New York and New Jersey, and the Transportation Systems Center of the U.S. Department of Transportation. Robert Skinner, Frank Koller, and John Yaney deserve particular credit for collecting and analyzing the data. We are all grateful to the MIT Civil Engineering Systems Laboratory for its financial assistance.

REFERENCES

1. DE NEUFVILLE, R., AND MIERZEJEWSKI, E., 1972. Airport Access Cost-Effectiveness Analysis, *Transportation Engineering Journal, ASCE*, Vol. 98, No. TE3, August, pp. 663–678.
2. DE NEUFVILLE, R., AND STAFFORD, J. H., 1971. *Systems Analysis for Engineers and Managers*, New York: McGraw-Hill Book Company.
3. DE NEUFVILLE, R., SKINNER, R. E., JR., AND KOLLER, P. F., 1971. A Survey of the New York City Airport Limousine Service: A Demand Analysis, *Highway Research Record 348*, pp. 192–201.
4. DOMENCICH, T. A., KRAFT, G., AND VALETTE, R., 1968. Estimation of Urban Passenger Behavior: An Economic Demand Model, *Highway Research Record 238*, pp. 64–75.
5. KOLLER, P. F., AND SKINNER, R. E., JR., 1971. *An Airport Access Demand Study*, M.S. Thesis, Department of Civil Engineering, MIT, Cambridge, Mass., June.
6. LAVE, C. A., 1969. A Behavioral Approach to Model Split Forecasting, *Transportation Research*, Vol. 3, pp. 463–480.
7. Port of New York Authority, Central Planning Division, 1968. Kennedy Airport Access Project: Travel Time and Cost Study, New York. Sept.
8. QUANDT, R. E., AND BAUMOL, W. J., 1966. The Demand for Abstract Transport Modes: Theory and Measurement, *Journal of Regional Science*, Vol. 6, No. 2, pp. 13–25.
9. RASSAM, P. R., ELLIS, R. H., AND BENNETT, J. C., 1970. The N-Dimensional Logit Model: Development and Application, Washington, D.C.: Peat, Marwick, Mitchell & Co.
10. WARNER, S. L., 1962. *Stochastic Choice of Mode in Urban Travel: A Study in Binary Choice*, Chicago: Northwestern University Press.
11. YANEY, J. C., 1972. *New York Airport Access Demand Study*, M.S. Thesis, Department of Civil Engineering, MIT, Cambridge, Mass., June.
12. YU, J., 1970. Demand Model for Intercity Multimode Travel, *Transportation Engineering Journal, ASCE*, Vol. 96, No. TE2, May, pp. 203–218.

11

The Survival of the Passenger Train: The Demand for Railroad Passenger Transportation Between Boston and New York*

Franklin M. Fisher

The nature of the demand for railroad passenger transportation is not a new subject. Indeed, there has been controversy about it between the railroads and the Interstate Commerce Commission, the railroads insisting that demand was not sensitive to price (inelastic), and the Commission generally taking the opposite view (10). The two chief existing empirical studies of this problem, by Dixon (4) and Aitcheson (1), have been accepted as showing the truth of the I.C.C.'s contention that the bulk of passenger demand is sensitive to price or elastic. Thus J.C. Nelson (9) states:

> ... the Depression evidence indicated that the coach travel demand was elastic while the parlor and sleeping car demand was inelastic. ...

> ... between cities such as New York and Washington, New York and Boston, and New York and Chicago, where good schedules can be maintained, low coach fares, if low enough and combined with attractive trains and courteous service, might easily bring about the return of considerable coach traffic to the roads.

In view of the results of the present analysis, the evidence on which such conclusions and policy recommendations rest seems at least somewhat open to question.

*Adapted extensively from Chapter 6 of *A Priori Information and Time Series Analysis*, North-Holland Publishing Company, Amsterdam, 1962. For the purposes of these case studies, which focus on modeling strategy, important sections of the original study, dealing with ways of knowing whether or not the statistical estimates are reasonable, have been omitted. The interested reader is encouraged to refer to the original text.

We here reexamine the problem, concentrating on the Boston–New York passenger run for 1929–1940 and 1946–1956. Our findings have a wider applicability than this, however, and it is easy to use them to forecast the long run trend in the demand for passenger traffic between those points. In view of the finding that passenger runs of the Boston–New York or New York–Washington length and population density are the only ones in which the railroads are not at a clear and real cost disadvantage (8), such a forecast sheds light on the ability of railroads to remain in the passenger business at all.

The techniques employed here illustrate that a priori information which is the a posteriori information of other studies may be the most valuable and convincing sort of outside knowledge, and that the comparison of separate studies using the same methods on independent sets of data yields the most convincing possible kind of result.

SHORT-RUN DEMAND FUNCTION FOR 1929–1940

Theory

The major difficulty with the earlier analysis by Dixon is that he estimates only the relationship between railroad fares and passenger traffic, and deals with the influence of other variables, such as personal income, only as an afterthought. There is, however, some excuse for such a procedure. The number of observations he had available is scarcely large enough, and multicollinearity was too high for any more sophisticated approach. In fact, we are unable to improve on his results directly using his data.

In the case of the Boston–New York passenger run, however, some data are available for 1929–1940. The obvious approach in this area, therefore, would be simply to perform a multiple regression of passenger traffic on whatever explanatory variables seem appropriate. Although we shall indeed use this approach, it would be a mistake to think that all our problems are thereby solved. As we have argued elsewhere (5), longer time series directly aid us in this way only if the additional observations are generated by the same economic mechanism that produced the data already possessed.

In the present instance, there seems reason to believe that the demand reaction in 1929–1934 differed considerably from that in 1935–1940. In general, this was the case because of the changing characteristics of the substitutes for rail transportation between Boston and New York. These are bus service, air travel, and transportation by private automobile.

Of these substitutes, perhaps bus service underwent the least change from the early to the late 1930s. It did improve, as better buses and better roads were built, but the change was probably not so striking for this railroad substitute as for the other two.

Automobiles gained from product improvements, and the most important highway development in the area, the opening of the first section of the Merritt Parkway, in July 1937, benefitted them, but not the buses. The full effects of the parkway system were probably not felt by the railroad until after World War II, but the increasing use of automobiles in the New England area was certainly a factor changing the substitution characteristics of the demand for passenger transportation by rail.

The airplane also provided a sharp difference in transportation opportunity between the early and the late 1930s. Air service between Boston and New York was already in existence at the start of the decade, but it was not until the introduction of the DC-3 in 1935–1936 that airplane travel began to be accepted as a safe, comfortable, and usual means of transport. Over the 1929–1940 period, airline passenger traffic between Boston and New York rose from 3986 passengers in 1929 to 18,417 in 1934 to 168,994 in 1940 (3). This development alone should make us wary of including the observations for the early 1930s with those of the later prewar period without any adjustment.

It is not easy to see just what adjustment should be made, however. We clearly have not yet the data, or the information, that would enable us to specify and estimate a model complete enough to account for these effects. On the other hand, there are clearly too many relevant variables which are not approximately constant to enable us to split the period in half. It is thus necessary to make a somewhat unhappy compromise and to use the entire period together with an arbitrary definition of a trend in the demand function. Fortunately, however, the results here can be checked, for it is easy to predict how the estimates for the prewar period should differ from those obtained below for the later demand curve.

The relevant variables for the demand function now in question are the prices of rail, bus, and air travel between Boston and New York; income; the per capita stock of automobiles; population; and a number of variables affecting the attractiveness of substitutes, such as airline safety records, transit times by different means, road quality, automobile service facilities, automobile repair records, and the like. Data on these last factors are largely unavailable, nor is it clear in what way they should enter the analysis. What we shall do is to assume that their relatively slow moving influence is eliminated by doing the analysis not on the annual data, but on the differences between sequential observations. Further, since it is the case that year-to-year percentage population changes were nearly constant, we may treat this variable in the same way.

The way in which the per capita stocks of cars enters is also somewhat complicated. The per capita stock of cars fell for most of the first part of the period and rose thereafter. Fortunately, however, yearly percentage changes in the per capita stock of cars were roughly constant over each half of the 1929–1940 period taken separately. We may therefore introduce a dummy variable taking on the value zero for all observations up to 1934–1935 and taking on the value of unity thereafter. This will roughly allow for the change in direction of movement of the automobile stock.

The introduction of our dummy variable has an additional advantage. It allows us to take account of some kind of shifts in the demand function, for it allows us to take care of a change in trend, that is, of a change in the attractiveness of substitutes, after 1935.

The data on railway passenger traffic, real personal income per capita, and rail, bus, and airplane fares are given for 1929–1940 in Table 11.1. The personal income figures are for New York, Connecticut, Rhode Island, and Massachusetts. The fares given are all one-way, and the railroad coach fare is used since coach traffic accounted for the bulk of the passengers carried.

Table 11.1 Railroad Passengers Carried; Personal Income; One-Way Rail, Bus, and Air Fares: 1929–1940

Year	Railroad Passengers Carried ($\times 10^3$)	Personal Income per Capita ($)	One-Way Real Fares (1947 dollars)		
			R.R.	Bus	Air
1929	1276	1672	11.3	5.46	40.6
1930	1179	1372	11.6	5.60	38.8
1931	971	1303	12.7	6.15	24.4
1932	741	1130	14.1	6.85	24.3
1933	623	1111	14.9	5.48	25.1
1934	728	1160	14.4	5.10	24.3
1935	803	1196	14.1	5.11	23.7
1936	1080	1326	10.2	5.85	23.2
1937	1207	1326	7.49	4.80	22.6
1938	1062	1263	8.41	5.97	21.2
1939	1126	1347	9.65	6.06	20.1
1940	1192	1421	8.15	4.92	19.9

Source: M. R. Colberg, W. C. Bradford, and R. M. Alt, *Business Economics*, Homewood, Ill., Richard D. Irwin, Inc., 1957.

Analysis

We assume that the underlying demand function can be approximated by an exponential relationship. Specifically, we take the number of railroad passengers carried, Q, to be a function of real personal income per capita, Y; and real one-way rail, bus, and air fares, P, B, and A, respectively:

$$Q = a\, Y^b\, P^c\, B^d\, A^f \qquad (11.1)$$

The exponents of Eq. (11.1) are, in this kind of model, identical to the elasticities of demand with respect to the associated variable.

These parameters were estimated by regression on the differences of the natural logarithms of the variables, a procedure that eliminates the effects of slow-moving, longer-run trends. Thus we obtain for the full 1929–1940 period:

$$\Delta \log Q = + 0.00532 + 1.68\Delta \log Y - 0.426\Delta \log P$$
$$\qquad\qquad\quad (0.34) \qquad\qquad (0.153)$$

$$+ 0.236\Delta \log B + 0.155\Delta \log A, \qquad R^2 = 0.916 \qquad (11.2)$$
$$(0.146) \qquad\qquad (0.159)$$

When the dummy variable, D, described above is added we obtain

$$\Delta \log Q = + 0.0282 + 1.83\Delta \log Y - 0.376\Delta \log P \qquad (11.3)$$
$$\qquad\qquad\quad (0.366) \qquad\qquad (0.158)$$

$$+ 0.290\Delta \log B + 0.191\Delta \log A - 0.0531 D, \qquad R^2 = 0.932$$
$$(0.151) \qquad\qquad (0.161) \qquad\qquad (0.0494)$$

In both cases we can have high confidence in the value of the income elasticity and of the correlation coefficient R^2. Technically, these are significant at the 1% level. The other coefficients are not significant. All coefficients have the sign that theory indicates they should have.

Both the differences and the similarities between Eqs. (11.2) and (11.3) are of considerable interest. The chief difference is obviously in the constant term, the unexplained yearly trend. When the dummy trend-change variable D is introduced, unexplained trend goes from well under 1% to almost 3% per year. Moreover, in view of the definition of D, this means that there was nearly a 3% increasing trend in passenger traffic from 1929 to 1935 but that, subtracting the coefficient of D from the constant term, from 1935 to 1940 there was a *negative* trend of nearly the same magnitude. This change in trend direction is just what we should expect in view of our earlier discussion of the growth in the per capita stock of cars in the late 1930s and the earlier decline thereof. This leads us to accept Eq. (11.3) as the better regression, a judgment that is fortified when we examine long-run influences more explicitly.

The similarities between Eqs. (11.2) and (11.3) allow us to draw some further conclusions. To begin, it seems strongly apparent that, in view of the size and significance of the income elasticity in both equations, we may immediately reject Dixon's conclusion "[that changes in business activity do] have some effect on passenger traffic, but a tremendous increase or decrease in such activity is required to produce a noticeable change in the regu-

lar fare traffic . . . " and that "variations with income level are not too striking." The evidence of our results is strongly the other way.

More important than the size of the income coefficient, however, is
the fact that all price elasticities are small. In particular, the demand for
railroad passenger transportation is seen to be inelastic with respect to railroad fares—the opposite of the I.C.C.–Aitcheson–Dixon conclusion. Finally,
given the magnitude of this last elasticity, the substitution elasticities with
respect to bus and air fares seem about the right magnitude. Note, in particular, that, as we should expect, the bus was the more important substitute
in the prewar period. The plausibility of all these results strengthens our
faith in our procedures.

LONG-RUN DEMAND FUNCTION

Controlling for a long-run trend allows one to estimate a short-run demand
function but hides the effects we may be most interested in for prediction
of future events. Accordingly, we now turn to the analysis of a long-run
demand function and analyze a positive trend in passenger traffic before, and
a negative trend after, 1935.

Theory

We argued above that by taking logarithmic first differences and including our dummy variable we roughly eliminated or controlled the effects
on railway passenger demand of the growth of population and the stock of
private automobiles. We now wish to investigate those effects. Of course,
we cannot do this in a refined way, for several reasons. In particular, we cannot avoid picking up influences associated with other variables, such as the
general improvement in air transportation. However, even a crude procedure
will be of interest, especially for purposes of comparison with similar results
for the postwar period.

As our measure of the stock of cars, we use total registrations of automobiles in New York, Connecticut, Rhode Island, and Massachusetts.
These figures, together with the stock of cars per capita and population for
the same region, are given for 1929–1940 in Table 11.2.

Our investigation of these long-run effects proceeds in the following
way. First, we estimate the parameters of Eq. (11.3) using the logarithmic
first differences of the data. This gives us an estimate of the various income
and price elasticities. We then use these estimates to calculate the total number of railroad passengers (as opposed to the changes in total) expected in
any year, working always with logarithmic data. We then try to explain the

Table 11.2 Total Automobile Registrations, Population, and Registrations per Capita: New York and Southern New England: 1929–1940

Year	Registrations ($\times 10^3$)	Registrations per Capita	Population ($\times 10^6$)
1929	2899	0.15522	18.678
1930	3056	0.15920	19.196
1931	3073	0.15837	19.405
1932	3052	0.15591	19.574
1933	3012	0.15272	19.725
1934	3027	0.15225	19.883
1935	3134	0.15620	20.062
1936	3311	0.16395	20.194
1937	3474	0.17164	20.241
1938	3480	0.17179	20.255
1939	3602	0.17772	20.267
1940	3753	0.18577	20.201

Source: F. M. Fisher, *A Priori Information and Time Series Analysis*, North-Holland Publishing Company, Amsterdam, 1962 (reprinted 1966).

differences, the residuals, between these predictions and the actual values in terms of longer-run effects. In other words, we use the a posteriori information of Eq. (11.3) as a priori information, adding a level to our statistical pyramid.

The specific process is basically as follows. We use the (not first-differenced) logarithms of the data to calculate, for each year 1929–1940, an expected value for the number of railroad passengers, also in logarithms, log Q from Eq. (11.3). For each of the subperiods with which the dummy variables were associated for 1929–1934 and 1935–1940, we then take the regression of the residuals, the differences between the actual and the predicted number of railroad passengers, on the logarithm of the stock of cars per capita. We use the stock of cars in per capita form to avoid the high degree of multicollinearity between the stock of cars itself and population.

If the stock of cars, per capita, were the only variable accounting for long-run trend in passenger traffic demand, this would suffice. However, as this is not the case, we adjust crudely for the effect of other factors, in the following manner. Forget for the moment all the statistical difficulties involved in the procedure of taking residuals as described (we cannot be better than crude here), and let R be the described calculated residuals, C the stock of cars per capita, F population, and V other long run variables. Let u be a random disturbance with the usual properties. Now, suppose that the "true" relationship explaining the residuals is

$$R = K_i + \alpha \log C + \beta \log F + \lambda V + u \qquad (11.4)$$

Further, take the regression of V on log F to obtain

$$V = K_i + \mu \log F \tag{11.5}$$

where the K_i, here as elsewhere, are constant terms of no interest. If the fit of Eq. (11.5) were very good, that is, if population and improvements of bus and air transportation are highly correlated, a rather good assumption in view of common trends, then the residual from Eq. (11.5) would have a relatively low variance and thus a low regression coefficient when regressed on log C. We may thus, to a first approximation, take it as independent of log C. This assumption will be particularly plausible if we agree to exclude from V variables relating to improvement of automobile facilities which are likely to be correlated with C, and to include their effects instead in the coefficient of C, α.

We can now obtain log F in terms of log C by regression:

$$\log F = K_i + \pi \log C \tag{11.6}$$

It is then easy to see that substitution of Eqs. (11.5) and (11.6) in Eq. (11.4) leads to:

$$R = K_i + [\alpha + \pi(\beta + \mu\lambda)] \log C \tag{11.7}$$

Therefore, the expected value of the estimate, a, of α obtained by regressing R on log C is

$$E(a) = \alpha + \pi(\beta + \mu\lambda) \tag{11.8}$$

Since a is observable and π can be estimated, we can thus get a rough estimate of α, given a value for $\beta + \mu\lambda$. Furthermore, even though precise estimates of $\beta + \mu\lambda$ are unavailable, we can still place limits on that magnitude, thus obtaining limits on α, the effect on railroad passenger traffic caused by long-run trends in automobile availability.

We may also estimate the nature of α on theoretical grounds. When the stock of cars is rising and more people are buying cars, we may presume that the persons who are buying cars are inclined to travel, and will also tend to substitute automobile travel for rail. Thus we may expect α to be negative as the stock of cars rises. But it is not the case that the reverse argument applies when the automobile stock falls, for here the probability of interest is the probability that a family which sells its car without replacement will become a train-riding family. That probability must be close to zero, however, for much the same reasons as we gave for supposing that a train-riding family was more likely to buy a car than a non-train-riding one. Particularly is this true if cars are sold without replacement: the incentive to sell will be low if a great deal of travel is contemplated, and families in straitened circumstances are unlikely to travel much. The former reason, indeed, is likely to be stronger here than in the previous case, since given the rather rapid depreciation in value of automobiles, the cost calculation involved in deciding to sell is likely to be in favor of retention if a lot of train travel is envisaged. We should thus expect families selling cars without replacement to add relatively little to train travel, other things being equal.

Moreover, this being the case, we are almost certain to observe a somewhat positive α for periods in which the car stock is slightly declining. A decline in the total car stock means that more cars are being sold than are being bought, not, generally, that no families are acquiring cars for other than replacement purposes. Since families acquiring cars for the first time can be expected to cut down on train travel, and since families selling cars without replacement cannot be expected to contribute much to train travel, the net effect should be downward, that is, in the *same* direction as the movement of the stock of cars. We should thus expect a zero or positive α for periods in which the stock of cars is declining.

Analysis

We are now nearly ready for the results. There is one difficulty of interpretation that remains to be discussed, however. It is not clear how the standard errors in the regression of R on log C are to be interpreted for significance tests. Aside from the problems involved in the fact that R is itself an estimate, it is unclear how many degrees of freedom are involved. Since we have argued that significance tests are not generally applicable anyway (5), this is not very disturbing, and we may continue to interpret standard errors as measures of goodness of fit only.

Estimating Eq. (11.7) for the 1929–1934 period, we obtain:

$$R = K_1 + 0.684 \log C, \qquad r^2 = 0.0144 \qquad (11.9)$$
$$(2.83)$$

Estimating Eq. (11.6) for the same period, we obtain:

$$\log F = K_2 - 0.719 \log C, \qquad r^2 = 0.336 \qquad (11.10)$$
$$(0.505)$$

Equations (11.8) and (11.9) give us estimates of $E(a)$ and π.

It remains for us to guess the value of $\beta + \mu\lambda$. Clearly, to begin, the sign of β should be positive. Other things being equal, population growth should have a favorable effect on railroad passenger traffic. Moreover, it seems plausible that β should be greater than unity—a growth in population means not only more people to travel, but also more people to whom to travel. In short, it seems reasonable to place β between 1 and 2 and probably closer to 1.

Second, λ can be expected to be negative since substitute quality was rising over our period. Since population was also rising, μ will be positive, so the product of the two parameters will be negative. It is impossible to say with any precision what the magnitude of that product should be, but it seems unreasonable to suppose that it is greater than β in absolute value, for that would mean that the gross effect of population growth on Boston–New York traffic, other things being equal, was negative. We shall probably be safe in taking $\beta + \mu\lambda$ as between zero and 2. This yields as limits on α for the 1929–1934 period by simple substitution in Eq. (11.7):

$$0.684 \leq \alpha \leq 2.12 \tag{11.11}$$

Thus α was positive, or, in view of the huge standard error in Eq. (11.9), zero, for the early 1930s, as predicted by our discussion.

Using the same procedure for the late 1930s, however, yields a far different result. Estimating Eq. (11.7) for 1935–1940, we obtain:

$$R = K_3 - 0.359 \log C, \quad r^2 = 0.428 \tag{11.12}$$
$$(0.208)$$

Estimating Eq. (11.6) for the same period, we obtain:

$$\log F = K_4 + 0.0354 \log C, \quad r^2 = 0.325 \tag{11.13}$$
$$(0.0254)$$

Again setting $\beta + \mu\lambda$ between zero and 2 (certainly closer to zero here than for the earlier period in view of the accelerated growth of substitute quality), we obtain new limits on α for the 1935–1940 period:

$$-0.359 \geq \alpha \geq -0.430 \tag{11.14}$$

so that the elasticity of railroad passenger demand with respect to the stock of cars appears to be on the order of -0.4 for the 1935–1940 period. This is, again, consistent with our theoretical discussion.

This result, together with the right sign on α for the early 1930s, strengthens our confidence in our model for railroad passenger volumes, Eq. (11.3). However, all our results will best be tested by applying similar methods to the data for the postwar period and comparing the resulting estimates with those already obtained.

PREDICTION FOR 1946–1956

We may now use our results to forecast the general postwar history of this passenger run and may predict the changes which we should expect to see in the demand function when the latter is estimated using the same techniques on postwar data. As these last predictions were substantially made before such data were secured, such an exercise was most valuable, as it allowed us to compare the results of two studies, carried out with the same methods on different sets of data. The consistency of such results is the best kind of evidence for the methods used.

Theory

Given our results, we should expect to see a very large increase in Boston–New York rail travel during and immediately after World War II, owing principally to increasing incomes and the decrease of the stock of cars per capita during the war, as well as to gasoline rationing. After the war, we should expect a decrease in Boston–New York travel by rail as the stock of cars increased rapidly and deferred demand for automobiles was satisfied. Both of these expectations are borne out.

After the war, we should expect certain changes to take place in the demand function. The improvement in airline travel should have produced a reasonably high positive coefficient for airplane fares. The elasticity of rail travel with respect to air travel may have become higher than that with respect to bus fare for the decade after the war, reversing the situation shown in Eq. (11.3), as increasing incomes and continued product improvement made air travel an increasingly attractive substitute.

Furthermore, we should expect changes in the short-run income elasticity. In the period already examined, when incomes were low, railroad transportation was a superior good—at least in the short run. With rising incomes, however, income elasticity probably fell as downward changes in income no longer induced so many travelers to ride the bus. Further, as income continued to rise, railroad travel between Boston and New York became something of an inferior good for those in high-income brackets as it became increasingly easy to afford air transportation. This effect probably also reduced income elasticity.

There can, of course, be no question about the change in price elasticity after the war. The steady improvement in substitute means of passenger transportation, especially the better roads and service facilities for automobiles, must have increased the price elasticity of demand.

Finally, we shall see that our analysis of the elasticity of demand with respect to the per capita stock of cars leads us to expect a substantial increase in that coefficient also.

We shall see that all these predictions are verified when we examine the results for the postwar period.

Analysis of Short-Run Demand Function

We turn now to the data for that period (Table 11.3). We begin with 1946 and end with 1956. The latter date is chosen as the end point because the opening of the Massachusetts Turnpike in 1957 and the completion of the Connecticut Turnpike in 1958 provided a large and discontinuous improvement in automobile and bus travel between Boston and New York, substantially changing the alternatives open to potential rail travelers.

Table 11.3 Railroad Passengers Carried; Personal Income; One-Way Rail, Bus, and Air Fares: 1946–1950, 1952–1956

			One-Way Rail Fares (1947 dollars)		
Year	Railroad Passengers Carried ($\times 10^3$)	Personal Income per Capita ($)	R.R. (cents/mile)	Bus ($)	Air ($)
1946	3090	1922	3.15	4.48	11.5
1947	2287	1719	2.92	4.51	10.7
1948	1914	1676	3.28	4.64	11.2
1949	1597	1654	3.43	4.69	12.6
1950	1458	1781	3.78	4.64	12.5
1952	1703	1821	3.42	4.56	12.0
1953	1649	1859	3.39	4.53	12.2
1954	1610	1857	3.23	4.31	11.6
1955	1540	1981	3.24	4.32	11.7
1956	1591	2081	3.30	4.54	11.5

Source: F. M. Fisher, *A Priori Information and Time Series Analysis*, North-Holland Publishing Company, Amsterdam, 1962 (reprinted 1966).

Data on through railroad passengers carried between the two cities were supplied by the New York, New Haven and Hartford Railroad, monthly for all years involved, with the exception of November 1950 through December 1951. This means that 1951 has to be excluded; the missing data for 1950 are estimated by assuming that travel in the missing 2 months was the same fraction of total travel for the year as was travel in the corresponding 2 months of 1949.

Data on rail fares on coach were also supplied by the New Haven, this time in the form of cents per mile. As the distance between Boston and New York has not changed, the results using such data are identical with those that would be obtained using total fare in view of our logarithmic form. As before, we use one-way coach fare as our variable. The yearly figures are constructed as weighted averages, the weights being the monthly traffic. Both here and in the series for bus and airline fares, the reduction in the federal transportation tax in 1954 from 15 to 10% was counted as a fare reduction. All fares are reported to include tax.

Our annual average money airline fares were constructed by linear interpolation from the one-way first-class fares charged by American Airlines. It is impossible to tell what allowance should have been made for the existence of round-trip and family-plan discounts (both here and for the railroad fare). However, provided that a reasonable constant proportion of traffic made use of such discounts, the effects of error here will not appear in any parameter of interest.

A more serious question is that of aircoach travel, which began in 1953 between the two cities and has grown considerably in relative importance since that time. Taking as representative all tickets sold by American Airlines for direct travel between the two cities, we find that coach passengers were only about 4.7 and 7.9% of all passengers in March and September 1956, although the figure had risen to about 21.0% by 1958. We thus feel justified in ignoring this problem in our data, although we clearly could not ignore it were we to use data from years after 1956.

Estimating the short-run demand function for the postwar decade, we obtain, as we did for Eq. (11.2):

$$\Delta \log Q = -0.0880 + 1.79 \Delta \log Y - 1.32 \Delta \log P$$
$$\quad\quad\quad\quad\quad (0.394) \quad\quad\quad (0.624)$$
$$+ 0.261 \Delta \log B + 0.641 \Delta \log A, \quad R^2 = 0.876 \quad\quad (11.15)$$
$$(0.859) \quad\quad\quad (0.581)$$

These results are very pleasing. Not only does every coefficient have the right sign, but comparison of Eq. (11.15) with Eq. (11.2) reveals that the point estimate of every coefficient has moved in the direction predicted on theoretical grounds. In particular, income elasticity has decreased very slightly and price elasticity increased appreciably; indeed, it appears that demand is now slightly price elastic. This result somewhat confirms the prevailing opinion, which however was based on the incorrect evidence of prewar results. Further, elasticity with respect to air fare has substantially increased and air transportation is clearly now a more important substitute than is the bus.

Finally, observe that the magnitude of the constant term in Eq. (11.15) indicates the presence of a more rapidly declining trend (nearly 10% per year) in the postwar decade than was the case in the prewar results. This is as we should expect, considering the rapid growth of the stock of cars after the war.

Analysis of Long-Run Demand Function

We now turn to the explanation of that trend, to the long-run demand function. The data for New York, Massachusetts, Rhode Island, and Connecticut on registrations of private automobiles, registrations per capita, and population are given in Table 11.4.

Table 11.4 Total Automobile Registrations, Population, and Registrations per Capita: New York and Southern New England: 1946–1950, 1952–1956

Year	Registrations $(\times 10^3)$	Registrations per Capita	Population $(\times 10^6)$
1946	3680	0.17803	20.671
1947	4055	0.19006	21.335
1948	4320	0.19663	21.972
1949	4576	0.20369	22.466
1950	5060	0.22593	22.395
1952	5456	0.24113	22.626
1953	5826	0.25163	23.154
1954	6025	0.25541	23.588
1955	6523	0.27528	23.697
1956	6774	0.28576	23.706

Source: F. M. Fisher, *A Priori Information and Time Series Analysis*, North-Holland Publishing Company, Amsterdam, 1962 (reprinted 1966).

Estimating Eq. (11.7), as before, yields

$$R = K_5 - 1.36 \log C, \quad r^2 = 0.959$$
$$(0.100)$$
$$(11.16)$$

Similarly, estimating Eq. (11.6) yields

$$\log F = K_6 + 0.267 \log C, \quad r^2 = 0.913$$
$$(0.00928)$$
$$(11.17)$$

Now, $\beta + \mu\lambda$ seems almost certain to be lower here than in the prewar period, because of road improvement. However, because of the relatively low magnitude of the estimate of π just obtained in Eq. (11.17) ($\pi = 0.267$),

the limits set on α are not very sensitive to our guess here. We may therefore continue to take $\beta + \mu\lambda$ as between zero and 2, remembering that the
true value is rather likely to be close to the former limit. Using Eq. (11.8),
this yields $-1.36 \geq \alpha \geq -1.90$ as our limits on α—a considerable increase
in absolute magnitude from the prewar period.

IMPLICATIONS FOR PASSENGER TRAINS

Now, what are the implications of all this for the future of the Boston–
New York passenger run? The picture seems bleak indeed for the railroad.
The increasing attractiveness of substitute means of transport will tend to
reduce traffic in the future, and it is doubtful whether future population
growth will offset this. The railroad thus seems certain to lose traffic, at
least relatively.

Indeed, the only hopeful sign in our results would seem to be the still
sizable coefficient of the income variable in postwar demand function, Eq.
(11.15). It would appear that the New Haven Railroad can hope to gain
passenger traffic from increases in personal income, if at all. But this hope
is forlorn. All our income elasticity estimates were estimates of the short-
run income elasticity with the stock of cars per capita constant. Our discussion also applied only to this case. Unfortunately for the railroad,
however, it is impossible to assume that the stock of cars is uninfluenced
by secular changes in income. Indeed, Chow finds that the elasticity of the
demand for the stock of automobiles with respect to income is greater than
2 (2). If this is the case, then in view of the high elasticity predicted and
found above of passenger traffic with respect to the per capita stock of cars,
the direct effects of a secular increase in real income will be more than offset
by its indirect effect through the stock of automobiles.

It follows that Boston–New York passenger traffic will not only fail to
increase because of secular rises in income, it will even decrease. Surely
there is little reason to believe that the effect of increases in the stock of cars
on this traffic will not be at least as great in the future as it has been in the
past. The continued improvement of the highway system, the opening of
the Massachusetts and Connecticut Turnpikes in particular, assures this,
and the inconvenience of traffic congestion does not seem likely to reduce
automobile usage in intercity travel, at least for the present.

Moreover, the picture is even darker for the railroads when we consider
the possible future of passenger traffic on other runs. Railroads appear to
be at a real cost disadvantage on all passenger runs other than those of about
Boston–New York length and population density (8). For shorter runs and
less dense population the bus, and for longer runs the airplane, seem to have
lower real costs. Coupled with our findings on the probable future of the
demand for passenger transportation on the Boston–New York run, which

probably apply in general to other runs of the same length, these facts strongly lead us to approve Interstate Commerce Commission Examiner H. Hosmer's 1958 statement that, without continued government aid, the railroads will be out of the passenger business, if not by 1970, as he states, then not much later (6). It is interesting to note that in 1971, eleven years after this study was completed, this is essentially what happened when the United States established Amtrak. As we said in 1960, it seems clear that the federal government must either subsidize or allow continued abandonment of passenger service.

Because of high costs, passenger service on most runs had already ceased to yield a profit by the late 1950s. Our results show that, to the extent that Boston–New York demand conditions are at all typical, there is little hope of relief on the demand side. The demand for railroad passenger transportation on that run is likely at best to remain stable in the long run as population and incomes grow.

Further, with a price elasticity so close to unity, lower rail fares cannot do terribly much to stimulate demand, although some relief is clearly possible. In particular, to the extent that our observed elasticity is an average of low-elasticity business demand and high-elasticity pleasure traffic, it is possible that experimentation with price discrimination, between different times of day, for example, may bring temporary relief. Even such gains as are achieved are likely quickly to disappear, however, as substitutes continue to become increasingly attractive.

ACKNOWLEDGMENTS

Many persons and institutions helped sustain me, intellectually and physically, during the research reported here. My greatest personal debt is to John R. Meyer. As my employer, my thesis director, and my friend he gave freely of his time. His advice and suggestions were invaluable.

REFERENCES

1. AITCHESON, B., 1941. *Preliminary Examination of Factors Affecting the Demand for Rail Passenger Travel*, I.C.C. Bureau of Statistics, Statement 4129, Sept.
2. CHOW, G., 1957. *Demand for Automobiles in the United States*, Amsterdam: North-Holland Publishing Company.
3. COLBERG, M. R., BRADFORD, W. C., AND ALT, R. M., 1957. *Business Economics*, Homewood, Ill.: Richard D. Irwin, Inc.
4. DIXON, W. J., 1941. *The Elasticity of Demand for Railroad Passenger Transportation*, Ph.D. Thesis, Yale University, New Haven, Conn.
5. FISHER, F. M., 1962 (reprinted 1966). *A Priori Information and Time Series Analysis*, Amsterdam: North-Holland Publishing Company.

6. HOSMER, H., 1958. End of Rail Travel by 1970 Foreseen, *New York Times*, Sept. 19, p. 1.

7. MAHAFFIE (Commissioner), 1949. Recommendations, *I.C.C. Reports*, Vol. 276, pp. 433, 488.

8. MEYER, J. R., PECK, M. J., STENASON, J., AND ZWICK, C., 1959. *The Economics of Competition in the Transportation Industry*, Cambridge, Mass.: Harvard University Press.

9. NELSON, J. C., 1959. *Railroad Transportation and Public Policy*, Washington. D.C.: The Brookings Institution.

10. Passenger Fares and Surcharges, 1936. *I.C.C. Reports*, Vol. 214, p. 174.

PART III

Optimization

12

Introduction

This part discusses the role that a set of mathematical techniques called *optimization* can play in the planning and management of large-scale systems, and then illustrates it through a series of seven case studies. As with the use of any type of analytic techniques within the context of complex problem solving, the focus for discussion falls not only on the various techniques available for analysis but also on the art of how such mathematical procedures are applied. Large-scale systems present considerable problems in terms of the number of decision variables and objectives, the stochastic variations in economic and physical conditions, and the representation of the systems being modeled. These issues must be acknowledged and addressed in a straightforward manner with proper attention paid to the particular important aspects of a given problem. No one procedure or series of procedures will be the panacea that solves all problems to the last detail. Rather, there is a sensitive tradeoff between the accuracy of the different techniques available and the reality of how we can represent any particular problem. For each such problem a decision must be made about which portions should be addressed with which techniques.

CONTEXT FOR OPTIMIZATION

We start with the premise that any system can be simulated in detail. That is, given a specific set of known parameters and fixed policy decisions for the system, it is possible with some degree of certainty to forecast precisely how the system will respond in terms of its objectives and measures of effectiveness. Simulations can vary from simple intuitive mental models all the way to sophisticated stochastic simulations on large computers. But they all have one thing in common: comparison of alternative configurations for the system can only be made by changing conditions, exercising the model over again, and comparing the results. For the average problem in large-scale systems, the possible combination of different levels of decision variables is so large that a detailed simulation of all alternatives would be extremely costly and often even physically impossible. What is needed is some guidance, either from judgment alone or from judgment aided by other analytic techniques, as to what is a good limited subset of the possible alternatives to be studied in detail.

Optimization can be of great use in the search for good alternatives to examine in detail. Optimization techniques are mathematical procedures for finding, among all possible feasible solutions to a problem, the alternative that best satisfies the stated objective for choice. There are two main differences between simulation and optimization. One is that the optimization process explicitly considers different levels of the decision variables, whereas simulation does not. This extra capability to look at possible alternatives at the same time must, however, have its price: in this case, the price is a reduced ability to model the system in detail. The second difference between simulation and optimization is that the optimization techniques require a specific mathematical abstraction of the objectives and constraints which, given the complexity of the original problem, may be quite superficial. Simulation, on the other hand, can be made as complex and detailed as time, money, patience, and need allow.

The message of this chapter and of the case studies is that optimization and simulation are tools that may be successfully used interactively rather than each standing alone as competing analytical techniques. The important issue is, then, not which is to be used but how they may be used in cooperation so that the strong points of each can contribute to the analysis of a problem.

The term *screening model* is used in this part to define a simple optimization model that works interactively with a more detailed simulation to guide analysis. In a sense almost all optimization models are screening models, for they are used to abstract and gain understanding about a more complex problem. To provide some idea of the alternative structural forms that a screening model might take, we describe the general structure of optimization and the mathematical techniques available for solving different types of problems. Emphasis is placed on the techniques applicable to the large, high-dimensional problems typical of systems planning and design, and on the important issues of how such techniques should be used.

The discussion focuses on how the tradeoffs in time and money invest-ment are to be made between optimization and simulation, while allowing each to address the type of question they can best handle. The issues of the greatest importance involve: how stochasticity will be dealt with, what ob-jectives are and how the conflict between them can be illuminated; how system interactions, such as those between supply and demand, are to be ap-proached; and how the dimensionality questions raised by different levels of decisions and decisions over time are to be best resolved. All of this em-phasis on analytic modeling must still, however, be considered in its proper perspective. Analysis in complex, ill-defined problem areas is done as an aid to judgment, not as a replacement for it. The analyst must keep firmly in mind that the audience for whom the work is being performed has many objectives. Good analysis recognizes the inherent conflicts without bias or prejudice and provides a better understanding of the system under considera-tion and more information for the decision process.

OPTIMIZATION TECHNIQUES

The growing and vigorous field of applied mathematics known as mathemat-ical programming is concerned with the solution of optimization problems. This involves finding the value of a set of decision variables that optimize a statement of objectives while not violating given mathematical statements of constraints on the levels those variables may take. This may be stated as:

Maximize (or minimize) $z = f(\bar{X})$ (12.1)

subject to the constraints

$$g_i(\bar{X}) \leq b_i, \quad i = 1, \ldots, m$$
$$\bar{X} \geq 0$$

(12.2)

where $\bar{X}$ = vector of the unknown decision variables, x_j, $j = 1, \ldots, n$

$f(\bar{X})$ = known objective function, which transforms specific levels of the decision variables to a single value, z, as a measure of evaluation in terms of the objectives of the problem

$g_i(\bar{X}) \leq b_i$ = ith constraint, a known function that relates levels of the decision variables to a specified bound, b_i, which must not be exceeded.

The optimization problem is to find the vector $\bar{X}$ that optimizes the objective function while satisfying the constraints.

As an example of an optimization problem, consider an automobile manufacturer who makes several models. He has limited resources in terms of steel, glass, rubber, plastic, production space, and labor and knows how much profit he can make from selling a given number of each model. His decision is how many automobiles of each model to build. The constraints of the problem show how much of each resource is used in producing each model and require that the limits on available resources not be exceeded. The objective function, in this case, will be to maximize profit where gain is measured by the amount of money to be obtained by selling so many of a particular model and cost is a function of the amount of materials, labor, and so on, expended in the manufacturing process.

To move from an original problem, such as that of the car manufacturer, to a mathematical form for solution, many assumptions must be made. These concern the forms of the constraints and objective function, and the relationships between the decision variables and the various functions. As a rule, the more involved these relationships are assumed to be, the more difficult it will be to solve the resulting optimization problem. Thus there is a strong interaction between the reality of the formulation and one's ability to solve the problem and obtain information about the decision. The choice of the structure of the optimization problems represents an implicit decision about the appropriate balance between complexity and computational flexibility.

Optimal Linear Programming

If one makes linear assumptions, that is, assumes that the objective function and constraints are linear functions of the decision variables, we have the special structure known as linear programming. Procedures are available, as software packages on all major computer systems, which will handle very large linear optimization models very quickly and with little trouble, and will provide a great deal of sensitivity information as well. For example, IBM's MPSX software system, available on the 360/370 series of computers, has a capacity of 16,000 constraints and an unlimited number of decision variables. Although it is difficult for the novice user to envision such a large problem, it does not take long to discover relatively simple problems which have an extremely large number of constraints and variables. Even more important, the software programs are relatively foolproof and can be used with little knowledge of computer programming. Further, they will generally always find a global optimal solution if an optimal solution exists. To give some idea of the efficiency of these packages, the author used MPSX on the IBM 370/165 for a problem with 650 constraints and 1300 variables and obtained the optimal solution and sensitivity analysis at a total cost of about $25.

Special structure in linear programming, if recognized, can be exploited to give even faster computational results. An entire class of network and transportation problems may be solved using special procedures, such as Ford and Fulkerson's Out of Kilter Algorithm, with far greater efficiency than with MPSX. However, recognizing structure and finding or writing a large enough computer code to handle such problems may prove to be difficult.

Nonlinear Programming

So far we have discussed the great computational ease of dealing with linear problems, but the reader may well be left with a nagging doubt about the validity of a completely linear model for any process he is concerned with. What happens in nonlinear modeling, since this seems to be a more realistic case? For nonlinear programming, the computational methods are based on the recognition of special structure and, for any comparable expense, involve fewer variables and constraints.

For the case where objective functions and constraints are nonlinear but separable, that is, where decision variables are raised to powers other than unity but do no interact between each other, a variation on linear programming known as *separable programming* is helpful. It uses linear approximations of the nonlinear functions and will provide global optimal solutions when maximizing concave functions or minimizing convex functions, and local optimal solutions otherwise. It is available on IBM equipment as part of the MPSX package, and on other systems. Several other optimization problems, such as quadratic programming, are similarly solved by transforming them to a linear programming form.

An important class of problems are those in which some or all decision variables take on integer values rather than any number in the continuum. Such a formulation often represents a problem more realistically. In particular, binary variables, which equal either zero or unity, are very useful in modeling logical decisions such as whether a project should be undertaken or not. Integer programming methods include branch-and-bound, group-theoretic methods and cutting-plane algorithms. The MPSX package previously discussed also solves integer programming problems. Its procedure is based on the use of branch-and-bound and the solution of a series of linear programming problems. Of necessity, solutions can only be obtained for integer programming problems smaller than those reported for linear programming. The time required to obtain solutions are not only much greater and thus more expensive but also depend heavily on the structure of the problem and the data.

For more highly nonlinear problems, methods such as search, geometric programming, and dynamic programming are available. These are most applicable to a series of problems that arise in technologic design. In technologic design, a particular well-defined process with interactive subsystems, such as a chemical process or a waste treatment facility, is to be optimized. Usually, in such cases the objective functions are highly interactive and non-linear, but the problem has relatively few decision variables and constraints. Thus their solution procedures involve methods most removed from linear programming. In search, complex nonlinear functions are manipulated to find points at which the first derivative vanishes, that is, points at which the necessary conditions for a local optimum are satisfied. Such methods include the Newton–Raphson, Golden Section, and Fletcher–Powell methods for unconstrained problems. These may be used for constrained problems by applying conversion methods such as that of Sequential Unconstrained Multipliers (SUMT). All the search procedures have convergence problems in some data cases, are extremely data dependent, and can be trapped by certain pathological cases.

Geometric programming, which requires a nonlinear objective function and nonlinear constraints of a very special form, finds a solution by operating on the dual formulation, which has a linear constraint set. Dynamic programming is a recursive optimization technique for solving nonlinear problems with special structure which permits decomposition of a problem. This structure is most often found in problems with multiple time steps. Dynamic programming works best with few constraints, but computation times are remarkably good when it can be applied. The main problems in its application is recognizing structures that permit its use, and in dealing with the increased dimensionality added by additional constraints.

A brief word should be said about what is to be done when all else fails. We have dealt with the interaction between simulation and optimization: in many cases simulation may be the only tool for application because a proper analytic formulation cannot be abstracted or solved. In such cases, intuition about the problem can be gained simply by exercising a simulation model. Alternatively, a heuristic procedure can be used. Such procedures consist of a set of rules for an algorithm that produces good solutions to a problem for which there is no way to demonstrate when the optimum is obtained. These procedures evoke "common sense" in searching for good solutions, and if anything is to be gained by this discourse it is the importance of common sense in the solution and application of optimization techniques.

A blind insistence on optimal solutions for a complex model that does not represent realistic situations can be worse than no analysis at all: it wastes time and destroys the confidence of the intended audience. Further, there is always the conflict between those who are interested in only the elegance of the mathematics and those who wish to solve a practical problem. Some of the cases that follow illustrate instances in which the most useful mathematical procedure is one that is quite crude from a mathematical viewpoint.

ADDITIONAL READING

Those interested in further references on optimization techniques themselves are offered the following suggestions. As general overall references, we recommend

HILLIER, F. S., AND LIEBERMAN, G. J., 1967. *Introduction to Operations Research*, San Francisco: Holden-Day, Inc.
WAGNER, H. M., 1969. *Principles of Operations Research*, Englewood Cliffs, N.J.: Prentice-Hall, Inc.

As regards linear programming

DANTZIG, G., 1963. *Linear Programming and Extensions*, Princeton, N.J.: Princeton University Press.
HADLEY, G., 1962. *Linear Programming*, Reading, Mass.: Addison-Wesley Publishing Company, Inc.
SIMMONDS, D., 1972. *Linear Programming for Operations Research*, San Francisco: Holden-Day, Inc.

The special procedures for solving network problems are presented by

FORD, L., AND FULKERSON, D., 1963. *Flow in Networks*, Princeton, N.J.: Princeton University Press.
FRANK, H., AND FRISCH, I., 1971. *Communication, Transmission and Transportation Networks*, Reading, Mass.: Addison-Wesley Publishing Company, Inc.

For integer programming, the reader may wish to consult

HU, T. C., 1969. *Integer Programming and Network Flows*, Reading, Mass.: Addison-Wesley Publishing Company, Inc.
PLANE, D., AND MCMILLAN, C., JR., 1971. *Discrete Optimization: Integer Programming and Network Analysis for Management Decisions*, Englewood Cliffs, N.J.: Prentice-Hall, Inc.

For a discussion of search and geometric programming

WILDE, D. J., 1964. *Optimum Seeking Methods*, Englewood Cliffs, N.J.: Prentice-Hall, Inc.
WILDE, D. J., AND BEIGHTLER, C. S., 1967. *Foundations of Optimization*, Englewood Cliffs, N.J.: Prentice-Hall, Inc.

A good treatment of the algorithms for dynamic programming is given by

NEMHAUSER, G. L., 1966. *Introduction to Dynamic Programming*, New York: John Wiley & Sons, Inc.

A discussion of recent developments in multilevel optimization is provided by

MESEROVIC, M. D., MACKO, D., AND TAKAHARA, Y., 1970. *Theory of Hierarchical, Multilevel Systems*, New York: Academic Press, Inc.

A comprehensive treatment of nonlinear programming is given by

ZANGWILL, W., 1969. *Nonlinear Programming: A Unified Approach*, Englewood Cliffs, N.J.: Prentice-Hall, Inc.

Finally, general textbooks on strategy for large-scale systems computations are

LASDON, L., 1970. *Optimization Theory for Large-Scale Systems*, New York: The Macmillan Company.
WISMER, D., ed., 1971. *Optimization for Large-Scale Systems with Applications*, New York: McGraw-Hill Book Company.

APPLICATION OF OPTIMIZATION—SOME STREET KNOWLEDGE

The sense running throughout this chapter should make clear our feeling about optimization and its role in planning, management, and design. It is a useful tool, but its successful application depends to a great extent on how it is used and on how the results are interpreted. This use is always one of suggesting good designs for further study, not as a direct design tool. The following issues and topics seem to check the major questions that each analyst must answer and which the new practitioner will, at one time or another, painfully have to confront.

Problem "Reality" and Problem Solution

There is a strong tradeoff between the accuracy of optimization techniques and the reality of the problem formulation. The optimization model is, at best, a simplified abstraction of a complex situation. Inclusion of nonlinear cost functions and constraints, references to the stochastic nature of the problem, multiple objectives, and many variables and constraints can, however, help make the model more realistic. Such niceties also make solution more cumbersome, if not impossible. Therefore, very early in problem formulation the analyst finds himself beginning to make simplifying assumptions. The extent of these assumptions depends very much on the problem under attack and the type of information the analyst wishes to obtain.

Simple models based on linear programming are easy to solve, and a great deal can be learned, at little cost, by performing enormous amounts of sensitivity analysis. In many planning problems, such simple analyses provide more than enough insight to motivate future detailed studies on specific parts of the problem. For other types of problems, which are of a more detailed level, the introduction into the model of nonlinearities and other realistic characteristics is absolutely necessary to gain any useful guidance. In some problems, finally, an optimization model either cannot be successfully abstracted or can only be solved approximately. The appropriate balance between the use of optimization and simulation and the right degree of complexity to incorporate into the screening model must, therefore, be chosen carefully according to the peculiarities of a given problem.

Some practitioners (let us refer to them as the analysis freaks) liken the practice of using a simplified model to obtain solutions to that of knowingly playing in a crooked poker game because it is the only game in town. Although the argument has some merit, it is obviously more comforting to have some answers than an elegant model that cannot be solved. The people who love complexity for complexity's sake will just have to live with the fact that there are whole classes of problems in which simple models make very valuable contributions. Cases that fit this mold are discussed below.

The Effect of Problem Type

The type of problem to be addressed does affect the degree of sophistication brought to the solution technique. For years optimization has been used in various aspects of planning and design, and it is interesting to see that people in different areas have developed very different techniques and viewpoints. One school of thought is that associated with chemical engineering and other aspects of technologic design, where interactive but well-defined components of a process must be put together to meet specifications and minimize costs. These practitioners developed techniques that solved extremely nonlinear problems with few variables and constraints. They use solution algorithms which are essentially "hill climbing," in which a search is made for the local hilltops in the response surface using a combination of insight, guile, and patience. The other main school of thought is that of the "planners," who deal with problems that usually have much more weighty structure. Their problems have thousands of variables and constraints but linear or nearly linear objective functions and constraints that will allow approximation by linear programming. It is in this type of problem, incidentally, that integer variables become needed. The case studies provide examples of both the planning and design types of problems.

Even within the large-scale problems faced by planners, there are still different levels of approach. There are many levels of decisions to be made about the configuration of any system. At the broadest policy level, resources must be allocated among competing technologies and operational strategies.

At the next level, specifics must be considered about how these policies will be divided along different lines of authority and what groups will produce what, and how. At the most detailed scale, the minute decision of how each individual component will be scheduled is of interest.

There is now no computationally feasible way to deal simultaneously with all these levels in systems planning and design. Although multilevel optimization has been successfully pursued by Meserovic and others, the problems we refer to have more dimensions and are more difficult to abstract. Our only possibility is thus to decompose the complete problem into a series of subproblems, one at each decision level. The effect of the other levels is represented by estimating parameters of their contribution to the level under consideration. A simple law of systems is that the optimal solution to a series of subproblems is not necessarily the optimal solution to the original problem. Nonetheless, proceed we must.

In general, each of these levels of planning and design requires quite different modeling efforts. In most cases, the broader the problem, especially for planning at the highest policy levels, the better the chance for a very simple model. It is only as one takes a more detailed view and must order all the myriad tasks that more complicated models are necessary. Different levels of problems may thus aptly require quite different degrees of sophistication in the model and the optimization procedure.

As a short aside, we mention briefly a particular syndrome that crops up occasionally among us all. Being more conversant in some techniques than in others, whenever we see a new problem it tends naturally to fit our favorite. This is the law of the hammer, in which technique is the hammer and every problem is a nail. This is mentioned here only in an effort to keep us all honest by continually asking if another way might not be better. In most cases there are several paths to Nirvana and they should all be considered at least. Of course this comment is applicable only to the reader's colleagues and never to himself.

Multiobjective Considerations

Although optimization is essentially a single objective technique, any problem worth analysis has more than one objective. The question of how to deal with more than one objective is much more than a question of how multiple objectives are handled mathematically. The discussion of that issue is the main thrust of Part IV. It is helpful at this point to emphasize that most problems do have multiple objectives. The hidden scent of other objectives is found throughout the case studies of this part. Sometimes they appear as constraints such as minimum service requirements or as demands that must be met. Other times they are put into the objective function with a conversion factor that changes their units to some common form such as dollars. After the discussion and case studies of Part IV the question of multiple objectives and their evaluation should be somewhat clearer but still a major consideration in problem formulation and solution.

Models of System Interaction

In the abstraction from the actual problem to an optimization model, some system interactions are easier to model than others. The most notable example of this is the interaction between supply and demand. As indicated in Chapter 5 and Part II, it is fairly easy to build a supply model. In optimization, it is similarly quite easy to construct a procedure that defines the least costly way of satisfying a given set of demands. But this optimization of supply presents some problems. First, the demands may represent different objectives and a conflict in resource use. If the demands are given externally and are not decision variables in the optimization, they may not receive the important attention they deserve. Second, there is a marked interaction between supply and demand in planning most large-scale systems. In transportation planning, for instance, the use of a service is distinctly influenced by the form, mode, and route of the facility provided. The difficulty arises from the fact that the interaction between supply and demand is extremely difficult to model in a convenient mathematical form.

In general, we do not model the complete interaction between supply and demand. Rather, we find ourselves relying on a model that solves a surrogate problem: What is the best way to satisfy a given demand? All the case studies in this part are of this form. They all attempt to define the best way to accomplish a specific function rather than allowing that function to be a decision variable. It will be a long time before information about both supply and demand is conveniently available in a form suitable for optimization. Meanwhile, substitute ways must be adopted to deal with the problem, and the sensitivity of the solutions to all assumptions pertaining to supply and demand must be particularly tested.

Role of the Analyst

The role of the analyst in the process of problem formulation and solution is delicate. The analyst must remember for whom the analysis is being done. In most cases in the public sector that audience has many different interests. There will always be conflict between different components of the problem; successful analysis helps to bring these issues out, not to hide them. The focus of the work should be as an aid to the decision, not the decision itself. The analyst's own biases should be submerged to present not a single "best" solution but a range of solutions representing combinations of interests that can serve as a framework for negotiation and collective involvement in the decision process.

CASE STUDIES

The seven case studies presented in Chapters 13 through 19 represent various aspects of the use of optimization in planning and design. Each should be read not only for the problems they address and the way they are approached but for the areas they do not cover. Remember the delicate tradeoff between simulation and optimization and note what has been left in each case to the simulation process. All the intricate aspects of each problem could not be presented in such a short amount of space, so the reader is directed to the references at the end of each chapter to further his knowledge in any particular area.

Chapter 13 is an example of a simple screening model applied to a quite complex planning problem: the provision of water supply in a large area at different time stages for the next 50 years. Notice the simplifying assumptions that allow the authors to use a completely linear model, which seems acceptable for the first broad-scale look at this planning problem. First they derive linear cost functions by guessing in advance the size of particular facilities even though they are decision variables. More important, they formulate a supply model even though this problem evidences aspects of supply–demand interaction. For each of several competing water uses, estimates of future use are made for the next 50 years which may or may not reflect changing preferences for water as this resource becomes limited or as technology changes. Would, for instance, agriculture get such a large share of the water in the year 2020 if the estimates were made to represent the competition for water at the time? This is obviously a case where sensitivity analysis on demand estimates and cost estimates is necessary. The very simple model form, in this case not only linear but a linear network, allows inexpensive manipulations to answer a whole series of "what if" questions about future conditions.

Notice also that the authors have not dealt with time by considering all time periods in one formulation and thus requiring facilities built in one time period to be available in the next. This would require a more extensive mixed integer programming model. Instead they took "snapshots" of the system over time; that is, they solved a separate problem for each time period without any assumptions of what had been constructed in earlier time periods. Here again we see the tradeoff between more realistic formulation and an easier solution made in favor of the latter. Stochastic variations are also not considered. This case illustrates the simple exercise of a screening model as a tool for examining broad policy issues and for getting a first estimate and understanding of future conditions so as to direct more detailed investigations in the future.

Chapter 14 presents an example of technologic design which, as mentioned above, is a natural problem for the use of a highly nonlinear optimization technique. The problem has few decision variables and constraints and can be solved by hand using a geometric programming technique. It is an example of small-scale problem solving where the decision variables are very detailed and the problem highly nonlinear. Notice the way the authors chose to deal with multiple objectives by quantifying the cost of flood damages and adding them to the cost of building the structure. The reader is directed to Part IV for further discussions on how to represent conflicting objectives in optimization models.

In Chapter 15 it is interesting to note that there are many ways that a location model might be built and to consider the way that the authors fit one to their problem. They assumed only a finite number of points for sources of waste, intermediate transfer points, and final disposal of wastes, rather than allowing any point on the plane to be a potential decision. This allowed a much easier formulation of the problem. An integer programming technique was used for solution, with the integer variables associated with whether or not a particular site was developed. This allowed a much more realistic representation of the economies of scale inherent in building facilities but increased the cost of the solution. It is also interesting that in a problem dealing with solid wastes where there are many different objectives, such as health and aesthetics, only the single objective of cost minimization is considered explicitly. Other objectives are explored implicitly through model manipulation and constraints. Notice also the scale at which the system was modeled—a broad policy level considering only the gross movement of material rather than the routing of individual vehicles in their daily tasks.

Chapter 16 deals with the question of when and where highway expansions should take place. The algorithm chosen, although it has a strong flavor of dynamic programming, is in fact a heuristic procedure which allows quick evaluations of a very complicated problem. Some important savings to be gained by using such heuristic procedures to sketch boundaries for feasible solutions are carefully outlined by the authors.

In Chapters 17 and 18 we take a closer look at the inclusion of stochastic elements in the optimization process. Chapter 17 describes the use of a static stochastic linear screening model to suggest both design and operating parameters for the management of the water resources in the Delaware River basin. A dynamic model is then used to schedule investment alternatives over time. In both cases simple models are used and the problem is decomposed. The authors make a strong case for the use of simulation here as a device for capturing the nonlinear and stochastic nature of the problem once the screening models have proposed good configurations. The objective function chosen is to maximize expected economic efficiency.

In Chapter 18 stochastic considerations are placed in the optimization model itself by using chance-constrained programming. Here the question of operating a multireservoir system with both water supply and recreational objectives is addressed by using different objective functions as criteria and comparing the results. The potential recreational benefits, which are an inverse function of the amount of seasonal drawdown for water supply, are in conflict with purely economic objectives. Two different objectives are studied. An economically efficient solution, which minimizes total drawdown regardless of where it occurs, is compared to a more equitable distribution, which minimizes the maximum drawdown. The authors have used fairly simple analysis techniques to do this. They have thus chosen, in comparison with the previous case, to trade the possibility of a more detailed study for the opportunity to explore multiple objectives more thoroughly. Both chapters subscribe strongly, however, to the use of simulation for further refinement of design.

Chapter 19 is an example of an optimization problem so complex that no screening model sufficiently realistic can be found. Instead, the authors must rely on a detailed simulation model and manipulate it to gain information about the system.

Thus the seven cases illustrate a range of models. We have very simple screening models; models that show greater realism by more accurately considering cost functions, stochastic behavior, and multiple objectives; and a complex problem that is handled by simulation alone. All the cases are representative of techniques for optimization in the planning, management, and design process, and all are in the spirit of analytic modeling to gain a better understanding of system performance.

13

Analysis of Water Reuse Alternatives in an Integrated Urban and Agricultural Area*

A. Bruce Bishop and David W. Hendricks

This chapter deals with the difficult question of choosing among alternatives over time to provide water supply for a complex urban and agricultural area. Using analytic techniques in a problem with as many intricate physical, social, political, and economic interactions associated with such a major policy issue requires a great deal of effort and concern to ensure that the essence of the decisions and choices is captured. What we attempt to show is that very simple models, which are readily available as computational tools, can go a long way, at relatively little expense in time or money, toward providing understanding of the system interactions and the dominant system alternatives under different assumptions of future conditions. With the results of such preliminary analysis, decision makers can see the important policy tradeoffs and sensitive areas where further investigations can be identified. These points are illustrated by a case study of water reuse in an area around Salt Lake City, Utah.

THE PROBLEM

The growing demands for existing water supplies and current water shortages emphasize the need for a comprehensive approach to the analysis

*Adapted extensively from Water Resources Systems Analysis, *Journal of the Sanitary Engineering Division* of the American Society of Civil Engineers, Feb. 1971, and from *Analysis of Water Reuse Alternatives in Integrated Urban and Agricultural Areas*, Utah State University, Logan, Utah, 1970.

and planning of integrated water reuse as well as a careful focusing on the nature of demand for water supply and its policy implications. Within the context of all water uses in a region, this case study is used to show a methodology for examining the contribution of water reuse to the total supply of water resources in a region.

The components of the water resource system, including both sources of supply and demand requirements, are shown in the matrix of Figure 13.1. Each row represents a different possible origin of supply, each column a different possible use of water. Both the sectors of water use and the categories of supply can be specified to any degree of refinement desired.

The matrix of sources of supply and sectors of demand for water depicts all possible combinations for satisfying the aggregate system demand with the aggregate available supply. Each element in the matrix represents a possible means of satisfying all or part of the demand requirements of a sector with all or part of the water from a given source. Category totals appear as marginals in the outer row and column.

In the past, water planning and management has been concerned mainly with the design and optimum operation of storage and distribution systems to regulate allocation of primary water supply to each use sector in both time and space. This approach is generally adequate when the primary water supply is in large excess of demands. However, in many areas the primary supply is no longer sufficient to meet the diversion requests of all users. Secondary and supplemental sources of water then become important, and water demands must be met by recycle and sequential reuse from secondary sources or by the development of supplementary supplies. This means that all combinations in the matrix of Figure 13.1 need to be considered for comprehensive planning of water utilization.

The purpose of this paper is to delineate the manner in which all permutations of water use can be explored and how the best alternatives can be selected. Specifically, the objectives are to formulate a conceptual framework for analyzing water reuse alternatives and to present a model for analyzing alternatives of sequential and recycle reuse in an integrated agricultural and urban environment. The function of the model is to determine the optimal allocations of water from each source of supply to each sector of use. The objective of the model can be either to minimize cost, which is the focus of this chapter, or to maximize net benefits. Some questions to be answered by the model are: Which origins of primary and secondary water supply might best be allocated to which use sectors, considering quantity and quality constraints and the objective of minimum costs? What should be the design capacities of wastewater treatment facilities, and when should they be phased into operation?

Since quality constraints may necessitate treatment of water before reuse,

Supply Origins \ Demand Destinations	Municipal	Industrial	Agricultural	Recreation Wildlife Hydropower	Systems Outflow	Annual supply availability by source
Primary supply — Surface water, Groundwater			{ Initial allocation of primary supply }			Outflow, Recharge
Secondary supply — Municipal effluent, Industrial waste, Agricultural return flow			{ allocation of water resources }			Municipal waste system outflow, Industrial wastewaters, Irrigation return flows
Supplementary supply — Imported water, Desalting of sea water			{ Allocation of supplementary supply }			Importation, Desalting
Total demands by use	Municipal diversion	Industrial diversion	Agricultural diversion	Miscellaneous diversion	Downstream outflow	Totals

Figure 13.1 Allocation alternatives for the water resource system.

three possible levels of treatment are considered in the analysis: conventional primary–secondary, tertiary, and desalting. To illustrate the use of the reuse model, it is applied to a specific metropolitan area.

WATER REUSE PLANNING AS A TRANSPORTATION PROBLEM

In searching for a model to represent the problem so that screening of alternatives could be attempted, it was decided to choose a simple model that could quickly provide insight and information about the system under consideration. To do this, an important issue must first be addressed: the relationship between supply of water for different uses and the demand for water by different uses. This is a true multiobjective problem in which different categories of demand represent contributions to different societal objectives. For example, water for agriculture may represent an economic objective, while recreational uses contribute most heavily to some quality of life objective. Thus, to consider a problem of maximizing net benefits for water use in which both the question of where supply should come from and how much should be supplied to each water use (i.e., where both supply and demand are decision variables) presents not only a problem of how such benefits are evaluated for each water use but of how benefits to different accounts are compared. This is the main theme of Part IV. For the time being, we will look simply at the question of the best way to supply a given set of demands. The estimation of the future demand for water presents a major set of assumptions about the future wishes of society and its preferences for the allocation of resources among competing uses. This needs to be done with much care. In fact, by evaluating different estimates of water demand, this conflict for resources among different objectives can be more clearly understood.

Once the assumption is made that the supply side of the issue will be the major focus of the model, an analytic framework can be quickly established in terms of the following optimization problem:

Minimize the cost to the region of transporting and treating water subject to the constraints that:

1. All given demands for water of a given quality in a demand category are met.
2. No source can supply more than is available.
3. Continuity in the physical system must be maintained, such that the amount that enters a particular node in the system equals the amount that leaves that node.

Such a problem statement immediately brings to mind the network problems in mathematical programming known as transportation or trans-shipment problems, which are readily solvable under certain assumptions.

The transportation problem is the problem of allocating a single commodity from given sources to meet given demands at least cost. It is formulated as follows. First, let a_i be the fixed amount of the commodity available at each of the i sources, and b_j be the fixed demand for the commodity at each of the j demand points. Let the variable x_{ij} be the amount of the commodity shipped from the ith source to the jth demand. Then, if c_{ij} is the unit cost of shipping the commodity from i to j, the objective function is to minimize the total cost, TC:

Minimize $$TC = \sum_i \sum_j c_{ij} x_{ij} \qquad (13.1)$$

subject to the constraints:

$$\sum_j x_{ij} = a_i, \qquad i = 1, 2, \ldots, m \qquad (13.2)$$

$$\sum_i x_{ij} = b_j, \qquad j = 1, 2, \ldots, n \qquad (13.3)$$

$$\sum a_i - \sum b_j \qquad (13.4)$$

and each $x_{ij} \geq 0$. Equation (13.2) indicates that the total amount shipped from each origin equals the supply available, Eq. (13.3) that the amount received at each destination equals the demand, and Eq. (13.4) that the total amount shipped must equal the total amount received. With each c_{ij} constant, the problem represented by Eqs. (13.1)–(13.4) is formulated as a linear programming problem with $m + n$ equations in $m \times n$ variables. Its optimal solution can be obtained through linear programming or the special algorithm for the transportation problem (3).

The transportation problem is well adapted to the problem of water reuse planning. Water from several origins or categories of supply must be transported to various destinations or sectors of use at minimum cost. Since the effluent from a sector can be made available for reuse in the system, sectors of use also become origins of secondary supply. A waste treatment plant, for example, may be the destination of municipal effluent, while at the same time it becomes an origin for treated wastewater available for reuse. The effluent from any sector can be allocated for use by another sector for sequential reuse, or it can be reallocated to the same sector by a recycle reuse of the water. The matrix of Figure 13.1, illustrating the concepts of water reuse, thus fits closely the format of the transportation problem.

The costs incurred in allocating water from any origin to any destination depend on the water quality of the source, the quality requirement at the destination, and the facilities required to transport and deliver the water from origin to destination. In adapting water reuse planning to the format of the transportation problem, the cost of "shipping" a unit amount of water from the ith origin to the jth destination is comprised of two components. Treatment costs are incurred when the quality of the source does not meet the requirements of the use sector, and the water must be treated to bring it to the required level of quality. Transportation costs arise from delivering the water from a given source to its point of use. The sum of these two components represents the cost, c_{ij}, between each origin and destination. The cost function thus guarantees that the quality constraints are fulfilled along with the quantity requirements.

THE AUGMENTED MODEL

Several additional considerations should be incorporated into the basic transportation model to make it applicable to the problems common to real water resource systems. Such problems include those associated with the operations of water treatment, desalting plants, and blending facilities; and limitations in the physical system which render some allocations infeasible. These considerations can be included without altering the basic structure of the "transportation model." This is done by augmenting the problem with additional sets of origins, destinations, and constraint equations, as described below.

Wastewater Treatment Operations

Water and wastewater treatment operations in the system include such facilities as primary and secondary treatment plants, tertiary treatment plants, and desalting plants. To include those in the model, the transportation matrix tableau is augmented by adding a column vector denoting the treatment facility as a destination, and a row vector which indicates that the facility is also an origin of treated water. The system balance is maintained by stipulating that the inflow to the plant as a destination must equal the outflow from the treatment plant as an origin. For example, in the problem structure shown in Figure 13.1, if the plant production is represented by row k, and the plant intake by column l,

$$\sum_j x_{kj} = \sum_i x_{il} \qquad (13.5)$$

When the treatment plants, either existing or proposed, are of a specified capacity, their outflow must be limited to the maximum permissible, a_k, and their inflow, likewise, to b_l. Thus:

$$\sum_j x_{kj} \leq a_k \qquad (13.6)$$
$$\sum_i x_{il} \geq b_l \qquad (13.7)$$

If, on the other hand, the optimal plant capacity and phasing of operations is to be determined, the flows through the plants are left unbounded. This allows the treatment operations to be used at the levels required for cost minimization. This approach can provide insight into either the optimum design capacities of proposed treatment facilities for a system or the best levels of operation of existing facilities.

Blending Operations

Blending operations, where water too high in total dissolved solids (TDS) to meet user quality requirements is mixed with a water low in TDS to produce an acceptable product, can be handled in a manner similar to that for treatment facilities. The capacity and production of the blending operation is established as in Eqs. (13.6) and (13.7), and the inflow–outflow balance is maintained by Eq. (13.5). One additional equation must be included to specify the blending ratio of the salty and the pure water. This is determined from the TDS of the water supply categories which can be allocated to blending. The general form of the equation is:

$$\sum (x_{ij})_{\text{salt sources}} = R \cdot \sum (x_{ij})_{\text{pure sources}} \tag{13.8}$$

where R is the ratio of pure water to salty water necessary to achieve acceptable quality.

Although Eq. (13.8) is an approximation, it is reasonably accurate for our purposes. A rigorous solution can be obtained, however. It requires the use of the salinity mass-balance equation for the blending operation. The allocative mix to the blending operation from the various origins having different salinity levels would be determined by means of a suboptimization known as the "refinery problem." This would consist of a least-cost allocation to the blending operation, the objective function, subject to the salinity mass-balance constraint.

Infeasible Allocations

In some cases limitations in the system may make it impossible or impractical to allocate water between some origins and destinations. It might be physically impossible to transfer water from a particular source to a particular user, or social and political constraints might prevent the use of water directly from a source to a user, for example the use of untreated municipal effluent for irrigation. Reasonable engineering judgments might also determine that some transfers were unnecessary. It might be recognized, for instance, that pure water from a groundwater source does not need wastewater treatment or desalting before use.

These limitations are recognized in the structure of the problem by assigning an unrealistically high cost to the element representing such an allocation. The high cost associated with that particular combination will prevent any allocation from taking place.

Salt Loading and Reuse Factor

In analyzing blending and desalting operations, it may be important under some conditions to determine the optimal system allocations based on maintaining the salt balance between each origin and destination. Where the TDS of the source water is too high, it could be reduced to acceptable levels by combining with a source of make-up water of better quality. Such a procedure might be used as an alternative to desalting brackish supplies or as a means of determining whether the operation of a desalting plant in the system should serve as a source of water for direct allocation or as make-up water for blending. The levels of sequential or recycle reuse that can be permitted for each use to maintain the salt balance can be determined in the manner suggested by Hendricks and Bagley (7).

Consumptive Use and Losses

The amount of consumptive use and losses in the system can be treated in the same manner as a class of transportation problems where the demands are less than the supply. To balance the amounts available and the demands, an additional destination is specified, with zero allocation costs to this destination. In the case of water reuse, this destination is identified as system outflow. Quality constraints can be imposed on the outflow, on effluents for example, by assigning a set of costs to the outflows. Water from other origins, already of sufficient quality, can be released to system outflow at no cost.

WATER REUSE MODEL OF AREA AROUND
SALT LAKE CITY, UTAH

A case study of the area around Salt Lake City, Utah, is presented to demonstrate how the model can be applied to an actual water reuse system and to show its utility and range of applications in water resources planning. The case study is intended to show methods and types of results. Therefore, the data used in the model, although based on the best sources available for this type of study, would require further refinement if the results were to be used for actual planning and decision making. The results obtained do indicate, however, trends and orders of magnitude.

Description of Study Area

The Salt Lake City area of north-central Utah is a major urban center in which varied uses of a limited water supply for municipal, industrial, agricultural, wildlife, and recreation purposes all lie in close proximity. The study area includes Salt Lake County and most of Davis County to the north. The three sources of primary water supply in the area are the Jordan River, which originates at Utah Lake and enters the Salt Lake Valley at the south and flows 45 miles northward to the Great Salt Lake; a number of streams flowing from the Wasatch Mountains on the east and some imported water from Deer Creek Reservoir; and groundwater. The quantity of water available from these sources is given in Table 13.1. Possible sources of imported water are not considered in this study.

Table 13.1 Water Available from Primary Sources

Source	Annual Acre-Feet
Jordan River	270,000
Wasatch streams	83,000
Groundwater	48,000
Total	401,000

Source: T. Arrow, *Groundwater in the Jordan Valley, Salt Lake County, Utah*, Utah State Engineer Water Circular 1, Salt Lake City, 1965.

The 1965 population of the area was estimated at 500,000. The trend is toward industrial development and continuing population growth. Mining and manufacturing are the principal industries of the area, but agriculture still plays a major role in the valley in terms of water use.

Projected Demands for Water

The projected demands, by each sector of water use, are summarized in Table 13.2. The table also contains projected secondary supplies, consisting of effluents or return flows from the municipal, industrial, and agricultural sectors, as well as the Farmington Bay Bird Refuge. The predicted consumptive use for each sector is simply the diversion minus the effluent or return flow. Return flows for 1980, 2000, and 2020 were made by using the same ratio of diversion to return flow as for the 1965 data.

Finally, the difference between the total primary supply and the total consumptive use is the system outflow, given in Table 13.3. This represents water flowing from the study area into the Great Salt Lake.

Table 13.2 Summary of Water Use Projections by Year (Acre-Feet × 10³)

Use	Demands for Diversion				Secondary Supply Effluent Return Flow				Consumptive Use			
	1965	1980ᵃ	2000ᵃ	2020ᵃ	1965	1980	2000	2020	1965	1980	2000	2020
Municipal	88	165	253	341	74	139	213	287	14	26	40	54
Industrial	124	204	290	376	87	143	203	263	37	61	87	113
Agricultural	270	224	209	194	103	86	80	74	167	138	129	120
Bird refuge	141	141	141	141	65	65	65	65	76	76	76	76
Total	623	734	893	1052	329	433	561	689	294	301	332	363

Source: A. B. Bishop and D. W. Hendricks, Water Resources Systems Analysis, *Journal of the Sanitary Engineering Division, ASCE,* Vol. 97, No. SA I, App. A; for footnote a values, O. L. Harline, *Municipal and Industrial Water Requirements, Utah Counties, 1960–2020,* and *Use of Water for Municipal and Industrial Purposes, Utah Counties, 1960–61,* Bureau of Business and Economic Research, University of Utah, Salt Lake City, July 1963.

Influent and Effluent Qualities

The differences in quality between water available from each supply origin and water required by each demand destination can be tabulated in a matrix of water quality differences. This provides the basis for determining the treatment costs necessary to match each source with its possible uses. The two quality criteria considered to be critical in this study are biological oxygen demand (BOD) and total dissolved solids (TDS). Both of these quality criteria must be satisfied in allocating water from a source to a user.

The matrix of Figure 13.2 indicates as marginals the nominal values for BOD in the primary and secondary water sources, and the BOD quality requirement for each water sector of use. The elements of the matrix indicate the amount of BOD that must be removed from a source of water in order to meet the quality requirements of a given use. If the BOD requirement is already met, this value is zero. Thus the matrix gives the initial and terminal BOD numbers between any origin and destination, and the required BOD removal.

The same type of information for TDS is given in Figure 13.3. The matrix indicates nominal values for the salt content of the water sources, and the allowable TDS for the water uses. The difference between initial and terminal values for TDS between any supply origin and demand destination indicates whether desalting or blending is required and provides a basis for estimating costs.

Water Costs

The water quality analysis given by the matrices of Figures 13.2 and 13.3 and the supply and demand quantities from Tables 13.2 and 13.3 provide the necessary information for specifying the treatment required, whether conventional, tertiary, desalting, and for estimating commensurate unit costs to make water available between all combinations of supply origin and use sectors.

Table 13.3 Primary Water Supply, Consumptive Use, and System Outflow by Year (Acre-Feet × 10³)

	1965	1980	2000	2020
Primary supply	401	401	401	401
Consumptive use	294	301	332	363
System outflow	107	100	69	38

Demand Destinations / Supply Origins		Uses			Treatments						Supply source quality (BOD mg/ℓ)
		Municipal	Industrial	Agricultural	Secondary	Tertiary	Desalting	Blending	Bird refuge	System outflow	
Primary supply	Surface water	20	0	0	0	0	0	20	0	0	20
	Groundwater	0	0	0	0	0	0	0	20	0	0
	Jordan river	20	0	0	0	0	20	20	0	0	20
Effluents for secondary supply	Municipal	300	270	0	280	300	300	300	280	280	300
	Industrial	300	270	0	280	300	300	300	280	280	300
	Agricultural	5	0	0	0	5	5	5	0	0	5
Treatments	Secondary	20	0	0	0	0	80	20	0	0	20
	Tertiary	0	0	0	0	N/A	0	0	0	0	0
	Desalting	0	0	0	0	0	0	0	0	0	0
	Blending	0	0	0	0	0	0	0	0	0	0
	Bird refuge	20	0	0	0	20	20	0	0	0	20
Required quality for use (BOD mg/ℓ)		0	30	300	20	0	0	0	20	20	

Figure 13.2 Matrix of water quality differences indicating required BOD removal (mg/liter).

Supply Origins \ Demand Destinations	Uses			Treatments						Supply source quality (TDS mg/ℓ)
	Municipal	Industrial	Agricultural	Secondary	Tertiary	Desalting	Blending	Bird refuge	System outflow	
Primary supply										
Surface water	0	0	0	0	0	N/A	0	0	0	300
Groundwater	0	0	0	0	0	N/A	0	0	0	300
Jordan river	300	0	0	0	0	790	300	0	0	800
Effluents for secondary supply										
Municipal	400	100	0	0	0	890	400	0	0	900
Industrial	400	100	0	0	0	890	400	0	0	0
Agricultural	500	200	100	0	0	990	500	0	0	1000
Treatments										
Secondary	500	200	100	0	0	990	500	0	0	1000
Tertiary	400	100	0	0	0	890	400	0	0	0
Desalting	0	0	0	0	0	N/A	0	0	0	10
Blending	0	0	0	0	0	N/A	0	0	0	500
Bird refuge	1000	700	600	0	0	1490	1000	0	0	1000
TDS influent required quality for use (TDS mg/ℓ)	500	800	900	2000	2000	10	500	3000	1500	

Figure 13.3 Matrix of water quality differences indicating required TDS removal (mg/liter).

Demand Destinations / Supply Origins		Uses			Treatments						Available source quantity (Acre–feet x 10^3)
		Munic–ipal	Indus–trial	Agricul–tural	Second–ary	Ter–tiary	Desalt–ing	Blend–ing	Bird refuge	System outflow	
Primary supply	Surface water	38	38	5	1000	1000	1000	38	5	0	83
	Groundwater	23	23	10	1000	1000	1000	23	10	0	48
	Jordan river	108	10ᵃ	5ᵃ	1000	30	59	1000	0	0	270
Effluents for secondary supply	Municipal[b]	135ᶜ	56	51	46	77	105	1000	46	46	74
	Industrial	115ᶜ	39	10	29	1000	88	1000	29	29	87
	Agricultural	108ᶜ	93	5	1000	30	59	1000	5	0	103
Treatments	Secondary	80	10	5	1000	33	59	1000	5	0	50
	Tertiary	1000	1000	10	1000	1000	1000	1000	0	0	100
	Desalting	15	10	5	1000	1000	1000	11	5	0	56
	Blending	0	0	5	1000	1000	1000	11	5	0	100
	Bird refuge	1000	1000	1000	1000	1000	59	1000	1000	0	65
Demands (Acre–feet x 10^3)		88	124	270	50	100	56	100	141	107	

Figure 13.4 Transportation model tableau indicating costs of transferring water from any source to any destination ($/acre–feet).

The cost data are presented in the "transportation tableau" of Figure
13.4. When applicable, the costs of pumping, transporting, and delivering the
water are included. Where allocations between a source of supply and a
demand sector are infeasible, a unit cost of $1000 per acre-foot is entered.
For the purposes of this study it was assumed that all costs are constant; they
are average costs. Actually, costs will vary with the quantity of water and level
of treatment required. This can be taken into account in the model analysis
through iterative solutions obtained by updating cost figures for sensitive
variables.

ANALYSIS AND RESULTS OF CASE STUDY

With the flexibility of the reuse model and the operations research techniques
used in conjunction with linear programming, a number of investigations can
be performed to evaluate water-reuse-system alternatives. Techniques of sensitivity analysis and parametric programming are available to evaluate the
effect of the constraints, the cost coefficients of the objective function, or the
coefficients of variables in the constraint equations which describe the structural relationships of the problem. Results of the case study are offered here
as examples of the types of analysis that can be performed and the information
and evaluations for planning decisions that can be derived.

A comparative study of optimal system configurations in general seeks
to answer the question: What is the optimal allocation pattern, given the
conditions for operating the system? For example, one may desire to know
the allocations at present levels of demand and the changes in the allocation
pattern as demands increase to projected levels in the future, assuming that
the primary water supply cannot be expanded. A similar study could be
carried out allowing for possible importation of supplementary water supplies.
Another important study might center on the operation of various types of
treatment facilities in the system, including conventional, tertiary, desalting,
and blending operations. This type of investigation would include optimal
allocation patterns assuming certain locations and capacities for various
treatment plants, and the design capacity and timing of construction of
possible treatment operations.

Plant Design Capacities and Construction Timing

This part discusses the results of the model investigation on plant design
capacity and construction timing. Figures 13.5 to 13.8 show the allocation
patterns for demand at 1965, 1980, 2000, and 2020 levels, assuming no expansion of the primary water supply.

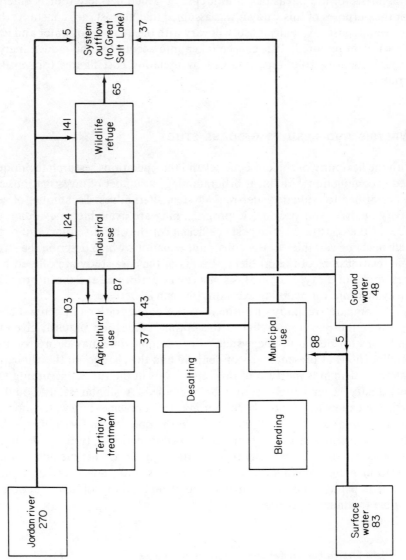

Figure 13.5 Year 1965 optimal allocation pattern (acre-feet × 10³).

In 1965 the optimal plan did not allocate water to any of the treatment and blending operations (see Figure 13.5). With 1965 supplies and demands, treatment of effluent water for reuse was not required. Its incorporation into the system would have increased costs. Municipal and industrial requirements were better satisfied from primary sources and agricultural requirements met by sequential use of effluent from municipal and industrial systems and recycling of irrigation return flow.

To satisfy the 1980 municipal requirements, however, the minimum cost allocation brings a tertiary treatment process into the optimal solution at a level of 34,000 acre-feet annually (Figure 13.6). The influent to the plant, water from Jordan River and irrigation return flow, with a TDS of 1000 mg/liter, is blended in a 1:1 ratio with the surface water supply of low TDS to produce a water acceptable for the municipal system. Hence careful consideration ought to be given to the use of a tertiary treatment and blending operation with a minimum capacity of 34,000 acre-feet for meeting expected demand in 1980.

Investigations for the year 2000 indicate that the 1980 trend would continue and that municipal demands would be met almost entirely by mixing tertiary-treated Jordan River water and irrigation return flow with surface water and groundwater supplies. Industrial requirements would be best satisfied by sequential use of municipal effluent, and agriculture demands by the use of industrial effluent (Figure 13.7).

Finally, at 2020 levels, as surface water and groundwater supplies are entirely used up in the blending operation, it appears that a desalting plant should be brought into the minimum cost basis to supply the additional blending water required to meet the municipal demand. This suggests, then, that the future water planning should include consideration of a desalting plant with minimum capacity of 39,500 acre-feet annually prior to 2020. Parametric analysis shows that the plant should be phased in just after the turn of the century because of the continuous nature of the demand function (Figure 13.8).

SUMMARY

The particular analysis described in the case study points up three useful aspects of the reuse model and the information it provides. First, it indicates how the primary, secondary, and recycle sources of supply should be allocated to satisfy demands at least cost. Second, given a constant water supply and the projections of future demands, it points up the types of treatment processes for water reuse which would meet demands most cheaply. Finally, the optimal allocation indicates as part of the solution the required capacities for treatment facilities and, through parametric analysis, the time at which they should be phased into the system.

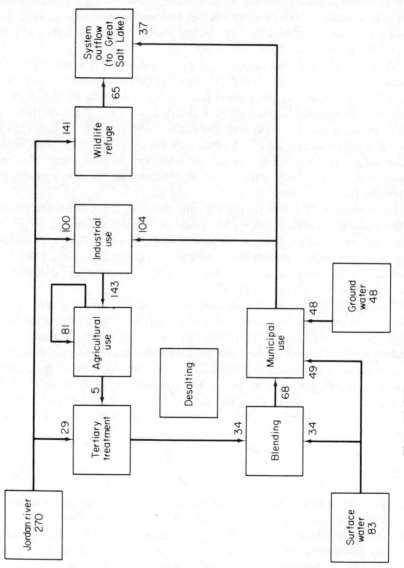

Figure 13.6 Year 1980 optimal allocation pattern (acre-feet $\times 10^3$).

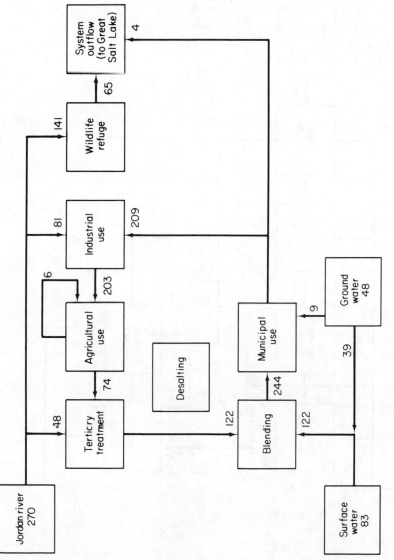

Figure 13.7 Year 2000 optimal allocation pattern (acre-feet $\times$ 10³).

187

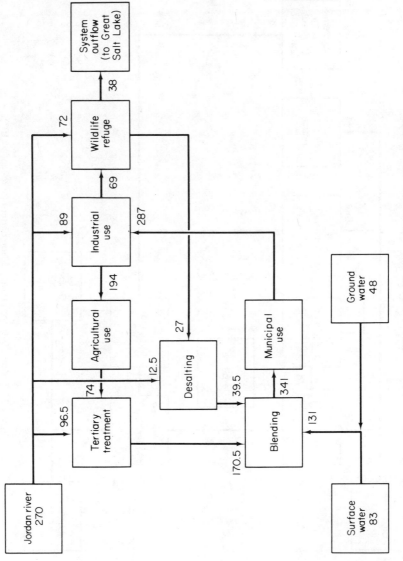

Figure 13.8 Year 2020 optimal allocation pattern (acre-feet $\times 10^3$).

ACKNOWLEDGMENT

This work was supported by the Utah Water Research Laboratory, College of Engineering, Utah State University, Logan, Utah.

REFERENCES

1. ARROW, T., 1965. *Groundwater in the Jordan Valley, Salt Lake County, Utah*, Salt Lake City: Utah State Engineer Water Circular 1.
2. BISHOP, A. B., AND HENDRICKS, D. W., 1971. Water Resources Systems Analysis, *Journal of the Sanitary Engineering Division, ASCE*, Vol. 97, No. SA1, Feb., pp. 41–57.
3. GASS, S. I., 1964. *Linear Programming—Methods and Applications*, New York: McGraw-Hill Book Company, 2nd ed.
4. HARLINE, O. L., 1963. *Municipal and Industrial Water Requirements, Utah Counties, 1960–2020*, Salt Lake City: Bureau of Business and Economic Research, University of Utah, July.
5. HARLINE, O. L., 1963. *Use of Water for Municipal and Industrial Purposes, Utah Counties, 1960–61*, Salt Lake City: Bureau of Business and Economic Research, University of Utah, July.
6. HAYCOCK, E. B., SHIOZAWA, S., AND ROBERTS, J. O., 1968. *Utah Desalting Study*, Salt Lake City: Utah State Office of Saline Water, Atomic Energy Commission.
7. HENDRICKS, D. W., AND BAGLEY, J. M., 1969. Water Supply Augmentation by Reuse, Banff, Alberta: *Proceedings of the Symposium on Water Balance in North America*, American Water Resources Association, June.
8. MILLIGAN, J. H., ANDERSEN, J. C., AND CLYDE, C. G., 1970. *Colorado River Allocation Study*, Logan, Utah: Utah Water Research Laboratory, Utah State University.
9. SMITH, R., 1969. *A Compilation of Cost Information for Conventional and Advanced Wastewater Treatment Plants and Processes*, Cincinnati: U.S. Federal Water Pollution Control Administration.
10. Utah State Department of Health, 1964. *Inventory of Municipal Wastewater Facilities in Utah*, Salt Lake City.
11. Utah State Department of Health, 1965. *Inventory of Industrial Waste Facilities*, Salt Lake City.
12. Utah State Department of Health, 1966. *Industrial Wastewater Facilities in Utah*, Salt Lake City.

14

A Cofferdam Design Optimization*

Farrokh Neghabat and Robert M. Stark

This chapter explores the use of nonlinear optimization in a highly technical question of design. A cofferdam is a temporary structure built to enclose an ordinarily submerged area to permit construction of a permanent structure on the site. Its specific function, which is to keep the construction site dry, and its short life (the project duration), contrast with the versatility and long life of most structures that concern civil engineers. The discussion here will be limited to circular cellular cofferdams, one of the types most commonly used for deep-water and large projects. The cellular cofferdam is a relatively simple steel structure filled with soil. It is constructed by driving straight web steel sheet piles into the ground to form a series of self-contained cells which are joined by short connecting arcs and filled with a suitable soil. Its analysis tends to be uncomplicated by many different types of failure modes.

Cofferdams function in a random environment characterized by fluctuations in surrounding water levels. If the water exceeds the height of the dam and floods the construction project, costs of repair, replacement, clean up, dewatering, and so on, are incurred besides the possibility of damage to the permanent structure. Increasing the height of the structure diminishes the chance of flooding but increases the cost of construction. Therefore, the design height is an appropriate balance between the costs of construction and of potential floods.

*Adapted extensively from A Cofferdam Design Optimization, *Mathematical Programming*, Vol. 3, No. 3; Dec. 1972, pp. 263–275.

The geometry of the cofferdam is a second aspect of its design. A variety of cell geometries is available to enclose the desired area. To be feasible, each of them must satisfy certain technological constraints so that the cofferdam will not overturn or slip on its foundation.

The conventional design of a cofferdam tends to be a trial-and-error procedure (7). A trial design is chosen from experience and analyzed for feasibility. In practice the designer manipulates this feasible design to take economic advantage of site conditions as they relate to the costs of the fill and steel sheet piles. The resulting design is feasible but neither necessarily, nor likely to be, optimal.

The analysis to follow integrates economic considerations into an expected cost objective subject to technological constraints. Geometric programming techniques are used to derive design values of the height, main and connecting cell diameters, and the cycle length.

THE MODEL

The formulation presented in this chapter pertains to circular cellular cofferdams that rest on rock and have no inside berms (Figure 14.1). The concept, however, may be applied to other types of cofferdams.

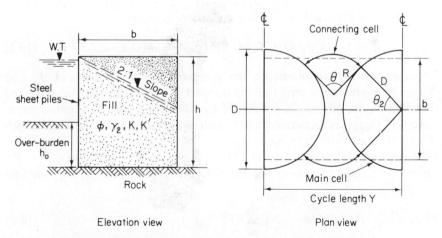

Figure 14.1 Plan and elevation of a circular cellular cofferdam.

Objective Function

The expected cost objective function consists of three components: the costs of fill, steel sheet piles, and the risk of flooding.

The cost of fill is proportional to the volume, $LY^{-1}Ah$, where A is the cycle area, that is, the area of a main and a connecting cell, L the total length of the project, Y the cycle length, and h the height of the dam from the river bottom. Using a proportionality constant, in terms of cost per unit volume, C_b, the cost $C_bLY^{-1}Ah$ reasonably represents the total cost of fill and associated costs of hauling, placing, and disposal.

The cost of the steel sheet piles is closely related to their weight. An estimate of this weight of the steel is obtained by multiplying the area density, ρ, of the sheet pile, ordinarily given in pounds per square foot, by the surface of the steel, $LY^{-1}Ph$, where P is the cycle perimeter. Letting C_s be a unit cost, the total cost of the steel, including associated costs of transportation, setting, driving, and removal of the sheet piles, can be represented by $C_s\rho LY^{-1}Ph$.

The cost of potential flooding is less easy to document (2, 3). It may be considered proportional to the number of floods, N, during the life of the dam and to the duration of the flood, $d(h)$. Both parameters may be random. The total expected number of flood days is simply $\bar{d}E(N)$, where $\bar{d}$ is the average duration of the flood corresponding to a mean height $\bar{h}$, estimated from hydrological data for the given site (4). Using C_f as a proportionality constant, in terms of cost per flood day, the expected cost of floods is represented by $C_f\bar{d}E(N)$. As an approximation, we may let $E(N) \approx t(C_1h - C_2)^{-1}$, in which C_1 and C_2 are nonnegative constants and t is the design life in years (2).

The total expected cost of the cofferdam is then:

$$Z = C_bL(Y^{-1}Ah) + C_s\rho L(Y^{-1}Ph) + C_f\bar{d}t(C_1h - C_2)^{-1} \qquad (14.1)$$

From elementary trigonometry, the geometric design variables A, P, and Y can be expressed in terms of the connecting cell radius R, the main cell diameter D, and the angle, θ, formed by connecting tees. For convenience, however, designers generally work with a rectangular section of length Y and average width b, chosen such that its area closely approximates that of an actual cell (Figure 14.1). Therefore, $A \approx bY$ and $P \approx 2(b + Y)$. Substituting these expressions in Eq. (14.1), the resultant objective function becomes:

Minimize $\quad Z = C_bLbh + 2C_s\rho Lh(bY^{-1} + 1) + C_f\bar{d}t(C_1h - C_2)^{-1} \qquad (14.2)$

Constraints

The cofferdam design must satisfy constraints to preclude failures (5). Slippage of the sheet piles on the river side and excessive interlock stresses at the joints are the foremost modes of failure. The cofferdams may also slide along the base and fail at the center line of the fill as a result of excessive vertical shear. For this analysis constraints are only formulated for slippage and interlock stress. A more comprehensive treatment of technological constraints is given in references 2 and 5.

The constraint on the slippage of the sheet piles is designed to maintain a factor of safety against failure which is sufficiently high. For this analysis it is assumed that a safety factor of 1.25 is adequate. The safety factor for slippage is the ratio between the moments resisting and causing failure (Figure 14.2) (5, 6). Taking moments about the toe of the cofferdam, T, yields:

$$\text{factor of safety} = \frac{b \tan \alpha (P_w + P_0)}{P_w h/3 + P_0 h_0/3} = 3\left(\frac{b \tan \alpha (\gamma_w h^2 + K'\gamma h_0^2)}{\gamma_w h^3 + K'\gamma h_0^3}\right) \quad (14.3)$$

where h, h_0 = height of cofferdam, and of the overburden

$\qquad P_w, P_0$ = pressure of water, and overburden

$\qquad \gamma_w, \gamma$ = unit weight of water, and submerged fill

$$K' = \text{Coulomb's coefficient} = \frac{\cos^2 \phi}{\left(1 + \sqrt{\dfrac{\sin(\phi + \alpha)\sin\phi}{\cos\alpha}}\right)^2}$$

$\qquad \alpha$ = friction angle between the fill and steel piling

$\qquad \phi$ = angle of internal friction

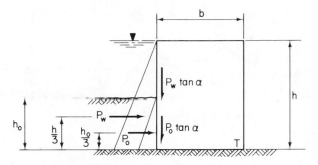

Figure 14.2 Forces causing and resisting slippage.

The constraint for slippage consequently becomes

$$C_3 h b^{-1} + C_4 h^{-2} b^{-1} - C_5 h^{-2} \leq 1 \quad (14.4)$$

where

$$C_3 = \frac{5}{12 \tan \alpha}, \qquad C_4 = C_3 \frac{K'\gamma h_0^3}{\gamma_w}, \qquad C_5 = \frac{K'\gamma h_0^2}{\gamma_w}$$

The computed maximum interlock stress, t_{max}, should not exceed the allowable stress, t_a, specified by the manufacturers of the steel sheet piles. The maximum tension occurs at the junction between the main cells and the connecting arcs (Figure 14.3) and is given by:

$$t_{max} = \frac{P_2 Y/2}{\cos(\theta/2)} \qquad (14.5)$$

where $P_2 = \frac{3}{4}(\gamma_w + K'\gamma)(h - h_0)$. The constraint for interlock stress then becomes

$$C_6 Yh - C_7 Y \leq 1 \qquad (14.6)$$

where

$$C_6 = \frac{3}{8}\left(\frac{\gamma_w + K'\gamma}{t_a \cos(\theta/2)}\right), \qquad C_7 = \frac{3}{8}\left(\frac{(\gamma_w + K'\gamma)h_0}{t_a \cos(\theta/2)}\right)$$

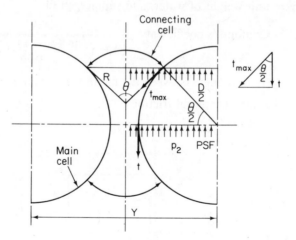

Figure 14.3 Maximum interlock tension.

OPTIMIZATION

The mathematical model described above may be solved using geometric programming (1,8). This technique requires that all terms in the objective function and constraint set be posynomial, that is, of the form $C_j \prod_{i=1}^{n} X_i^{a_{ij}}$, where $C_j > 0$ is a given constant, $X_i > 0$ are decision variables, and a_{ij} are given constants. Because of the power of the technique in certain types of problems, it is often advantageous to convert nonposynomials to equivalent posynomial form. Such is the case here.

The first term of the objective function, Eq. (14.2), is a posynomial; the second term can be rearranged to achieve posynomial form; but the last requires more manipulation. Use of a common technique replaces $(C_1 h - C_2)$ by another variable, say H, and introduces the constraint:

$$C_1^{-1} h^{-1} H + C_1^{-1} C_2 h^{-1} \leq 1 \qquad (14.7)$$

All terms in the objective function are then posynomials.

For design purposes, shallow overburden of only a few feet deep may, as a rule, be neglected without appreciably affecting the design of the cells. When there is no overburden, that is, when $h_0 = 0$, the model becomes:

$$\text{Minimize} \quad Z = (C_b L)bh + (2C_s \rho L)hbY^{-1} + (2C_s \rho L)h + (C_f \bar{d}t)H^{-1} \quad (14.8)$$

subject to

$$
\begin{aligned}
C_3 h b^{-1} &\leq 1 \quad \text{(slippage)} \\
C_6 h Y &\leq 1 \quad \text{(interlock tension)} \\
C_1^{-1} h^{-1} H + C_1^{-1} C_2 h^{-1} &\leq 1 \quad \text{(change of variable)} \\
h, b, Y, H &> 0
\end{aligned}
\qquad (14.9)
$$

This formulation leaves a highly nonlinear optimization problem which is quite difficult to solve. What makes geometric programming so appealing is that if all the conditions are met for its use, the dual problem can be written and it will prove to be much easier to attack. For this problem the dual is to maximize

$$d(\mu) = \left(\frac{C_b L}{\mu_{01}}\right)^{\mu_{01}} \left(\frac{2C_s \rho L}{\mu_{02}}\right)^{\mu_{02}} \left(\frac{2C_s \rho L}{\mu_{03}}\right)^{\mu_{03}} \left(\frac{C_f t \bar{d}}{\mu_{04}}\right)^{\mu_{04}} C_3^{\mu_{11}} C_6^{\mu_{21}}$$
$$\left(\frac{C_1^{-1}(\mu_{31} + \mu_{32})}{\mu_{31}}\right)^{\mu_{31}} \left(\frac{C_1^{-1} C_2 (\mu_{31} + \mu_{32})}{\mu_{32}}\right)^{\mu_{32}} \qquad (14.10)$$

subject to

$$
\begin{array}{ll}
\mu_{01} + \mu_{02} + \mu_{03} + \mu_{04} = 1, & \text{normality condition} \\
\mu_{01} + \mu_{02} - \mu_{11} = 0, & \text{orthogonality condition for } b \\
\mu_{01} + \mu_{02} + \mu_{03} + \mu_{11} + \mu_{21} - \mu_{31} - \mu_{32} = 0 & \text{\textit{"}} \quad \text{\textit{"}} \quad \text{\textit{"}} \quad h \\
-\mu_{02} + \mu_{21} = 0, & \text{\textit{"}} \quad \text{\textit{"}} \quad \text{\textit{"}} \quad Y \\
-\mu_{04} + \mu_{31} = 0, & \text{\textit{"}} \quad \text{\textit{"}} \quad \text{\textit{"}} \quad H
\end{array}
$$

All μs are greater than zero.

Notice that the constraints are linear equality constraints with three more variables than equations. If the number of dual variables were equal to the number of dual constraints, which is equivalent in the primal problem to the number of posynomials being equal to the number of primal variables plus one, the solution would be found simply by solving a set of simultaneous linear equations. That case is known as the zero-degree-of-difficulty case. The case under consideration has three degrees of difficulty and will require some manipulation of the linear constraint set to find the optimum solution.

The dual variables can be expressed in terms of μ_{01}, μ_{03}, and μ_{32} by

$$\mu_{02} = \mu_{21} = \tfrac{1}{4}(1 - 3\mu_{01} - 2\mu_{03} + \mu_{32})$$
$$\mu_{04} = \mu_{31} = \tfrac{1}{4}(3 - \mu_{01} - 2\mu_{03} - \mu_{32}) \qquad (14.11)$$
$$\mu_{11} = \tfrac{1}{4}(1 + \mu_{01} - 2\mu_{03} + \mu_{32})$$

The optimizing values of the dual variables μ_{01}^*, μ_{03}^*, and μ_{32}^* are obtained by first substituting Eq. (14.11) into the dual objective, Eq. (14.10), then setting partial derivatives of $\ln d(\mu_{01}, \mu_{03}, \mu_{32})$ with respect to each dual variable equal to zero, and, finally, solving the resulting three equations (14.12) simultaneously:

$$\frac{4C_8\mu_{01}^2}{\mu_{03}} + 3\mu_{01} + 2\mu_{03} - \mu_{32} = 1$$

$$4C_9\mu_{03}\sqrt{\frac{\mu_{32}}{\mu_{01}}} + \mu_{01} + 2\mu_{03} + \mu_{32} = 3 \qquad (14.12)$$

$$\frac{4C_{10}\mu_{01}\mu_{32}}{\mu_{03}} + \mu_{01} + 2\mu_{03} - 3\mu_{32} = 3$$

where

$$C_8 = (C_sC_b^{-1}p)^2\,(\gamma_w + K'\gamma)\,[8t_a\,\cos(\theta/2)]^{-1}$$
$$C_9 = [\,(5C_bLC_f\bar{d}t)\,(12C_2\,\tan\alpha)^{-1}\,]^{1/2}\,(2C_s\rho L)^{-1}$$
$$C_{10} = (24C_1C_s\rho\,\tan\alpha)\,(5C_2C_b)^{-1}$$

The optimal values of the remaining dual variables are obtained from Eq. (14.11) by substitution of μ_{01}^*, μ_{03}^*, and μ_{32}^*.

The corresponding optimum of Z, $Z^* = \min Z = \max d(\mathbf{\mu})$, is the global optimum in view of the posynomial objective. This optimum is used to relate the dual variables to the design variables as follows:

$$\mu_{01}^* = (C_bL)bh(Z^*)^{-1}, \qquad\qquad C_3hb^{-1} = 1$$
$$\mu_{02}^* = (2C_s\rho L)bhY^{-1}(Z^*)^{-1}, \qquad C_6hY = 1$$
$$\mu_{03}^* = (2C_s\rho L)h(Z^*)^{-1}, \qquad\qquad C_1^{-1}h^{-1}H = \mu_{31}^*(\mu_{31}^* + \mu_{32}^*)^{-1} \qquad (14.13)$$
$$\mu_{04}^* = C_f\bar{d}tH^{-1}(Z^*)^{-1}, \qquad\qquad C_1^{-1}C_2h^{-1} = \mu_{32}^*(\mu_{31}^* + \mu_{32}^*)^{-1}$$

This is an important result for the design of the circular cellular cofferdam under the assumptions discussed earlier. In a simple manner, the design can be-explored for sensitivity to various parameters, particularly to the flood damage assumptions in the objective function.

In the case of deep overburden, the solution for $h_0 = 0$ should be modified by considering the effect of constraints of Eqs. (14.4) and (14.6). The model would now contain higher degrees of difficulty which one could attempt to solve using existing geometric programming codes. Alternatively, one might try a trial-and-error procedure.

NUMERICAL EXAMPLE

Consider the design of a circular cofferdam resting on rock foundation with no overburden and berm. The cofferdam will be 800 ft long and has an anticipated life of 1 year. It is to be constructed of 15-in., MP 101 sections of steel sheet piling with an area density of 28 psf and an allowable interlock tension of 8000 psi. The cost of purchasing, setting, and driving the sheet piles into the bedrock is estimated at $163 per ton. This unit cost includes the expenses of removing the piles and their salvage value upon completion of the project. The cost of fill material, including its haul, placement, and finally its removal and disposal, is estimated at $5.67 per cubic yard. The cost of flooding is estimated at $8000 per day. For the site in question, hydrologic data yield a mean sample height which is exceeded, on the average, 5 days during the year. The data on the annual flooding frequencies (Table 14.1) provide an expression for the expected number of floods in terms of the design height of the dam by using a least-squares fit:

$$E(N) = (0.8h - 33.3)^{-1}$$

Table 14.1 Expected Number of Floods for Various Heights of Dam

Height of Dam, h (ft)	Expected Number of Floods, $E(N)$
40	4
42	2
44	1
46	0.5
48	0.25
50	0.15
52	0.10

Other parameters include

$$\gamma = \text{unit weight of submerged material} = 65 \text{ pcf}$$
$$\alpha = \text{friction angle between fill and piling} = 21°50'$$
$$\phi = \text{angle of internal friction} = 28°50'$$
$$K' = \text{coefficient of active earth pressure for fill} = 0.29$$
$$\theta = \text{angle between tees} = 60°$$

The simultaneous equations for μ_{01}, μ_{03}, and μ_{32} are, by substitution into Eqs. (14.12):

$$\frac{0.636\,\mu_{01}^2}{\mu_{03}} + 3\mu_{01} + 2\mu_{03} - \mu_{32} - 1 = 0$$

$$0.504\,\mu_{03}\sqrt{\frac{\mu_{32}}{\mu_{01}}} + \mu_{01} + 2\mu_{03} + \mu_{32} - 3 = 0 \qquad (14.14)$$

$$\frac{2.004\,\mu_{01}\mu_{32}}{\mu_{03}} + \mu_{01} + 2\mu_{03} - 3\mu_{32} - 3 = 0$$

Solving simultaneously yields $\mu_{01}^* = 0.52$, $\mu_{03}^* = 0.25$, and $\mu_{32}^* = 1.73$. The remaining dual variables using Eqs. (14.11) are $\mu_{02}^* = \mu_{21}^* = 0.17$, $\mu_{04}^* = \mu_{31}^* = 0.06$, and $\mu_{11}^* = 0.68$. These optimal values of the dual variables indicate, even before the optimal dimensions of the cofferdam cell are known, how the relative proportion of costs should be allocated. In this case, 52% of the total cost should be spent on fill material ($\mu_{01}^* = 0.52$), and 42% on the steel sheet piles ($\mu_{02}^* + \mu_{03}^* = 0.42$). The cost of the expected flooding should account for only 6% of the total cost ($\mu_{04}^* = 0.06$).

The minimum total cost, Z^*, is obtained by calculating its equivalent, the maximum value of $d(\mu)$, from Eq. (14.10). By simple substitution this yields $Z^* = \$633,900$.

The optimum dimensions of the cofferdam cell are calculated from Eqs. (14.12) using Z^* and optimal weights μ^*. The optimum height of the cofferdam, h^*, is derived either from the third term in the objective function or the second term of the third constraint of Eq. (14.9). Using the former:

$$2C_s\rho L h^* = \mu_{03}^*\,Z^* \qquad (14.15)$$

so the optimal height is:

$$h^* = (0.25)(633,900)\left[\frac{2(163)(28)(800)}{2000}\right]^{-1} = 43.4 \text{ ft}$$

The optimal width, b^*, is calculated from the first term of the objective function, Eq. (14.9). Thus:

$$C_b L b^* h^* = \mu_{01}^* Z^* \tag{14.16}$$

so the optimal width is:

$$b^* = 0.52(633,900)[(5.67)(800)(43.4)/(27)]^{-1} = 45.2 \text{ ft}$$

The optimum cycle length, Y^*, can then be obtained from the second term of the objective function of Eq. (14.9). This gives:

$$(2C_s \rho L h^*)b^*(Y^*)^{-1} = \mu_{02}^* Z^* \tag{14.17}$$

or, alternatively, from the calculation for h^*:

$$(\mu_{03}^* Z^*)b^*(Y^*)^{-1} = \mu_{02}^* Z^* \tag{14.18}$$

so the optimal cycle length is:

$$Y^* = \frac{\mu_{03}^*}{\mu_{02}^*} b^* = \frac{0.25}{0.17}(45.2) = 66.4 \text{ ft} \tag{14.19}$$

Using elementary geometric relations, the optimal diameter of the cofferdam cell, D^*, and the optimal radius of the connecting cell, R^*, can be calculated, as suggested by Figure 14.1 (2). One obtains:

$$D^* = 57.5 \text{ ft}$$
$$R^* = 16.2 \text{ ft} \tag{14.20}$$

CONCLUSIONS

Structural engineers are members of one of the better developed technologies and oversee an enormous investment. Only very infrequently are infeasible structural designs encountered. Yet optimal designs are of comparable scarcity. A growing literature, dating from about the mid-1960s, attests to a recognition of the potentialities for optimization and reliability in structural analysis and design.

Important research results have emerged and have been implemented in structural design. However, there seem to be three main impediments to more substantial progress. First, the failure modes of structures can be difficult to define. Second, familiar structures have long lives and are often more versatile than originally imagined. Third, the complexity of structural analyses limits the generality with which optimization techniques can be utilized.

The cofferdam has important advantages as a candidate for research into the optimization of structural design. Its failure modes are distinct, its life is transient, its failure rarely results in loss of life, and the appropriate structural analyses are tractable. The preceding analysis of a circular cofferdam illustrates these observations.

The expected-cost objective function and the associated constraints explicitly account for relevant economic and technological considerations and permit the extensive sensitivity analyses which are required. The geometric programming formulation results in three simultaneous equations whose solution is related to the design height and design geometry. Geometric programming is well known for the insights it can provide into the optimal allocation of resources among the component costs. This information provides valuable shortcuts to experience, and reduced calculation for an improved design. The procedures developed here promise to relieve much of the tedium and expense of conventional trial-and-error methods, besides providing an improved design.

Sensitivity analyses of parametric information are feasible and, of course, a needed source of insight. This appears to be particularly required in this instance because cost information appears not to be well developed either for cofferdam design or for the effects of flooding. In part this lack of cost information reflects the transient character of the structure, the geographical variation, and even the resources of the contractor. The development of design procedures such as the one presented here may stimulate better assessment and documentation of costs. It is a step toward greater standardization of cofferdam design, and the core ideas appear applicable to the design of other flood control structures.

ACKNOWLEDGMENTS

Financial support during the research period was provided by an NSF research grant and by the Civil Engineering Department of the University of Delaware. The authors wish to express their appreciation to the Bell Telephone Laboratories for their assistance in the final preparation of this work.

REFERENCES

1. DUFFIN, R. J., PETERSON, E. L., AND ZENER, C., 1967. *Geometric Programming*, New York: John Wiley & Sons, Inc.
2. NEGHABAT, F., 1970. *Optimization in Cofferdam Design*, Ph.D. Dissertation, University of Delaware, Newark, Del., June.
3. NEGHABAT, F., AND STARK, R. M., 1970. *Optimum Cofferdam Height*, preprint, American Society of Civil Engineers National Meeting on Transportation, Construction Division, July.

4. STARK, R. M., AND NICHOLLS, R. L., 1972. *Mathematical Foundations for Design, Civil Engineering Systems*, New York: McGraw-Hill Book Company.
5. TENG, W. C., 1962. *Foundation Design*, Englewood Cliffs, N.J.: Prentice-Hall, Inc.
6. TERZAGHI, K., 1945. Stability and Stiffness of Cellular Cofferdam, *Transactions of American Society of Civil Engineers*, Vol. 110, pp. 1083–1119.
7. WHITE, L. W., AND PRENTIS, E. A., 1950. *Cofferdams*, New York: Columbia University Press, 2nd ed.
8. WILDE, D. J., AND BEIGHTLER, C. S., 1967. *Foundations of Optimization*, Englewood Cliffs, N.J.: Prentice-Hall, Inc.

15

Location Models: A Solid Waste Collection Example*

DAVID H. MARKS and JON C. LIEBMAN

This chapter investigates a problem that is common to a great many engineering projects, that of determining the number, type, and location of facilities that should be built to meet perceived demands. The use of optimization in the analysis of such a problem has increased rapidly in the past decade, with the availability of more sophisticated computational equipment and with the development of mathematical techniques necessary to deal with certain fixed charges, that is, economies of scale, characteristics of the problem which make it decidedly nonlinear. The algorithm used here is one of integer programming using the technique of branch-and-bound (see reference 11 for a detailed description). The computational formulation and solution procedures represent a very different optimization approach from those in the preceding chapters. However, it is important to note that we have chosen the main emphasis to be the type of problem that may be analyzed in this fashion and the sort of questions that one would investigate with it. The methods of analysis are not panaceas for pouring out "optimal" solutions, since the real world with its immense complexity defies exact analogs. The results are regarded as having contributed to the engineers' understanding of the options and interactions in the system: as an aid to judgment, not a replacement for it. This chapter briefly discusses location problems in general and applies mathematical programming to a public sector

*Adapted extensively from *The Journal of the Urban Planning and Development Division, Proceedings of the American Society of Civil Engineering*, Vol. 97, No. UP1, April 1971, pp. 15–30.

problem: the analysis of alternatives for the collection and disposal of solid wastes in the short run.

Given that we are interested in a model that will allow the investigation of the question of the location of facilities, the literature is full of many competing model formulations and solution techniques. The review article by Marks et al. (6) provides a complete delineation of such approaches. For the present discussion, however, the models can be differentiated into two classes: those in which the tradeoffs are and are not commensurate. Optimization requires a single objective to be maximized or minimized with all other system considerations appearing explicitly or implicitly as constraints. For most private-sector problems, such as locating warehouses or distribution facilities, and for some public-sector problems, this objective is to minimize economic costs, and the main tradeoff is between facility costs and transportation costs.

Most of the major algorithmic and computation work has been directed toward this problem, for it is well defined and represents a means of expressing alternatives in units pertinent to the direct interests of the decision makers, specifically, reduced costs to meet established goals. There is a class of public-sector problems where the tradeoff between construction and service is no longer in commensurate units. The location of fire and ambulance emergency equipment, for instance, trades the cost of facilities against the value of time in reaching the demand. These problems are usually approached by surrogate methods, where time or distance is used as a measure of effectiveness subject to budgetary limitations. The formulation and solution of these problems are consequently more involved. In this case study, the system has strong cost objectives, and a model of the first type is appropriate.

SCOPE OF THE SOLID WASTE PROBLEM

The solid waste system, as defined, includes the generation of waste at a source, its collection and transport, and its disposal. It is a large, expensive, and vital system, with peculiarities and problems that are in urgent need of analysis. The yearly cost of operating all such systems in the United States was reported by Black (2) to be $3 billion in 1964. This amount makes it one of the foremost public works costs, ranking after educational, highway, welfare, and fire and police protection expenditures. The amount of waste generated in the United States from all sources approaches 125,000,000 tons/yr. The indications are that these costs and volumes will continue to rise drastically (3).

A main difficulty in an analysis is the determination of some measure of effectiveness for the service provided. In this case the service provided is a public good: the entire community suffers in decreased values associated with public health and aesthetics when even one of its members refuses to dispose of his solid wastes properly. A measure of the value of a public system is often difficult to come by. For the solid waste system it can be taken as the degree of service offered to the customer in terms of the frequency of collection, the types of wastes removed, the locations from which waste is collected, and the general level of satisfaction shown by the consumers in order to provide the incentive for everyone to dispose of waste material properly.

Management to achieve a desired level of service at minimum cost is the goal, then, of the system operator who must study the various control alternatives available to improve the efficiency of the system. Cost analyses by Ludwig and Black (3) reveal that 85% of the solid waste system cost is due to collection and only 15% to disposal. This does not necessarily mean that the only approach to the problem is through improvements in collection. Despite the fact that most analysts consider that waste loads are given and they begin designing a system from that point, it would seem that the collection costs could be decreased by eliminating the wastes before they require collection and transport.

Some ideas have already been advanced for improving collection. Prominent among these is the building of transfer facilities within a city at which collection vehicles could discharge their loads to special-purpose transport vehicles, and thus return to their collection jobs sooner. The suggested transport vehicles have ranged from large tractor-trailers to barges and railroad cars. This is the concept we explore and have applied to Baltimore as a case study.

FACILITY LOCATION MODEL

We first establish a model of the transfer-station location problem. Then a solid waste collection system is analyzed to see how the optimal cost, number of transfer stations, and location pattern change with changing parameters and additional constraints. The model of the transfer process is based on a model developed by Marks (4, 5) for determining which intermediate nodes should exist in a transshipment network where there is a fixed charge cost function and a capacity constraint for each potential node. For the problem of locating transfer stations for the solid waste collection system, the sources are the solid wastes to be collected, the sinks are disposal points for the solid waste, and the intermediate points are the potential transfer stations. No attempt has been made to model the collection process within each collection area, so that only transport from the collection area to final disposal with or without transfer is considered. The mathematical model is given below.

There is a set of p sources of waste, K, with amount S_k at each source. In addition, there is a set of n disposal points, J, for the waste, each with an upper bound on the capacity of D_j. A set of m intermediate facility sites, I, has been suggested as transshipment points between the sources and sinks. Each proposed facility has a fixed charge, F_i, a variable unit cost, V_i, which is a linear function of the amount shipped through the facility: and a capacity, Q_i. The problem is to find which facilities should be built and which sources and sinks each facility serves, so that the total cost of facilities and transshipment is minimized.

In mathematical form this problem becomes the following:

Minimize $$\sum_i F_i y_i + \sum_i \sum_j c_{ij}^* X_{ij}^* + \sum_i \sum_k c_{ki}^{**} X_{ki}^{**} \qquad (15.1)$$

subject to feasibility constraints. In Eq. (15.1) $y_i = 1$ if the ith facility is built, 0 otherwise; X_{ij}^* is the flow of material, and c_{ij}^* is the unit cost of transferring this material from facility i to a disposal point j; X_{ki}^{**} is the flow of material, and c_{ki}^{**} is the unit cost of transferring this material from source k to intermediate point i.

The first feasibility constraint is that all wastes must be collected:

$$\sum_i X_{ki}^{**} = S_k \qquad (15.2)$$

Second, all flows that enter a facility are expected to leave it:

$$\sum_j X_{ij}^* = \sum_k X_{ki}^{**} \qquad (15.3)$$

Also, the flow that enters a transshipment facility must either not exceed its capacity, if it exists, or equal zero:

$$\sum_i X_{ki}^{**} \leq Q_i y_i \qquad (15.4)$$

Likewise, flows to each disposal point cannot be expected to exceed its capacity:

$$\sum_i X_{ij}^* \leq D_j \qquad (15.5)$$

Finally, the flows cannot, of course, be negative:

$$X_{ij}^*, X_{ki}^{**} \geq 0 \qquad (15.6)$$

and $y_i = 0$ or 1, as indicated before.

Several additional constraints or side conditions might also be of interest. A budget constraint on the number or cost of the facilities to be built, or a mutual exclusivity constraint which allows a particular facility to exist only if another does not, might be included. The algorithm suggested for solution of this problem leads to optimal results with such added constraints and is presented by ReVelle et al. (9).

The model was applied to Baltimore using readily available data when they existed and best estimates when such data were not available. The model was used for screening purposes to show where sensitivity occurs and, therefore, where investment of time and money should be made to obtain better data.

BALTIMORE'S SOLID WASTE COLLECTION SYSTEM

The city of Baltimore, Maryland, had a population in 1960 of just over 1 million people, a land area of 80.3 square miles, and physical characteristics that ranged from high-density urban slums to verdant, low-density residential subdivisions. Solid waste collection and disposal as well as general sanitation activities are carried out by the city's Bureau of Sanitation of the Department of Public Works. For purposes of logistics and supervision, the city is subdivided into five autonomous districts, the northwestern, the northeastern, the central, the eastern, and the western. As of 1965, the Bureau employed 1271 persons, owned approximately 340 vehicles and pieces of equipment, and operated on an annual budget of over $7,000,000 (1). Collection accounts for almost 85 % of the department's operation, and the predominant collection operations are mixed refuse collection and street cleaning.

The Bureau collects mixed refuse from residential and noncommercial sites twice a week, once on either Monday, Tuesday, or Wednesday, and once on either Thursday, Friday, or Saturday. As collection in the earlier part of the week represents the pickup of 4 days' accumulation of wastes, the crew size is one driver and three laborers. In the later part of the week, the crew size is reduced to one driver and two laborers. Two types of vehicles are generally used for collection. They are both equipped with compaction equipment and have capacities of 20 cubic yards and 13 cubic yards of compacted solid wastes each. The larger vehicle is thought to be more efficient than the smaller and is assigned to the general tasks, while the smaller is saved for alleys and other size-restricting jobs.

In 1965 the Bureau collected 336,893 tons of mixed refuse, 31,395 cubic yards of ashes, and 216,622 cubic yards of street dirt. The collection of the mixed refuse from residences involves 92 vehicles, each with three daily route assignments per week, for a total of 276 routes. The city owns two incinerators for the disposal of wastes, and the collection vehicles take the solid waste directly from the collection area to the incinerator. The Reedbird incinerator has a daily capacity of 600 tons, and the Pulaski incinerator has a daily capacity of 800 tons. The location of these facilities is shown in Figure 15.1.

Several alternatives were considered for transfer sites. These are also shown in Figure 15.1. Sites A, B, C, D, E, and F are sites where transfer may

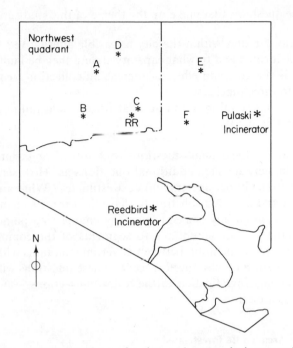

Figure 15.1 Proposed transfer sites and present incinerator sites, Baltimore, Md.

take place to larger vehicles. Site RR is a site for possible transfer of the waste to railroads.

Analysis for Baltimore

Analysts involved in studying, and planning for large-scale public systems such as the solid waste collection and disposal systems must be concerned with both the long and the short range. Not only must they study the evolution of new technology and methods that may dramatically change alternatives and options in the future, but they must be concerned with the system as it now exists. The ensuing description is directed toward the analysis of the system within the short-range picture, in order to give some feeling for the alternatives that are available for immediate introduction. These alternatives may be physical changes, such as the building of new structures; policy changes, such as the modification of the manner and character of the services and rules of operation of the system; or changes in the methods for carrying out certain objectives of the system. Short range alternatives falling into each of these categories make up the prominent issues that require analytical attention.

The three questions that make up the theme of this study are:

1. Are transfer sites within the city a feasible alternative? If so, where should they be located and to what capacity should they be built?
2. What is the cost and effect of increasing collection frequency from twice a week to three times a week?
3. Under what conditions and at what price would rail haul become a feasible alternative for the city?

Subordinate to these main questions, but still of great interest to the analyst and planner, are such additional questions as: How sensitive is the system to variations in parameters and cost estimates? What are the effects of the political and aesthetic constraints that might force a change from the solution most economically efficient from the engineering point of view to one that is perhaps more acceptable to segments of the community? Are there advantages of cooperation between governmental units within a region where such cooperation does not now exist? These questions will be subject to some preliminary investigation in the following sections, using the large-scale facility location model.

Collection Area To Be Investigated

The area chosen for detailed study was the northwest division of the city of Baltimore, which has also been studied by Truitt et al. (10). It is necessary to subdivide this area further for the analysis. A convenient way to do this is to consider each census tract as a subarea. The exact population was, then, accurately known from the decennial census, and data on housing density were based on estimates made from census data by the city of Baltimore, Department of Planning, as reported by Truitt et al.

The density of housing is important because it gives some indication of the speed of collection within a subarea. It would be suspected that the more closely spaced the pickup points are, the faster the collection process may be in terms of pounds per hour collected. Based on a statistical study made by Truitt et al., the population density in this work was classified either as low (less than 10 housing units per net acre) or high.

Estimation of Present System Costs

A most important aspect of the study is the establishment of some benchmark against which the results of changing the system may be measured to determine their efficiency. With the present system, wastes are collected twice a week, and published figures on the cost of running it are available (1).

Therefore, computation of its cost serves not only as a benchmark but also as a check on the validity of the model.

The yearly cost of collection by the city from the northwest division for 1965 was $722,000. The model reports $16,664 per week for collection and disposal, of which $12,666 per week, or $658,632 per year, are collection costs. The difference between the two estimates is thus 9.7%, which may be explained by the fact that the 1960 population estimates were used in the model, while the actual cost figure represents the cost of collecting from the 1965 population of the same region. As the northwest division of the city, particularly at its outermost fringes, is gaining in population, it would be expected that the model estimates would be slightly lower.

Estimation of Costs for Collection Three Times a Week

An estimate was made of the total cost of three-times-a-week collection without transfer stations, including disposal costs, with a resulting total system cost of $17,399 per week. When compared to the twice-a-week benchmark of $16,664 per week, this shows only a 4.4% increase in cost to increase the collection frequency. However, this estimate is really only a lower bound on the actual cost of three-times-a-week collection and thus should be viewed with care. There were several assumptions made in the calculation of the cost of three-times-a-week collection that should be clearly stated.

The first assumption is that the routes for the vehicles will be changed from those of the twice-a-week collection, to ensure that each truck is close to capacity before it makes the trip to the transfer or disposal point. Indeed, as there is no instrumentation on a vehicle to indicate to the driver the accumulated load, routes are usually assigned in terms of definite areas, with the driver often returning to the disposal point when he reaches a certain place in his route. Because less waste is generated in a given area for three-times-a-week collection, as there has been less time since the last collection, the vehicle must travel farther to get a completed load. The average load on a vehicle, an important factor in the calculation of costs, may consequently drop, especially if routes are improperly designed. Should the average load for the three-times-a-week collection drop by 10%, the overall cost difference between twice and three times a week would rise 8%. Likewise, should the average collected weight drop by 25%, the costs would rise 14%. Thus extreme care must be taken in designing the routes to make sure that enough load is picked up before transportation. In particular, careful attention must be paid to the time it takes to cover the routes. Most operations work under a no-overtime constraint, which means that at a certain time the truck will return, regardless of its present load.

A second assumption that has been made in these computations is that the rate of waste generation is the same for twice a week as for three-times-a week collection. However, work by Quon et al. (7) based on actual observations in a controlled experiment in Chicago shows increases in total waste on the order of 30 to 50% when the collection frequency was increased from once- to twice-a-week. Analyses were made to test the effect of increases in waste on the cost of three-times-a-week collection. These runs show that the estimation of the difference between the twice- and three-times-a-week collection is quite sensitive to the assumption made about waste generation (Table 15.1). Great care should, therefore, be taken in estimating the difference in cost between the twice- and three-times-a-week collection.

Table 15.1 Effect on Costs of Different Assumptions About Increases in Waste Loads for Three-Times-a-Week Collection with No Transfer

Increase Over Twice a Week Collection (%)		Total Cost ($/week)
Waste	Cost	
0	4.4	17,399
5	9.6	18,270
10	14.8	19,132
15	20.0	19,998

In measuring and comparing differences between individual runs when both have the same collection frequency, however, these assumptions have no serious effect on the calculation because the same assumptions appear in both. The remainder of the runs made with a collection frequency of three-times-a week were all made with the assumption that the average truck load does not decrease from twice-a-week collection and that the total waste generated does not change.

Desirability of Transfer Stations

With the best data available for the various cost parameters and for the cost of facilities, a series of analyses were carried out with transfer stations of different capacity to see if they would lower the cost of collection. In every case they did. The results of the analyses also give some interesting insights into the system. Runs were made for collections of both two and three times a week, and for four different-sized transfer stations, ranging from 40% of the estimates of weekly waste load, 600 tons/week, to over 100%. The results of all these runs are presented in Figure 15.2. For alternatives with a capacity of 900 tons/week and more, the cost function for the higher-capacity stations is more favorable and quite flat, indicating little sensitivity to size within that

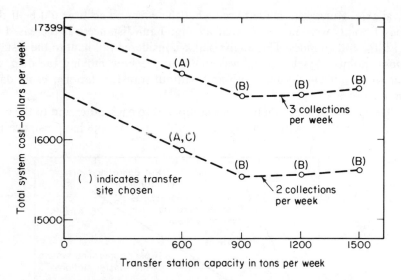

Figure 15.2 Total system cost versus size of transfer station.

range. More important, for each solution only one transfer site was chosen and for all six cases that site was B. The building of only one transfer station indicated that some waste load was still going directly to the final disposal point without transfer. Also, the facilities with 1200 and 1500 tons/week in capacity were used at less than capacity, with about 40% of the waste in both cases going directly to disposal. Later runs showed that this proportion was very sensitive to the location of the final disposal points. The cost for the three-times-a-week collection with transfer is comparable with that of the present two-times-a-week collection without transfer.

For the smaller-capacity (600 tons/week) stations, a different alternative, A, was chosen and in one case two alternatives, A and C, were chosen. In general, the difference between the solution with transfer stations and the one without is in the range of 4 to 7%, with the two-times-a-week collection frequency allowing a better decrease. The absolute difference between the transfer and nontransfer is a rough measure of how much the weekly costs of the facility could increase before transfer is no longer favorable.

Effect of Increased Haul Distance to Final Disposal

The present average haul distance from the northwestern division to disposal at the Pulaski incinerator was reported by Truitt et al. (10) to be 8 miles one way; this was verified by the present authors. As the city expands, the tendency is to move the waste disposal farther out, both to avoid creating a nuisance to nearby neighbors, and because unused sites near the city are unavailable. Runs were made to see what the effect on costs would be if the disposal points were moved so that average haul distance was increased to 10, 12, 14, and 16 miles. This analysis also considered rail haul to the distant disposal points. A set of runs was made to see how moving the disposal point would affect a system with and without transfer stations, using data from reference 8.

The results of these analyses are compared to each other and to the cost of the rail haul alternative in Figure 15.3. As the average haul distance in-

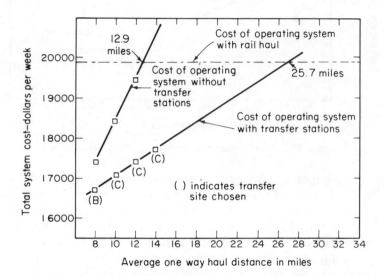

Figure 15.3 Transfer facilities become more desirable as haul distance increases (collection frequency = three times a week).

creases, transfer becomes more and more favorable. Further, combined use of trucks for collection and rail for long-distance hauling eventually dominates all truck operations both with and without transfer facilities. Without transfer, rail haul becomes feasible at 12.9 miles average haul distance. If transfer is allowed, rail haul is not a suitable alternative until the average haul distance is almost 26 miles. This indicates that as the pressure mounts to move disposal farther out from the city, a transfer facility system would preclude rail haul for some time, if the relative costs of the alternatives would remain the same.

Further, the location chosen for all transfer stations 10 or more miles out was the same, C. This indicates that the site selection in this case is quite stable. Also, as soon as the disposal point has been moved only 2 miles, all waste in the system goes through the transfer facilities and none is taken directly to the disposal point. This is the reason that C was chosen for this case while B was chosen for the previous cases with the present location of facilities. The initial choice between B and C, with a haul distance of 8 miles, favors B only slightly, and when the conditions are changed to cause more load to go through a facility, C becomes the best alternative.

Figure 15.4 shows how the twice-a-week collection system would react

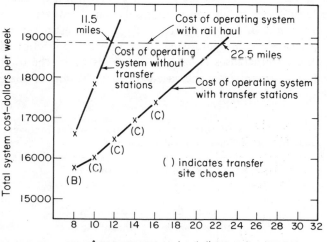

Figure 15.4 Transfer facilities become more desirable as haul distance increases (collection frequency = twice a week).

under the same changes in haul distance. The results are quite similar, with C being chosen as the transfer site and the rail-haul alternative becoming feasible at a slightly smaller haul distance than in the previous case.

Thus it would appear that the questions concerning transfer and rail haul are intimately linked. Under the present conditions, rail haul is not an economical alternative. At a cost of $18,791 per week for the present system, it represents a 12.8% increase in cost over the system with no transfer and an 18.3% increase over the system with transfer. However, if haul distances should increase over time, rail haul would become a desirable alternative in a fairly short time if no transfer is established. If transfer facilities were built, they would reduce costs to the extent that rail haul would not be economical for a considerable time. As for transfer itself, it does exhibit some, but not extraordinary, savings over the present system without transfer. However, for future conditions, these savings will increase, thus making transfer even more favorable.

Sensitivity to Other Parameters

Other analyses were made to show the sensitivity of the system to estimates of the various parameters. As the estimate of the waste in the three-times-a-week collection has been shown to have an effect on the solution, several additional runs were made to study this situation. Trials were made for three-times-a-week collection frequency with and without transfer facilities and for an increase in the waste load of 5%, 10%, and 15%. The results are shown in Table 15.2. The cost of the systems with a transfer station is even more sensitive than the nontransfer solution. However, in each case, the site selected was B, indicating that only the cost estimate, not the site chosen, is sensitive to this parameter. Twice-a-week collection was tested for its sensitivity to increases in waste in six trials where the total waste to be collected was progressively increased by 60%. Results of these runs are shown in Figure 15.5.

Table 15.2 Effect on Costs of Different Assumptions About Increases in Waste Loads for Three-Times-a-Week Collection

Increase over Twice-a-Week Collection with No Transfer (%)

Waste	Cost With Transfer	Without Transfer	Site Selected
0	4.4	6.4	B
5	9.6	11.4	B
10	14.8	16.6	B
15	20.0	21.5	B

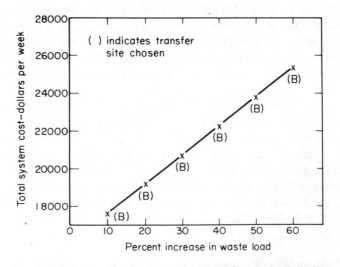

Figure 15.5 System cost is a linear function of waste loads (collection frequency = twice a week).

As for three-times-a-week collection, the cost of the system varied linearly with the waste load. This was because of the linear nature of the cost function after a facility was built. The same site, B, was also chosen for the transfer point for every solution. Even when the total waste was almost double the transfer-station capacity, it was not optimal to build a second facility at a different site.

The sensitivity of the solution to the speed of the transfer vehicle from the transfer facility to the disposal point was investigated. Present estimates place this speed at 16 mph because this is the approximate speed in traffic of the collection vehicles, and the transfer vehicles are larger tractor-trailers which must operate under largely the same traffic conditions. An increase in the traffic speed from 16 to 30 mph would bring about a decrease of about 7.5% in the cost of the system. However, it would be expected that the cost of the remedial measures necessary to bring about this increase in speed would more than outweigh any savings in transportation costs.

An attempt was also made to see how collection rates affected the solution. The collection rates were varied in the analyses by -10%, $+10\%$, and $+20\%$. Increasing or decreasing the collection rate by 10% leads to about a 5% change in the cost of the system. This falls into the range of being somewhat sensitive. Thus care should be taken in estimating these rates. Further, work rules that tend to increase the collection rate without too great an increase in cost, as well as inexpensive time-saving devices, would also bear some investigation.

The effect of changing the estimates of the costs of the facilities is also of interest. This gives an idea as to how carefully cost estimates have to be made for preliminary studies. Runs were made to show how fixed cost increases of 25%, 50%, and 100% would affect the solution. At about a 75% increase in fixed costs, transfer facilities are no longer feasible, but up to that point the same facility was chosen each time. A rough estimate as to how much the costs of the facilities could increase could have been found from earlier runs simply by looking at the absolute difference in cost between the transfer and nontransfer solutions. If the increase in the cost of the facility exceeded this difference, a transfer would no longer be economical.

Political, Aesthetic, and Regional Constraints

Many times, constraints occur which preclude the use of an alternative that would otherwise seem to be the best from an economic point of view. A site may be excluded, for example, either because it may be more useful for other municipal functions or, politically, because neighborhood groups oppose a facility on the site. Alternatively, regional constraints may not allow for cooperation among different political subdivisions.

Consider first an example where a particular site is not allowed to be used. The results of the analysis to this point show that B is the best transfer site. Now consider what occurs when this site is excluded from the set of alternatives. In the case of the twice-a-week collection, site A is chosen and the increase in the cost is about $60 per week. For the three-times-a-week collection, the site chosen is site C, and the difference is about $110 per week. Both of these changes are minuscule compared to the total cost of operating the system. This indicates that there may be readily available alternatives to the best site that will not raise the cost of the solution significantly. Conversely, it implies that optimum location is not the predominant issue and that the analyst ought to become concerned with feasible locations. In any case, the cost of satisfying additional constraints which prohibit certain sites is not particularly high, and the decision maker has a great deal of leeway in choosing between alternatives when other criteria must also be considered.

The effect of regionalization will be treated only briefly in this example, as neither the time nor the data were available for assembling an extremely large regional case. However, the question can be looked at implicitly by assuming that the northwestern division is a unified region, and asking how much the costs would increase if it were subdivided into two noncooperating regions. Suppose the region were divided into two areas so that the northern areas could use only transfer sites A, D, and E, and the southern areas could use only sites B, C, and F. In this case, transfer facilities are still economical, but only at an extremely slight advantage over no transfer at all. This would suggest that regionalization would mean increased savings. Although this analysis was not extended to the entire city, it might be expected that each of the administrative divisions would not have a transfer facility, but only two or three would be built.

Costs of Analysis

Sixty runs of the computer model were executed to obtain the above results. This required 45 minutes of computer time on the IBM 7094. At commercial rates of $500 per hour, this represents a cost of only about $375 for machine time.

SUMMARY OF EXPERIMENTAL ANALYSIS

The following general statements about the optimal configuration of the waste collection system for Baltimore can be made, based on the rough analysis with estimated data for the northwestern division. These findings would need verification with better cost data before they could be viewed as completely reliable.

First, it appears that building a transfer facility is economically desirable and would result in an annual saving, for 1969 conditions, of about 6% in the total cost of operating the solid waste collection system. This saving would be expected to increase in the future, as the system expands and grows. The most favorable size of the facility to be built is the largest plant that could be useful, the one with 500 tons/week capacity. This plant offers substantial capacity for future demands at minimal additional cost to the system. At this size, some of the waste material will be taken to transfer points and some will still be taken directly to the disposal point.

A rail-haul alternative would cost about 12% more to operate than the present system and about 18% more than the system with a transfer facility. Haul distance would have to increase an average of from 3.5 miles to 5 miles for rail haul to be favorable without transfer, and from 14 miles to 18 miles to be favorable with transfer.

The cost of changing the frequency of collections from twice to three times a week is difficult to assess because of the sensitivity of this calculation to estimates of how much the waste load might be. The added cost might range from 4.5% if there is no increase in the waste load, up to 21.5%, with a 15% increase in the waste load. All these figures are based on the assumption that the routes for the vehicles under three-times-a-week collection be redesigned for efficiency purposes. It would be strongly advised that more studies of the waste load question in this region be carried out before any firm decision be made on changing collection frequency.

Although differences in assumptions make estimating the cost of some configurations of the system difficult, the preferred site for a transfer facility is remarkably insensitive to change in the parameters of the problem. Regardless of the collection frequency, site B is the best site for the transfer facility. If for some reason site B could not be used, it would cost little extra to use the alternative sites available.

The cost of the system shows some sensitivity to collection rates, which indicates that care should be taken in measuring them and an investigation of some means of altering them will be of some help. Also, the costs are sensitive to the size of the region being serviced: it is better to form cooperative regional groups than to operate independent districts.

CONCLUSIONS

A facility location model has been used to screen preliminary data in order to learn something about the issues of whether and where to locate solid waste transfer stations in the city of Baltimore, Maryland. Use of this model has indicated that a transfer facility would be economically desirable, but that only a few, rather than many, facilities should be built. The cost estimates, although not the choice of location, are, however, very sensitive to estimates of load increases. Thus investigation of the effect of collection frequency on the generation of waste loads is of great value. In all, the use of the model with available data and a small amount of computer time has generated a great deal of information about the solid waste collection system.

ACKNOWLEDGMENTS

This work was supported by the Department of Health, Education, and Welfare, Environmental Control Administration, through Grants 5TO1 UI 01049 and 1RO1 UI 00828. In addition, we wish to thank Charles S. ReVelle of The Johns Hopkins University for his many helpful comments and support.

REFERENCES

1. *Baltimore, Department of Public Works*, 1966. Annual Report for the Year Ending Dec. 31, 1965, Baltimore.
2. BLACK, R. J., 1964. The Solid Waste Problem in Metropolitan Areas, *California Vector Views*, Vol. 11, Sept.
3. LUDWIG, H. F., AND BLACK, R. J., 1968. Report on the Solid Waste Problem, *Journal of the Sanitary Engineering Division, ASCE*, Vol. 94, No. SA2 (Proceedings Paper 5909), April, pp. 355–370.
4. MARKS, D. H., 1969. *Facility Location and Routing Models in Solid Waste Collection*, Ph. D. Thesis, The Johns Hopkins University, Baltimore.
5. MARKS, D. H., LIEBMAN, J. C., AND BELLMORE, M., 1970. Optimal Location of Intermediate Facilities in a Trans-Shipment Network, paper presented at the 37th National Operations Research Society of America Meeting, Washington, D.C., April.
6. MARKS, D. H., REVELLE, C. S., AND LIEBMAN, J. C., 1970. Mathematical Models of Location: A Review, *Journal of the Sanitary Engineering Division, ASCE*, Vol. 96, No. UP1 (Proceedings Paper 7164), March, pp. 81–93.
7. QUON, J. E., TANAKA, M., AND CHARNES, A., 1968. Refuse Quantities and Frequency of Service, *Journal of the Sanitary Engineering Division, ASCE*, Vol. 94, No. SA2 (Proceedings Paper 5917), April, pp. 403–420.
8. Refuse Disposal via Railroads, *Public Works*, July 1968.
9. REVELLE, C. S., MARKS, D. H., AND LIEBMAN, J. C., 1970. Private and Public Sector Location Models, *Management Science*, Vol. 16, No. 11, July, pp. 692–707.
10. TRUITT, M. M., LIEBMAN, J. C., AND KRUSE, C. W., 1969. Simulation Model of Urban Refuse Collection, *Journal of the Sanitary Engineering Division, ASCE*, Vol. 95, No. SA2 (Proceedings Paper 6508), April, pp. 289–298.
11. WAGNER, H. M., 1969. *Principles of Operations Research*, Englewood Cliffs, N.J.: Prentice-Hall, Inc.

16

Cost–Effectiveness Evaluation
of Freeway Design Alternatives*

ADOLF D. MAY AND JOHN H. JAMES

This chapter deals with the selection of the optimal series of freeway design improvements in a network where the effects of decisions are interactive and measures of effectiveness must be established. A heuristic and an optimal algorithm, both sensitive to the sequential nature of the decision process and thus dynamic programming in nature, are specified for the formulated problem. Then a case study on a California freeway is presented to show the use of such techniques.

THE PROBLEM

Congestion is now encountered during weekday afternoon peak periods on the northbound roadway of the Eastshore Freeway (I-80) in the San Francisco Bay area. Through driver complaints and preliminary engineering studies, the California Division of Highways has recognized this section of roadway as a problem. It has, therefore, requested an evaluation of possible freeway improvements. While other alternatives are possible, and have been discussed, they have been specifically interested in the possibilities of adding a lane or lanes along portions or along the entire length of the study section. The question of what additions are most effective is the problem addressed by this analysis.

*Adapted extensively from *Bay Area Freeway Operations Study—Part I: The Freeway Model* and *Part II: On the San Francisco–Oakland Bay Bridge*, University of California, Institute of Transportation and Traffic Engineering, Berkeley, Calif., Aug. 1970.

The system to be studied will be defined in terms of geographic boundaries and time limits. The system is limited to an approximately 10-mile section of the northbound roadway and associated on-and-off ramps of the Eastshore Freeway (I-80). The period of time to be studied is limited to 3:45 to 6:15 on a typical weekday afternoon.

The system is divided into subsections. A subsection is defined as a portion of the roadway over which demand and capacity remain constant. There were, consequently, 30 subsections for this section of roadway. Subject to constraints that are defined later, 1, 2, or 3 lanes may be added to each subsection. The total number of possible combinations of lane additions, that is, the total number of alternatives to be explored, is between 2^{30} and 3^{30}.

Detailed traffic data are available for the section of the freeway to be studied. They consist of counts of the number of vehicles entering various on-ramps and destined for various off-ramps, by 15-minute periods. In addition, the design characteristics of each subsection are available as well as selected relationships between the speed and flow of traffic. These data are described in detail in project reports (1, 4). Knowledge of these flows permits one to evaluate the performance of the freeway system for any configuration of lanes.

STRUCTURE OF THE PROBLEM

The structure of this problem has some quite particular features which have important consequences for the possible forms of analysis. These are worth discussing in detail. They not only characterize this situation, but a broad category of design problems, including such varied facilities as a railroad main line with its intermediate yards and spurs; a ship channel with varying cross sections and locks; and an automatic baggage and cargo handling system for an airport.

Basically, the freeway system can be thought of as a sequence of facilities, represented by each of the subsections, designed to process or service a variable flow of arrivals. As the arrivals are discrete, and are processed for their own particular characteristics, each subsection acts essentially as a service facility serving a queue. The whole system behaves like a series of queues, being processed through a series of service facilities.

Two important features of the system derive from these characteristics. First, there is the nonlinear behavior associated with each service facility or, in this case, subsection of the freeway. As is known from queuing theory, the delays that are associated with any service facility are closely associated with the degree of saturation or congestion. In queuing theory generally, we speak of saturation in terms of ρ, the ratio of the rate of arrivals to the maximum possible rate of service. As saturation increases, that is, as ρ increases from zero toward unity, the delays increase very rapidly, approximately:

$$\text{delays} \approx \frac{K}{1 - \rho} \tag{16.1}$$

In highway design, we refer to the volume/capacity ratio instead of ρ, but the effect on delays is similar. The net result for all such facilities is that improvements in the system, defined in terms of decreased delays, are a highly nonlinear function of investments in capacity.

The second important characteristic is the highly interactive nature of the series of queues. This leads to a convex objective function (for a maximization problem) to which it is extremely difficult to apply conventional optimization procedures. Because of the highly nonlinear nature of Eq. (16.1), it is important to reduce the degree and duration of saturation where it might occur. In a heavily loaded facility it is, therefore, sometimes important to try to prevent traffic from reaching critical points at critical moments. It may thus be desirable to deliberately introduce delays into the system at carefully selected points to improve the overall performance of the system. This characteristic has already been documented for other systems, such as lock operations on a barge canal, and is, in fact, the principle behind some of the deliberate metering of traffic at on-ramps of some congested highways at rush hour. Conversely, it may well be counterproductive to remove bottlenecks and reduce delays at some points, because this may increase the total delay in the system. Thus, we cannot assume that the objective function is concave. Nor, of course, can we assume that the value of improvements in one subsection is independent of the improvements in other subsections.

As a result of these factors inherent in this problem, we cannot use the most common optimization techniques. Linear programming is out, and so is any kind of piecewise linear programming resulting from the convexity of the objective function. Nor can we proceed, in the general case, to implicit enumeration techniques such as dynamic programming, because of the interdependence between the subsections.

Yet the problem is, indeed, huge. The number of alternatives, just for this limited problem, is between 2^{30} and 3^{30}, that is, of the order of 10^{10}. Even if we had a computer that could evaluate 1000 alternatives every second, it would take us several months of continuous calculations to consider all possibilities. As it actually took about 4 seconds to examine any configuration of lanes, we must look for approximative methods that will give us improved, if not optimal, solutions at a reasonable cost.

THE MODEL

In model formulation it is necessary to choose measures of effectiveness that represent the major objectives of the system and to relate them in a meaningful fashion. For the purpose of improving the freeway capacity, the measures of effectiveness chosen were the annual amount of money spent for construction and maintenance on the one hand, and the annual passenger-hours saved by the construction. The basic problem to be solved was formulated in words as: Find the lane additions that will

Maximize passenger hours saved

subject to the constraints that

the resulting freeway is a feasible design, and
a given budget constraint for the additions is not violated

Mathematically, we proceed as follows. Let X_j be the number of lanes added to subsection j, and let $\bar{X} = (X_1, X_2, X_3, \ldots, X_k)$ be a collection of lane improvements. Define $G(\bar{X})$ as the annual effectiveness of the lane arrangement $\bar{X}$, and $C(\bar{X})$ as the annual cost of lane arrangement $\bar{X}$. The objective, given an investment C, is to find the optimal lane arrangement $\bar{X}^*$ such that

$$G(\bar{X}^*) = \max G(\bar{X}) \tag{16.2}$$

subject to $C(\bar{X}) \leq C$

The model still requires a definition of the effect of lane changes in the subsections, $G(\bar{X})$, of the cost function, $C(\bar{X})$, and of feasible highway designs. We turn first to the description of system changes.

A deterministic predictive model (FREEQ, pronounced "free queue") was developed for the analysis of freeway systems (4). Essentially, given the design features of a freeway and the on-ramp to off-ramp flow demands for the study period, the FREEQ model determines the total passenger time expended in the system. Knowing the total delays associated with the base case of no new construction, we can therefore calculate $G(\bar{X})$.

For a complete description of FREEQ and for previous applications of the model, the reader is referred to the original reports (1, 4). The basic assumptions of the FREEQ model are as follows. First, traffic is treated as a compressible fluid, wherein an individual vehicle is regarded as an integral part of the flow and is not considered individually. Second, traffic demands are assumed to remain constant and do not fluctuate over any given time interval; they are expressed, for a given subsection, as a step function over the entire time period under consideration. Third, traffic is considered to propagate downstream instantaneously, once it is loaded onto the freeway, unless there are capacity constraints. Finally, the capacities of subsections, including weaving sections and merging points, are estimated using the standard methods defined in the *Highway Capacity Manual* (3). The FREEQ model is quite efficient. For example, it took up 34K words of computer core storage, used about 4 seconds of central processor time on the CDC 6600 for an evaluation of a 10-mile directional freeway for $2\frac{1}{2}$ hours of real time, at a cost of only $1.50.

Assuming the construction cost of widening a subsection is linear with respect to the number of lanes added, the annual cost of improvement, $C(\bar{X})$, may be taken as

$$C(\bar{X}) = R \sum C_j X_j \tag{16.3}$$

where C_j is the initial cost of widening subsection j by one lane and X_j is the number of lanes added to the existing subsection. Assuming that the additional annual maintenance cost of new lanes is a fixed percentage, M, of the initial cost of the improvement, the scaling factor, R, is

$$R = CRF + M \tag{16.4}$$

where CRF is the capital recovery factor:

$$CRF = \frac{r(1 + r)^n}{(1 + r)^n - 1} \tag{16.5}$$

and r is the interest rate and n the economic life of the improvement. For the following analysis, r was taken equal to 6%, n to 10 years, and M to 0.5% of the construction cost.

The highway design is limited by feasibility in the following manner. The alternatives open to the decision maker are the lane arrangements made by adding 0, 1, 2, ... lanes to each subsection. A lane-improvement plan is feasible if it is consistent with good highway design practice. This means that the number of lanes downstream of an off-ramp must either be equal to or be one less than the number of lanes upstream of the off-ramp. Likewise, the number of lanes downstream of an on-ramp must either be equal to or be one more than the number of lanes upstream of the on-ramp. Finally, only one lane can be added or dropped at a subsection boundary, and the maximum permitted number of lanes in a subsection is 6. These constraints can be represented by a network (Figures 16.1 and 16.8) and a simple dynamic programming algorithm can be used to generate feasible lane arrangements automatically.

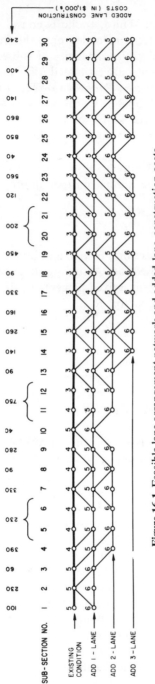
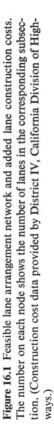

Figure 16.1 Feasible lane arrangement network and added lane construction costs. The number on each node shows the number of lanes in the corresponding subsection. (Construction cost data provided by District IV, California Division of Highways.)

Figure 16.8 Feasible lane arrangement network based on minimum design which results in no congestion.

ANALYSIS

If all feasible lane arrangements were known along with their cost and time savings evaluation, they could be plotted on a cost–effectiveness diagram, with the X axis being the annual cost and the Y axis being the annual savings in passenger hours. Consider the alternatives depicted in Figure 16.2, where each point represents the result of a particular lane arrangement.

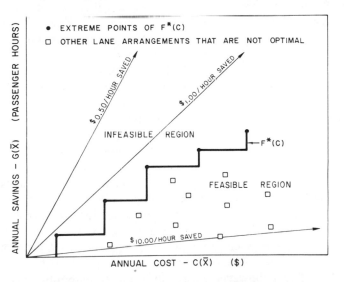

Figure 16.2 Example of an optimal cost–effectiveness curve.

It is easy to see that certain alternatives dominate others; that is, for a given budget, one alternative has a better saving than all others with the same budget. These points for different budgets are represented in the figure by darkened points; dominated alternatives are shown by square dots. The locus of the darkened points define the optimal cost–effectiveness curve $F^*(C)$. Each of these lane arrangements is optimal and the collection is referred to as the set S^*. The problem becomes to design a procedure that efficiently identifies the locus and character of the curve and its component points or, at least, some approximation of it.

The decision maker can utilize the cost–effectiveness diagram to determine the course of action to be taken in one of three ways. One way is to specify the amount of investment C, and to select the extreme point of $F^*(C)$ having a cost equal to or less than C. The slope of the line from the origin to this extreme point is numerically equivalent to the number of passenger-hours saved per dollar invested and its reciprocal is the cost–effectiveness in dollars invested per passenger-hour saved. Another approach is for the decision maker to specify the maximum value to be inputed to improvement, for example $2.00 per passenger-hour saved, and to select the extreme point of $F^*(C)$ which lies on or slightly above the intercept of the cost–effectiveness curve and the sloping line from the origin which corresponds to the specified maximum cost–effectiveness ratio. The third method is to specify the minimum acceptable annual savings in passenger-hours and to select the extreme point of $F^*(C)$ having an annual savings in passenger-hours which will first satisfy that goal.

If the objective function were concave and if the effects of investments in each subsection were independent, the cost–effectiveness function could be easily found by marginal analysis. That is, we could simply find which might be the most effective lane to add first, which second, and so on. This requires a minimal computational effort. A maximum of 90 evaluations (30 subsections times a maximum of the 3 lanes it is possible to add) would be required.

Unfortunately, because the feasible region is nonconvex and the subsections are interdependent, this marginal analysis procedure does not guarantee that we find the optimal cost–effectiveness curve, $F^*(C)$. In fact, in situations such as these, marginal analysis is frequently known as the myopic rule because, by focusing on the value of immediate improvements, one may short-sightedly be led into suboptimal arrangements. Yet, although we know that marginal analysis does not guarantee that we find $F^*(C)$, we do not have to assume that it pathologically always drives us away from it. And thus, because we need some way to approximate $F^*(C)$ without looking at all 2^{30} or 3^{30} combinations, we will use marginal analysis to approach $F^*(C)$. We will also, however, check our results for suboptimality by various ways.

Before we may proceed, however, we must deal with a further problem. This has to do with the fact that we may not build all the lane improvements at once, but may choose to plan for a sequence of construction. We cannot use the general optimal cost–effectiveness curve, $F^*(C)$, for this because, although we can build the lane arrangement corresponding to any point on $F^*(C)$, it may be impossible to pass from this point to higher points without destroying part of the freeway. In short, because of the interdependence of the performance of the subsections, each lane arrangement defining a point on $F^*(C)$ is not necessarily a subset of higher points. Yet we wish our solution to have this characteristic so that the improvement programs can be staged over time.

To overcome this difficulty, we define a sequential investment cost–effectiveness curve, $F_{CE}(C)$, such that each point on the curve corresponds to a lane arrangement which is a subset of the next more costly point. Here lane arrangement A is a subset of lane arrangement B, if the set of improved subsections corresponding to A is a subset of the improved subsections corresponding to B. It is reasonably easy to generate $F_{CE}(C)$, and we shall take it as one reasonable approximation, subject to sensitivity analyses, to the solution of our general problem.

A straightforward way to define the sequential investment cost–effectiveness curve is as follows. Start with the lane arrangement corresponding to the existing freeway; call this arrangement $\bar{X}_0$. Make a FREEQ run for this lane arrangement and note the subsection(s) where the volume–capacity ratio is equal to unity. These subsections will be designated as the bottlenecks corresponding to lane arrangement $\bar{X}_0$. We know, in general, that the addition of lanes at these points will be most effective. For each of these bottlenecks define a lane arrangement $\bar{X}_i$, such that the ith bottleneck and only those subsections needed to make a feasible lane arrangement are widened. For each $\bar{X}_i$, a FREEQ run is made to obtain the effectivness $G(\bar{X}_i)$. The most cost-effective improvement is $\bar{X}_A$, defined as the $\bar{X}_i$ for which $[G(\bar{X}_i) - G(\bar{X}_0)] / [C(\bar{X}_i) - C(\bar{X}_0)]$ is maximum. The process of finding $\bar{X}_A$ for a given $\bar{X}_0$ is referred to as a stage in the marginal analysis process.

For purposes of identification, the most cost-effective lane arrangement for a stage is labeled by adding an A to the label of $\bar{X}_0$; the second most cost-effective arrangement is labeled by adding a B; and so on. These steps are repeated from stage to stage by considering $\bar{X}_A$ of a stage as $\bar{X}_0$ of the next stage. That is. suppose that lane arrangement $\bar{X}_A$ is built and then ask the question: What is the next most cost-effective lane arrangement? The answer to this question yields $\bar{X}_{AA}$. The process is continued until the FREEQ runs associated with some lane arrangement $\bar{X}_{A...A}$ results in no bottlenecks. The cost–effectiveness curve $F_{CE}(C)$ is defined by connecting each of the points labeled only with A's by a step function.

Cost–Effectiveness Analysis

The sequential investment analysis described above was used to calculate the cost–effectiveness curve $F_{CE}(C)$ for the northbound Eastshore Freeway. The successive stages of improvements from the existing freeway conditions to noncongested freeway conditions are shown in block-diagram form in Figure 16.3 and in cost–effectiveness form in Figure 16.4.

The block diagram of Figure 16.3 begins with the existing freeway conditions as shown in stage 0. In stage I there are are four bottlenecks. The first bottleneck can be removed by adding a lane to subsections 5 and 6; the second bottleneck can be removed by adding a lane to subsections 20 and 21; the third bottleneck can be removed by adding a lane to subsection 25; and the fourth bottleneck can be removed by adding a lane to subsections 20, 21, 22, 23, 24, and 25. The most cost-effective improvement plan would be adding a lane to subsections 20 and 21 and has a marginal cost–effectiveness ratio of $0.15 per passenger-hour saved. The process is continued until at the completion of stage XIII the freeway is not congested, and the final

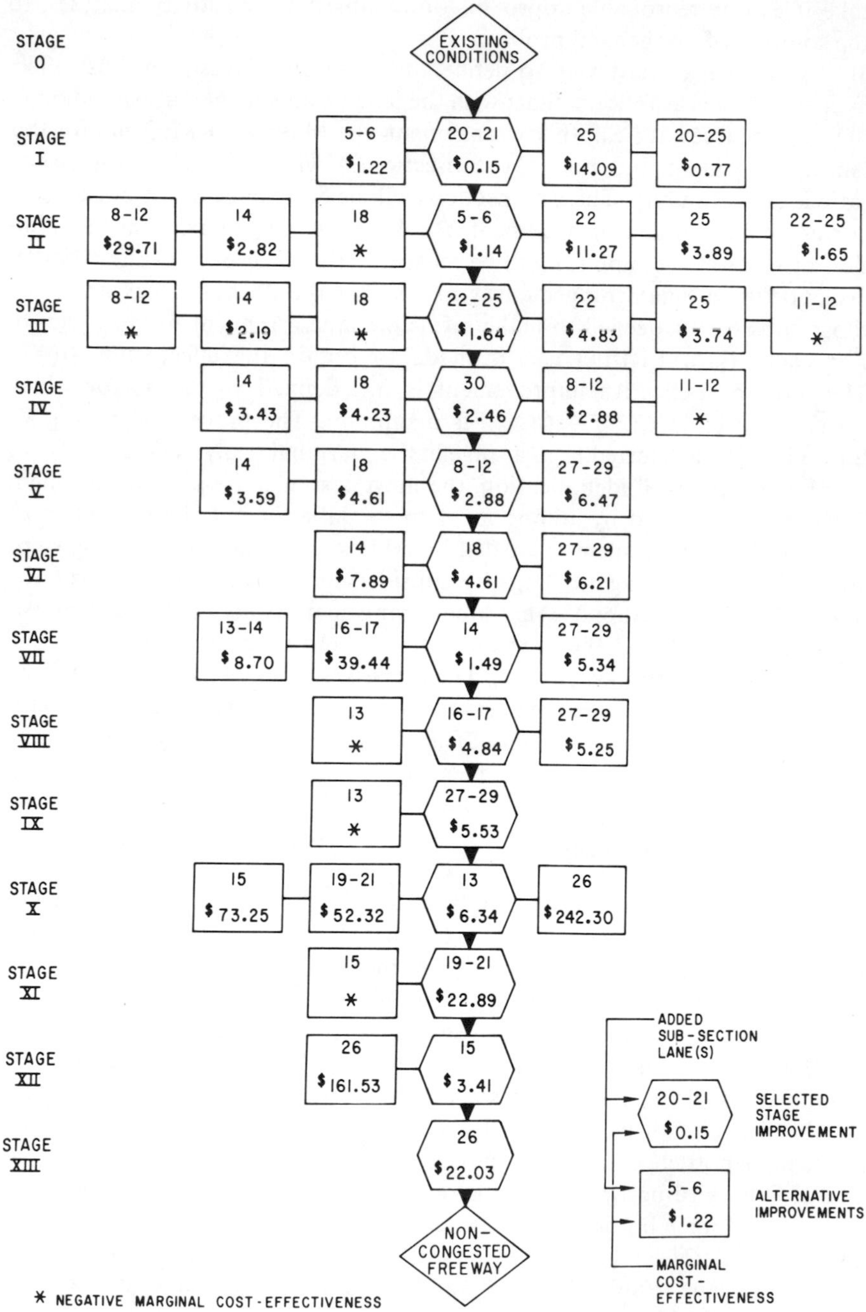

Figure 16.3 Block diagram of most cost–effective lane arrangement method applied to northbound Eastshore Freeway (I-80).

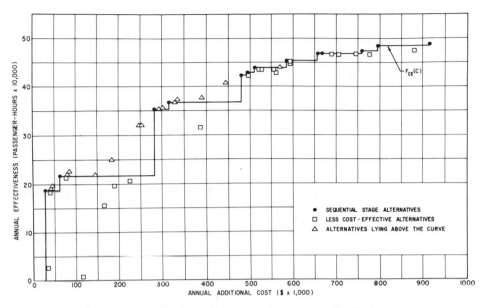

Figure 16.4 Cost–effectiveness diagram of most cost–effective lane arrangement method applied to northbound Eastshore Freeway (I-80).

marginal cost–effectiveness ratio is \$22.03 per passenger-hour saved. At each stage the most cost-effective alternative is selected and is included as a subset of all further alternatives.

The results of this investigation of the various improvement plans are shown in cost–effectiveness form in Figure 16.4. Each point on this diagram corresponds to one of the alternatives shown in Figure 16.3, and represents a particular improvement which has a specific increased annual cost and savings, in passenger-hours, when compared to the existing freeway conditions. There are a total of 48 alternative improvement plans, and they fall into three groups. The first group consists of the solid dots, which lie on the $F_{CE}(C)$ enclosure curve. These are the 13 optimal sequential-stage alternatives. The second group is denoted by the square dots, which represent the alternatives which are less cost effective and lie below the $F_{CE}(C)$ enclosure curve.

The alternatives in the third group, denoted by triangular dots, deserve particular attention. They represent alternatives which lie above the $F_{CE}(C)$ enclosure curve. As such, they indicate that the sequential investment cost–effectiveness curve, $F_{CE}(C)$, is, as we have already discussed, not equivalent to the global cost–effectiveness curve $F^*(C)$. They do not, however, invalidate the concept of the $F_{CE}(C)$ curve, since these designs which lie above $F_{CE}(C)$ have certain characteristics which limit their usefulness. On the one hand, these alternatives are improved intermediate steps between the stages shown in Figure 16.3 and thus may become the optimal selection if the ultimate budget falls between stages. On the other hand, if the ultimate available budget either is equal to the cost of a particular stage or permits the ultimate development of all stages, the optimal selection remains on the $F_{CE}(C)$ enclosure curve. In other words, the "triangular dot" alternatives do not lead to nor are they subsets of later alternatives lying above the $F_{CE}(C)$ enclosure curve.

Sensitivity Analyses

As the cost–effectiveness analysis we used only leads to an approximation of the optimum solution set S^*, it is essential that we explore the goodness of this approximation by sensitivity analyses using alternative procedures. Two underlying philosophies were maintained in these additional analytical procedures. First, it was assumed to be always more cost-effective to add a lane to a bottleneck subsection before considering adding a lane to a nonbottleneck subsection. Second, it was assumed desirable to employ the sequential investment approach, that is, to select a sequence of alternatives in which each stage is a subset of all following stages.

The sequential investment cost–effectiveness curve, $F_{CE}(C)$, was essentially generated by a marginal analysis procedure. At each stage the most cost-effective alternative was selected, and all future analyses branch from that alternative only. Because of the nonconcavity of the objective function, it is entirely possible that this marginal analysis is myopic and that a sequence which is ultimately optimal branches from an alternative which is not the most cost-effective at some stage. We can thus consider alternative procedures for automatically generating sets of desirable alternatives.

An alternative intuitive way to solve the sequential investment problem is as follows. Proceed as in the marginal analysis method, but for each stage select the next improvement plan on the basis of which lane arrangement reduces delays the most, rather than on the basis of which is most cost-effective. This method was used to calculate a different cost–effectiveness curve, which we can call $F_E(C)$, for the northbound Eastshore Freeway. The successive stages of improvement from the existing freeway conditions to noncongested freeway conditions are shown in block-diagram form in Figure 16.5 and in cost–effectiveness form in Figure 16.6.

The block diagram of Figure 16.5 is similar in form to Figure 16.3, except the procedure in the sequence is to always select the most effective lane arrangement rather than the most cost-effective lane arrangement. Stage 0 represents existing conditions and the sequence continues until stage XI is completed and no congestion exists along the freeway. At each stage the most effective alternative is selected and included as a subset of all further alternatives.

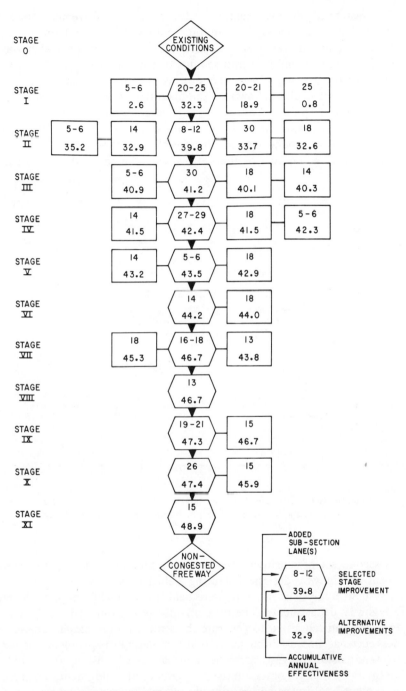

Figure 16.5 Block diagram of most effective lane arrangement method applied to northbound Eastshore Freeway (I-80).

The results of this investigation of the various improvement plans are shown in cost–effectiveness form in Figure 16.6. Each point on this cost-effectiveness diagram represents a particular improvement which has a specific increased annual cost and savings in annual passenger-hours when compared to the existing freeway conditions. There are a total of 31 alternative improvement plans shown on the diagram, of which 11 points lie on the $F_E(C)$ curve, 17 points above the $F_E(C)$ curve, and three points below the $F_E(C)$ curve. In comparison with the results obtained using the most cost-effective lane-arrangement method, this method appears to be less desirable because of the large number of alternatives which lie above the $F_E(C)$ curve.

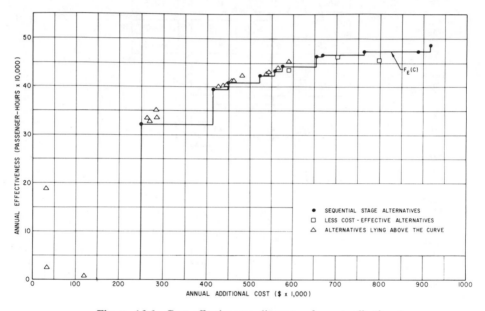

Figure 16.6   Cost–effectiveness diagram of most effective lane arrangement method applied to northbound Eastshore Freeway (I-80).

A second set of sensitivity analyses were based on the experience and judgment of highway designers. During February and March 1972, three 2-day systems analysis workshops were held for the California and Nevada state highway departments. The participants were grouped into three-man study teams and requested to generate the sequential investment cost–effectiveness curve as already discussed. Upon completing the problem, the study teams were asked to investigate other alternatives which they felt might lie above the $F_{CE}(C)$ curve, based on their experience and judgment. The results of these additional investigations are plotted in Figure 16.7, which includes the results of the previous investigations as well as the $F_{CE}(C)$ and the $F_E(C)$ curves. Only a few alternatives were found to lie above the $F_{CE}(C)$ curve.

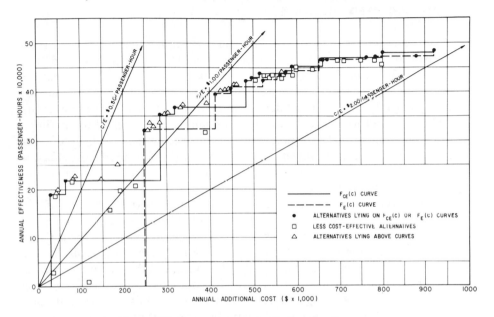

Figure 16.7 Summary of cost–effectiveness diagrams.

SUMMARY

The sequential investment cost–effectiveness analyses proved to be a practical method of calculating an approximate solution to the problem of optimum expansion of congested freeways. Numerical experience seems to indicate that the resulting $F_{CE}(C)$ curve is nearly optimal. Further, the method is simple and the solution required only 48 runs out of a possible 2^{30} to 3^{30} alternatives.

From a practical point of view, two specific redesign alternatives were identified. These alternatives are attractive because they have low cost per unit of effectiveness, they have low construction costs, and their improvement is contained in all later extended improvement plans. Adding a lane to subsections 20–21 is estimated to cost $200,000 and will result in a cost-effectiveness of $0.15 per passenger-hour saved. Adding a lane to both subsections 20–21 and 5–6 is estimated to cost $430,000 and will result in total cost effectiveness of $0.28 per passenger-hour saved. Both alternatives are now under active consideration by the California Division of Highways.

DYNAMIC PROGRAMMING SOLUTION FOR NONCONGESTED FREEWAYS

A more exact solution is available for evaluating freeway design alternatives for noncongested sections of freeway. This is considered to be less practical because the normal procedure is to consider only alternatives that eliminate or minimize congestion. However, there are two reasons for examining it in detail. First, it demonstrates how mathematical programming techniques (in this case, dynamic programming) can be utilized to obtain an exact solution to real transportation design problems. Second, it provides results which can help evaluate the assumption that it is more cost-effective to add a lane to a bottleneck subsection before considering adding a lane to a nonbottleneck subsection.

The key characteristic of this problem which permits the use of the dynamic programming technique is the independence between the effect of alternative improvements. The absence of bottlenecks, in which the arriving volume exceeds the available highway capacity at a given time, means that the flow is not being metered so as to alter the loads, and thus the performance, of subsections downstream. Under these conditions, adding a lane to one subsection does not change the cost effectiveness of adding a lane to another subsection. This greatly simplifies the problem and permits an exact solution.

The starting point for the dynamic programming analysis is the minimum freeway design which results in no congestion (Figure 16.8). To reach this stage from the existing design, one lane is added to subsections 5–6, 8–19, and 22–30 and two lanes are added to subsections 20–21. These improvements are estimated to cost \$6,520,000 (annual cost of \$918,500) and would reduce the total passenger-hours expended per afternoon peak period from 5025 to 3068 passenger-hours.

Mathematically, the objective, given an investment C, is to find the optimal lane arrangement $\bar{X}^*$ such that

$$G(\bar{X}^*) = \max\ G(\bar{X})$$

subject to $\qquad\qquad\quad C(\bar{X}) \leq C$ $\qquad\qquad\qquad\qquad$ (16.6)

where, as before, $\bar{X}$ is a feasible lane change, $G(\bar{X})$ the time savings, $C(\bar{X})$ the annual cost of the alternative, and C the budget constraint. Also, as before, the alternative plans for lane additions are limited to those consistent with good practice with respect to lane balance and maximum number of lanes. This considerably reduces the number of alternatives to be considered. By varying C from 0 to an upper limit of C, the optimum cost-effective enclosure curve for the set of extreme points $F^*(C)$ can be determined. In the example problem that follows, the upper limit of C is set at \$500,000.

The analysis was carried out using a variable increment dynamic programming process similar to the one described by de Neufville and Mori
(2). As usual, a significant number of possible alternative improvement
schemes can be eliminated during the analysis so that the combinatorial
problem is greatly reduced and the computations become feasible. The
costs and effectiveness of the optimal sequence of improvements are shown in
Table 16.1. They are also plotted on the cost–effectiveness diagram of Figure
16.9, which includes the curve $F_{CE}(C)$ given earlier in Figure 16.4.

Table 16.1 Summary of Cost–Effectiveness Calculations for Noncongested Freeway

Alternative (Addition of Lanes to Subsections)	Construction Costs ($ X 10³)		Peak-Hour Savings (Passenger-Hours)		Cost Effective-ness ($/Pas-senger-Hour)
	Initial	Annual	Daily	Annual	
18	90	12.7	1	250	50.80
1	100	14.1	2	500	28.20
14	140	19.7	3	750	26.20
14, 18	230	32.4	4	1000	32.40
1, 14	240	33.8	5	1250	27.00
1, 14, 18	330	46.5	6	1500	31.00
1, 14, 27	380	53.5	7	1750	30.60
1, 14, 18, 27	470	66.1	8	2000	33.10
1, 14, 16, 27	540	76.0	9	2250	33.80

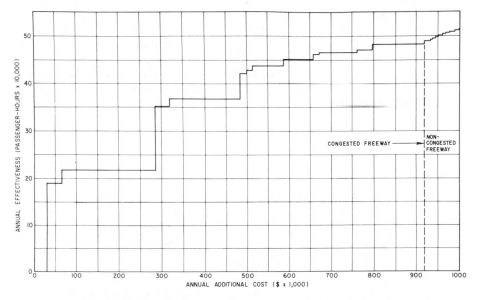

Figure 16.9 Cost–effectiveness diagram for congested and noncongested northbound Eastshore Freeway (I-80).

The results do indicate that the earlier stages of improvement when the freeway was congested were most cost-effective. However, the later stages of improvement when the freeway was congested were less cost-effective than improvements on the noncongested freeway. In the light of these results, the earlier assumption that it is always more cost-effective to add a lane to a bottleneck subsection before considering adding a lane to a nonbottleneck subsection is questioned, particularly if the subsection bottleneck results in minimum congestion and is very expensive to eliminate. On the other hand, on heavily congested freeways, the assumption appears to be applicable.

ACKNOWLEDGMENTS

The financial support and continued cooperation of the California Division of Highways and of the U.S. Department of Transportation are gratefully acknowledged.

REFERENCES

1. AIDOO, J. F., GOEDHART, R. W., AND MAY, A. D., 1970. *Bay Area Freeway Operations Study—Final Report, Part III: On the Eastshore Freeway (I-80) Northbound.* Berkeley, Calif.: University of California, Institute of Transportation and Traffic Engineering, Aug.
2. DE NEUFVILLE, R., AND MORI, Y., 1970. Optimal Highway Staging by Dynamic Programming, *Transportation Engineering Journal, ASCE,* Vol. 96, No. TE1, Feb.
3. Highway Research Board, 1965. *Highway Capacity Manual,* Special Report 87, Washington, D.C.: National Academy of Science–National Research Council.
4. MAKIGAMI, Y., WOODIE, L., AND MAY, A. D., 1970. *Bay Area Freeway Operations Study—Final Report, Part II: On the San Francisco–Oakland Bay Bridge.* Berkeley, Calif.: University of California, Institute of Transportation and Traffic Engineering, Aug.

17

The Combined Use of Optimization and Simulation Models in River Basin Planning*

HENRY D. JACOBY AND DANIEL P. LOUCKS

In this chapter, a particularly complex water resource system is analyzed using both optimization and simulation, not as competing techniques but as interacting partners in the search for good management policies.

Simulation models of river basins are being used by an increasing number of public and private agencies in the planning and management of water resources (5, 7, 8). The popularity of this type of analysis is due in part to the extreme flexibility of simulation models and the relative ease with which they may be developed. Although seemingly endless hours may be spent writing and debugging computer programs, analysts who work with these models develop a healthy respect for the complexities of river basin configuration and management practice that can be incorporated, and the speed with which many years of basin operation can be analyzed. And, no doubt, simulation techniques prove useful because their application can be accomplished with a thorough knowledge of the basin, which is essential for any such analysis, and with experience in manipulating computer languages. Other specialized skills, such as facility with the mathematics of optimization, are not essential.

Simulation analyses have one difficulty, however, particularly if the basin offers a large number of design alternatives. In most basin simulations, the analyst himself must choose the physical design to be studied in a particular computer run—where the design might include the size, location, and timing of dams, canals, irrigation systems, municipal and industrial water supplies, power stations, and pollution control works. The computer model

*Extensively adapted from *Water Resources Research*, Vol. 8, No. 6, Dec. 1972; pp. 1401–1414.

simply simulates the performance of the system chosen under a set of expected hydrological circumstances and system operating policies. By means of successive runs with alternative combinations of physical facilities and operating rules, the analyst searches for the system with the highest present value of net benefits. Unfortunately, if the basin is at all complex the number of possible designs can be extremely large. Methods must be used to "screen" the full range of alternatives and identify a subset to be analyzed in detail by simulation.

The Delaware River basin, which is introduced below as a sample case, exemplifies the problem. We discuss a simulation model of the basin that includes 35 reservoirs, of which only six have fixed capacities. Ignoring the variables defining reservoir operating rules, targets for water supply, recreation and power, and hydroelectric capacities—and assuming only two possible sizes for each new reservoir—a complete evaluation of alternatives requires 2^{29}, or over 500 million, simulation runs! If engineering judgment is used to eliminate 99 % of these possibilities as clearly inferior (an unlikely feat), and if computation time per run is held to 1 minute (also unlikely), it still requires over 10 years of computer time to evaluate the remaining 1 % of the possible configurations. Random sampling of designs may provide statistical estimates of the optimum values of the decision variables, but the formulation and execution of such sampling procedures is far from a trivial task.

The screening process may be approached in several ways. The most common way is to utilize engineering and economic judgment, supported by manual calculations. Or, it is possible to construct highly simplified and flexible basin simulation models that can perform a quick survey of alternative designs. Finally, preliminary screening may be carried out with the use of analytical optimization models. It has long been recognized that a combination of analytical optimization and simulation models offers the promise of significant economies in the analysis of large river basin systems (2). Application of both of these techniques in the analysis of the same river basin is, however, relatively rare, although some examples are known (1, 12).

In this chapter we report the results of a study designed to explore the difficulties and advantages of the joint use of these two types of models. Specifically, it addresses the issue of whether relatively simple optimization models, which are both mathematically tractable and computationally feasible for large systems, can effectively screen out the less beneficial alternatives and leave only a few that merit further study. Put another way, can the different models, when applied to the same basin, be made to give comparable values of the criterion function, and will there be a correspondence between the designs and operating policies formulated by an analytical model and those highly valued according to simulation analysis?

The answer to this issue is not obvious. In order to gain the great power of analytical models in selecting high-valued designs from a wide range of alternatives, sacrifices must be made in the reality of the system description. Often it is not possible to incorporate many of the important complexities that characterize the hydrology of a large river basin and the interaction among its physical components. That these simplifications are necessary, of course, means that one cannot hope to find the true global optimum design for a complex system by means of an analytical model. It will be sufficient if analytical models can be shown capable of the screening task—leaving the verification and improvement of the resulting designs and operating policies to the more detailed approximations that are possible with simulation. Our results indicate a great potential for analytical models in this context.

THE DELAWARE RIVER BASIN

An excellent case example for investigating the joint use of these models is provided by the work of Hufschmidt and Fiering (6) on the use of digital computer simulation for evaluating alternative water resource structures and operating rules, and a revised version of this work is used. The Delaware River basin is shown in Figure 17.1; over 22 million people and a major segment of the nation's industry are located within its boundaries. In addition, water from the basin is diverted to metropolitan New York and New Jersey and serves an additional 15 million people. The basin waters are heavily utilized now, and there is a large number of prospective water management projects. Following Hufschmidt and Fiering, our study was limited to the development of surface waters; groundwater management was not taken into account. The uses explicitly considered were recreation, flood control, hydroelectric power, and water supply. Considerations of water quality were introduced through a set of limitations on minimum flow and wastewater treatment in the major water supply areas.

A schematic representation of the basin and the existing and potential reservoirs, hydroelectric power plants, and major water supply areas is presented in Figure 17.2. As shown in the figure, the projects considered in this study included 35 reservoirs, 9 run-of-river and 12 variable-head hydroelectric sites, and 4 major water supply areas in the basin itself.

The benefit, loss, and cost data incorporated into both the simulation and screening models were expressed as continuous functions of inputs and outputs. In general, these functions were estimated from data found in numerous reports on the Delaware basin. Loss functions were used to evaluate the consequences of departures from planned target outputs. Many of these targets—for example, water supply quantities, lake levels in recreation areas, and firm power outputs—were unknown variables, as were all the reservoir and hydroelectric plant capacities that did not exist at the time the data were collected. Details on the derivation of these functions can be found in Hufschmidt and Fiering (6).

Figure 17.1 The Delaware River Basin.

Two assumptions were made regarding recreation and power. It was assumed that growth in the demand for power in the area far exceeds the potential development of hydroelectric power. Thus any economically feasible peaking power will presumably find a ready market. And it was assumed that the development of recreational facilities at any site in the basin will not affect the demand for recreation at other sites. Outdoor recreation demands in the basin have been estimated to be far in excess of the development potential of all reasonable feasible storage projects (14).

Both historical and synthetic monthly streamflows were used. The distribution of unregulated flows at each gauge site in any particular month was assumed to remain the same each year. The actual monthly streamflow at each project was assumed to equal the regulated flow at the site plus the

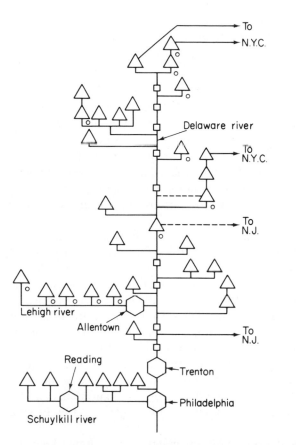

To
N.Y.C.

Delaware river

To
N.Y.C.

To
N.J.

Lehigh river

Allentown

To
N.J.

Reading

Trenton

Philadelphia

Schuylkill river

LEGEND

△ Reservoir site

▢ Run-of river hydro-
   electric site

∘ Variable–head
   hydroelectric site

⬡ Major water supply
   area

—— Existing diversion

---- Potential diversion

Figure 17.2 A schematic representation of the Delaware River
Basin and the existing and potential reservoirs, hydroelectric plants,
and major water supply areas.

random but correlated unregulated interflow that enters the basin between
the flow-regulating sites immediately upstream and the project site itself. The
probability distributions of unregulated flows were derived from historical
records. Synthetic streamflows supplemented the historical flows and were
used extensively in the simulation model. The time of flow from the uppermost
site to the mouth of the Delaware was assumed to be well within the monthly
periods defined by both types of models.

Within the simulation models, short-duration flood flows were randomly
generated and routed through the river system. The analytical screening
models did not explicitly incorporate flood flows themselves but considered
flood control storage as a variable whose value was dependent on the functional
relationships between the costs of reservoir storage capacity and the expected
downstream benefits of flood control.

SCREENING AND SIMULATION MODELS

Simulation Model

Because a thorough description of the Delaware simulation model is given by Hufschmidt and Fiering (6), only a brief summary needs to be presented here. The input data for the simulation model are divided into three components: permanent data, design data, and simulated unregulated streamflows at each of 25 gauging stations. The permanent data specify the parameters that are not decision variables. This information includes the economic and physical functional relationships between the various components of the river basin system.

The design data specify the values of the decision variables that can be determined by basin managers. They include the capacities of reservoirs and hydroelectric power plants and the targets for recreation, hydroelectric energy production, and water supply. In addition, and very importantly, the design data define the operating rules for determining the reservoir releases and allocations of water to different uses. The operating rules are a function of all the other design and economic parameters of the system and are described in Hufschmidt and Fiering (6) and in Maass et al. (9).

The streamflow data, consisting of a 50-year monthly sequence of unregulated synthetic flows for each of the 25 gauging stations in the basin, were used in applying the model. The synthetic flows were originally generated by a multivariate model developed by Thomas and Fiering (13) and by Fiering (3, 4).

For our screening-simulation study, estimates of the annual benefits, losses, and costs as functions of the input and output parameters were projected for the years 1980 and 2010. Using these parameters in the preliminary screening models, several alternative design configurations were selected on the basis of their potential for achieving maximum net benefits in those years. Then the preferred designs were simulated for a period of 50 years, using the same annual benefit, loss, and cost functions. By routing 50 years of monthly synthetic streamflows through the basin, estimates were obtained of the average and standard deviations of the annual net benefits that could be expected in the years 1980 and 2010 for each alternative.

By conducting a static analysis prior to a more comprehensive dynamic screening and simulation study, it was possible to reduce significantly the number of decision variables that had to be considered. It was assumed that those alternative designs eliminated in the static analyses would not need to be reconsidered when examining questions of project scheduling and sequencing. [This is, of course, not necessarily true, and the reader may wish to compare

this approach with the one taken by May and James in Chapter 16.—*Eds.*]
Thus, by using static analyses to obtain snapshots of the most economically
beneficial river basin systems in the years 1980 and 2010, the subsequent
dynamic analyses could concentrate on how best to develop those systems
over time.

Screening Model

To be useful, the analytical model must include a number of decision
variables: long-term or over-year storage, together with short-term or within-
year storage requirements; reservoir and hydroelectric power plant capacities;
targets for recreation storage, firm power, and water supply; and flood storage capacities. It also must contain the same economic functional relationships
that are included in the simulation model. Finally, the model must be adaptable to a dynamic economic analysis as well as to static or steady-state application.

Optimization techniques do not yet exist that can simultaneously satisfy
all the above conditions. After experiments with several nonlinear models and
algorithms, linear programming was chosen for the task of preliminary screening. It was chosen primarily because of its relative efficiency, compared to
nonlinear methods, in examining numerous design and operating policy
variables. Nonlinear functions were made piecewise-linear. Concave cost and
loss functions required several trial solutions using linear approximations
before a reasonably accurate solution could be obtained. Separable programming algorithms would have eliminated the need for some trial solutions had
they been available at the time of this study.

Preliminary calculations indicated that deterministic linear programming
models, using mean monthly flows, are not capable of capturing the significant
variations in the over-year and within-year flows that contribute to the need
for water storage in the basin. To achieve useful results, it was necessary to
develop stochastic linear programming models. These models incorporate
some of the hydrologic risk inherent in the Delaware system by defining more
than one possible streamflow and more than one possible reservoir storage
level in each period within a year. Thus a variety of discrete flows and reservoir volumes, and their corresponding probabilities of occurrence, were
considered in the determination of reservoir and power plant capacities, the
various supply targets, and the parameters of the reservoir operating rules.

In order to actually carry out screening calculations, of course, it is necessary to introduce simplifying assumptions and limitations that are not necessary in the simulation program. To show how this is done, the next two
sections describe the essential components of the static version of the

stochastic model used in this particular application and summarize the procedure used to extend the static results of the dynamic problem. It is not possible to present the models in full detail here; a more comprehensive discussion of these types of models is given by Loucks et al. (8).

STATIC SCREENING MODEL

We now develop a static screening model adaptable to many river basins, and which we apply to the Delaware River.

Objective Function

At each site s, let B_s^R, B_s^F, B_s^P, and B_s^W be the annual gross benefits for recreation, flood control, hydroelectric power, and water supply, and let C_s be the annual costs of all facilities. The objective of the model is to maximize the expected total annual net efficiency benefits derived from all sites:

$$\text{Maximize } E[\textstyle\sum_s (B_s^R + B_s^F + B_s^P + B_s^W - C_s)] \qquad (17.1)$$

Specification of Flows and Storage Volumes

The Delaware simulation model considers 12 monthly periods in each year. To reduce computation time and costs, the programming models divided each year into only six periods, defined both by similar hydrologic conditions and the importance of recreation in the summer months (Table 17.1).

Table 17.1 Periods Defined in Static Screening Model

Period	Months
1	February, March, April
2	May, June
3	July
4	August
5	September, October
6	November, December, January

To define more than one flow in each of the six periods, the range of unregulated flows at each gauge site in each period is divided into four intervals, each denoted by the subscript i or j. These intervals are defined so that the transition or conditional probability, P_{ij}^t, of an unregulated flow in interval j in period $t + 1$, given an unregulated flow in interval i in period t, is the same for each gauge site throughout the basin. Making the P_{ij}^t the same for all gauge sites simplifies the calculation of the various flows and storage volumes that are needed to obtain the expected net efficiency benefits.

Once the intervals of flow are defined, the unconditional steady-state probabilities, P_{it}, of each flow at any time can be derived either from historical and simulated records, or from a knowledge of the transition probabilities, by solving the following set of equations:

$$P_{j,t+1} = \sum_i P_{it} P_{ij}^t, \qquad j = 1, 2, 3; \ t = 1, 2, \ldots, 5 \qquad (17.2)$$
$$\sum_i P_{it} = 1, \qquad\qquad t = 1, 2, \ldots, 6 \qquad (17.3)$$

In the above and all following equations, if t is the last period in the year ($t = 6$), then $t + 1$ is the first period in the year.

The unregulated flows at each site within the basin at any time, I_{it}^s, can be defined by a linear function of the flows at any gauge site s', $G_{it}^{s'}$. The flows at each gauge site for each interval are selected so that they, together with the steady-state probabilities P_{it}, maintain the first three moments of the probability distribution of the historical unregulated flows in each period. The unregulated flow can then be taken as:

$$I_{it}^s = \frac{\text{watershed area upstream of site } s}{\text{watershed area upstream of gauge site } s} \cdot G_{it}^{s'} \qquad (17.4)$$

The total expected flow Q_{it}^s at each site at any time equals the uncontrolled flow I_{it}^s at that site, less the difference between the uncontrolled flows and the expected releases $R_{it}^{\bar{s}}$ at all control sites $\bar{s}$ immediately upstream, and less any net withdrawals $A_{it}^{\bar{\bar{s}}}$ at sites $\bar{\bar{s}}$ between the control sites $\bar{s}$ and the site s:

$$Q_{it}^s = I_{it}^s - \sum_{\bar{s}} (I_{it}^{\bar{s}} - R_{it}^{\bar{s}}) - \sum_{\bar{\bar{s}}} A_{it}^{\bar{\bar{s}}} \qquad (17.5)$$

Equation (17.5) is sufficiently accurate if the uncontrolled flows throughout the basin are relatively highly cross-correlated, that is, if they are all in the same interval i throughout the basin. For the purposes of preliminary screening, but not of simulation, it can be assumed that this is so. It is then not necessary to examine all possible combinations of flows at all sites upstream of any site s in order to calculate the expected flow, Q_{it}^s.

So that we may consider several possible volumes of reservoir storage at each reservoir site, the subscripts k and l are introduced. Just as the subscripts i and j designate intervals of flow, the subscripts k and l define intervals of volume of reservoir storage, not actual values. Unlike the known flow intervals i and j, however, the intervals of storage volume k and l are unknown. Corresponding to the known flows at the gauge sites $G_{it}^{s'}$ or the unknown expected flows Q_{it}^s for each interval i, are unknown volumes of reservoir storage, S_{kt}^s, for each interval k.

Based upon these definitions, each reservoir release, R_{kilt}^s, is specified by a simple continuity equation. It is the difference between the sum of the initial storage volume S_{kt}^s and the expected inflow Q_{it}^s and the final storage volume $S_{l,t+1}^s$:

$$R_{kilt}^s = S_{kt}^s + Q_{it}^s - S_{l,t+1}^s \qquad (17.6)$$

The expression for reservoir release can be simplified by assuming a functional relationship, $l = l(k, i, t)$, between the subscripts k, i, and l in each period t. In this study, l was taken equal to the integer portion of $(k + i)/2$ for all reservoir sites and all periods t. The specification of l does not define the actual storage volumes, S_{kt}^s and $S_{l,t+1}^s$, or the associated flows, Q_{it}^s and R_{kilt}^s; these remain unknown decision variables. The specification of l does, however, permit a reduction in the number of continuity equations: since l is uniquely defined by k and i for each period t, Eq. (17.6) need only be defined for all k, i, and t and R_{kilt}^s can be redesignated R_{kit}^s. This specification of l also makes it possible to compute the joint probabilities P_{kit} associated with reservoir storage volumes, inflows, and releases, even though these are unknown variables.

Given the assumptions that, at any time, the uncontrolled flows throughout the basin are in the same interval i and that the functional relationship for l holds, it is possible to show that the only combination of storage volumes S_{kt}^s that can eventually exist simultaneously at all reservoir sites is that in which the subscript k is the same. The total flow at any site can thus be expressed as a function of the initial interval of reservoir storage k and interval of unregulated flow i, and Eq. (17.5) becomes:

$$Q_{kit}^s = I_{it}^s - \sum_s (I_{it}^s - R_{kit}^s) - \sum_{\bar{s}} A_{kit}^{\bar{s}} \qquad (17.7)$$

The continuity equation, Eq.(17.6), at each reservoir site also can be written to incorporate the inflow Q_{kit}^s expected for any interval of storage k:

$$R_{kit}^s = S_{kt}^s + Q_{kit}^s - S_{l,t+1}^s \qquad (17.8)$$

Finally, the steady-state joint probabilities P_{kit} of various storage volumes and flows at each reservoir site can be calculated from a knowledge of the transition probability, P_{kilj}^t, of having storage volume within interval l and a flow within interval j in period $t + 1$, given storage volume within interval k and a flow within interval i in period t. Since the function for l and the transition probabilities of flow, P_{ij}^t, are assumed the same at each reservoir site, the transition probabilities are simply:

$$P_{kilj}^t = \begin{cases} P_{ij}^t & \text{if } l = l(k, i, t) \\ 0 & \text{otherwise} \end{cases} \qquad (17.9)$$

From a knowledge of the transition probabilities P_{kilj}^t, the steady-state joint probabilities P_{kit} can be calculated in the same way the steady-state probabilities of flow intervals i or j were found, that is, by solving Eqs. (17.2) and (17.3). The only difference is that the subscript i is replaced by k and i, and the subscript j is replaced by l and j:

$$P_{l,j,t+1} = \sum_k \sum_i P_{kilj}^t P_{kit} \tag{17.10}$$

$$\sum_k \sum_i P_{kll} = 1 \tag{17.11}$$

Just as the steady-state flow probabilities P_{it} do not depend on the intervals defined by the subscripts i and j, the steady-state joint probabilities P_{kit} do not depend on the intervals represented by the subscripts i, j, k, and l. The storage volumes S_{kt}^s or releases R_{kit}^s at each reservoir site s remain continuous unknown variables. Their values are influenced, but not determined, by the joint probabilities P_{kit}.

Specification of Objective Function and Constraints

Five sets of equations are used to define the expected net benefits derived from particular values of the storage volumes and flows. The net benefits were associated with each of the elements of the objective function specified in Eq. (17.1): recreation, flood control, hydroelectric power, water supply, and costs. Constraints are also associated with these activities and are defined concurrently.

The benefits derived from recreation can be assumed to depend on the level of reservoir storage used for planning the development of recreation facilities, and on losses associated with deviations from that level. To define the planned level of storage and the variations that might occur, each period's initial reservoir storage, S_{kt}^s, is set equal to the seasonal target storage volume, T_s^R, less any deficit D_{kt}^{Rs}, or plus any excess E_{kt}^{Rs}:

$$S_{kt}^s = T_s^R - D_{kt}^{Rs} + E_{kt}^{Rs} \tag{17.12}$$

The variation in the number of visitor-days for any particular planned level of storage can be specified by a fraction v_{st} for each period t at each site s ($\sum_t v_{st} = 1$). Estimating the value per visitor-day as \$1.50 (15) and considering the number of visitor-days expected for any target storage level and deviations from it, the recreation benefit and loss functions, $\beta_s^R(T_s^R)$ and $L_{st}^R(D_{kt}^{Rs}, E_{kt}^{Rs}, T_s^R)$, respectively, can be derived. Piecewise linear approximations of these functions were then used to define the expected benefits from recreation:

$$E[B_s^R] = \beta_s^R(T_s^R) - \sum_t \sum_k \sum_i P_{kit} v_{st} L_{ts}^R(D_{kt}^{Rs}, E_{kt}^{Rs}, T_s^R) \tag{17.13}$$

The expected flood control benefits in each period t can be defined as a function of the equivalent flood storage capacity, ES_t^s, just upstream of any potential damage site. This equivalent capacity can be expressed as the actual flood storage capacity, SF_t^s, if any, plus a piecewise linear function specifying the storage capacity at site s that will reduce the peak of a standard project flood by the same amount as a flood storage capacity of SF_t^s or ES_t^s at the next upstream control site, $\bar{s}$. The expected annual flood control benefits are, then, the sum of the expected flood storage benefits, $\beta_{st}^F(ES_t^s)$ for each period t:

$$E[B_s^F] = \sum_t \beta_{st}^F(ES_t^s) \tag{17.14}$$

In this connection, the unknown reservoir capacities K_s are permitted to range from zero to a maximum feasible level, K_s^{max}, based on hydrologic and economic criteria. The minimum total capacity at each reservoir site is assumed to be the sum of each period's maximum storage volume, $S_{\bar{k}t}^s$, for $\bar{k} = \max\{k\}$, and flood storage capacity SF_t^s. Thus, at any site, the reservoir capacity is defined by the constraints:

$$S_{\bar{k}t}^s + SF_t^s \leq K_s \leq K_s^{max} \tag{17.15}$$

The production of hydroelectric power at any site depends on the installed generating capacity, H_s; the flow through the turbines; the average storage head H_{klt}^s, which is a function of the initial and final storage volumes in each period; a constant k_s for converting the product of flow, head, and plant efficiency into energy; the number of hours in each period, h_t; and the set of plant factors F_t assigned to hydroelectric stations in the basin for the different periods of the year. The plant factors indicate the amount of firm energy a hydroelectric plant is expected to contribute to the power grid to which it is connected. If the firm power target for each period is some fraction η_{ts} of the annual target, T_s^P, each hydroelectric plant will be expected to produce $h_t F_t \eta_{ts} T_s^P$ in firm energy. The total energy actually produced at each site will include any deficit in firm energy, D_{kit}^{Ps}, plus any excess energy E_{kit}^{Ps}.

The reservoir release R_{kit}^s provides an upper bound on the amount of energy that can be produced if all the water is routed through the turbines:

$$h_t\, F_t\, \eta_{ts}\, T_s^P - D_{kit}^{Ps} + E_{kit}^{Ps} \leq k_s R_{kit}^s H_{klt}^s \tag{17.16}$$

To incorporate this constraint into the linear programming model, trial solutions with assumed heads are required. These and other approximations for maintaining linearity were made so that if any project were not in the solution, it could safely be assumed that it would not appear in an optimal solution; therefore, it would be dropped from further consideration. This is the essence of the screening process.

Plant capacity H_s also constrains the total energy that can be produced:

$$h_t F_t \eta_{ts} T_s^P + E_{kit}^{Ps} \le h_t H_s \qquad (17.17)$$

and nonnegativity is required:

$$h_t F_t \eta_{ts} T_s^P - D_{kit}^{Ps} \ge 0 \qquad (17.18)$$

Benefit and loss functions for hydroelectric power, $\beta_{st}^P(\eta_{ts}T_s^P)$ and $L_{ts}^P(D_{kit}^{Ps}, E_{kit}^{Ps}, T_s^P)$ respectively, are based on the market value of firm energy and dump energy and the economic loss if a deficit in the firm energy occurs. The expected benefits from power are derived from these functions:

$$E[B_s^r] = \sum_t \beta_{st}^P(\eta_{ts}T_s^P) - \sum_t \sum_k \sum_i P_{kit} L_{ts}^P(D_{kit}^{Ps}, E_{kit}^{Ps}, T_s^P) \qquad (17.19)$$

To calculate water supply benefits it is necessary to define the variables representing annual water supply targets T_s^W, and their known proportion allocated to period t, α_{st}. The water allocated at each supply site s is, then, limited to the total water supply available, Q_{kit}^s:

$$\alpha_{st}T_s^W - D_{kit}^{Ws} \le Q_{kit}^s \qquad (17.20)$$

where D_{kit}^{Ws} is the deficit in water supply at any site. The withdrawal at each site, A_{kit}^s, can be assumed to be an estimated portion, γ_{st}, of the water supply allocation that would be consumed:

$$A_{kit}^s = \gamma_{st}(\alpha_{st}T_s^W - D_{kit}^{Ws}) \qquad (17.21)$$

The alternative cost method can be used to derive each benefit function associated with the water supply target, $\beta_s^W(T_s^W)$. The study by Russell et al. (11) of the drought in the northeastern United States during the mid-1960s provides data for the derivation of the short-run loss functions, $L_{st}^W(D_{kit}^{Ws}, \alpha_{st}T_s^W)$. The expected water supply benefits are, then:

$$E[B_s^W] = \beta_s^W(T_s^W) - \sum_t \sum_k \sum_i P_{kit} L_{ts}^W(D_{kit}^{Ws}, \alpha_{st} T_s^W) \qquad (17.22)$$

The costs of the facilities are comprised of capital and operating costs. At each site, s, where applicable, capital cost functions are defined for total reservoir storage capacity, $C_s^S(K_s)$; recreation facilities, $C_s^R(T_s^R)$; and power plant capacities, $C_s^P(H_s)$. Assuming a constant capital recovery factor, CRF, for all projects, the annual costs attributed to capital costs equal:

$$(CRF) [C_s^S(K_s) + C_s^R(T_s^R) + C_s^P(H_s)] \qquad (17.23)$$

The annual operation, maintenance, and repair, OMR, costs are assumed to be proportional to the capital costs of the facilities for reservoirs and recreational developments, and are considered a function of the installed capacity, H_s, for power stations. Letting O_s^S, O_s^R, and $O_s^P(H_s)$ denote the annual OMR costs for reservoir capacity, recreation, and power, the total annual cost C_s at each site s equals:

$$C_s = (CRF + O_s^S) C_s^S(K_s) + (CRF + O_s^R) C_s^R (T_s^R) + (CRF)C_s^P (H_s) + O_s^P (H_s) \qquad (17.24)$$

Many of the above functions and inequalities can be combined, thus reducing considerably the number of variables to be defined for the screening model. These procedures are not carried out in this discussion, so the assumptions made during the development of the model can be emphasized.

Additional constraints in the model include those required for the piecewise linearization of nonlinear benefit, loss, and cost functions. The majority of these constraints are simply upper bounds on the variables defining each linear segment. Such bounds can be incorporated into a special bounds section for more efficient operation by many of the more recent linear programming algorithms. In this respect, the algorithm actually used to solve the model for the Delaware River was relatively inefficient.

DYNAMIC SCREENING MODEL

Once the static screening and simulation analyses are complete, the optimal scheduling of the completion dates of the projects selected by the static analyses can be determined by a relatively simple and unsophisticated dynamic model. Its objective is to define the sequence of project completion dates that maximize the present value of expected annual net benefits, considering the construction and operation of these projects and the operation of existing structures.

As is the case in any river basin system, the benefits derived from each project depend on many of the other existing structures and their coordinated operating rules. Such dependence precludes the use of some rather efficient methods for dynamic analyses (10). The dynamic model we used for the Delaware requires estimates of the annual net benefits associated with each possible combination of proposed projects, that is, for each state of the system. The static simulation model is used to estimate the expected benefits for each state for each 5-year period from 1970 to 2010.

To simplify the simulation task, it can be assumed that at the end of each 5-year period prior to 1980 and 2010, each of the desired projects to be completed by 1980 and 2010 would either be fully operational or nonexistent. Knowledge of the projects that should be operational by 1980 and 2010 considerably reduces the number of combinations of projects that have to be simulated. Construction staging alternatives, changes in nonstructural alternatives and transient operating policies associated with the filling up of reservoirs, and so on, are ignored in this dynamic screening analysis. Also, this analysis does not consider any constraints on funds required for project construction. Budget limitations, if known, could reduce even further both the number of project combinations and the time required to define the expected annual net benefits associated with each system state.

The dynamic model developed for the screening of scheduling alternatives in the Delaware basin is solved using a backward-looking algorithm. The variable B_{dy} is defined as the maximum present value of the expected total net benefits up to and including year 2010 given the state of the design d (i.e., a specific combination of projects) for the river basin system in year y. What is desired is the value of this maximum expected net benefit variable as of the initial year, say 1965, and the states of the river basin system that are required from 1970 to 2010 in order to achieve this value.

Let the function $B_y(d)$ denote the expected annual net benefits from design state d in year y. The annual expected net benefits associated with any particular state are assumed to remain constant over the 5-year period, that is, from year y to $y + 4$. The desired states of the river basin in years 1980 and 2010 and the annual net benefits expected in those years were known from the static screening and simulation analysis. Proceeding from year 2010 and its desired state, D, and discounting at an annual rate of r, the maximum total expected net benefit in year 2005 is calculated for each state d ranging from the desired state in 1980 to the desired state in 2010:

$$B_{d,\,2005} = \theta B_{2005}(d) + \phi B_{D,2010} \tag{17.25}$$

where discount constants $\theta = \sum_{i=0}^{4} (1 + r)^{-i}$ and $\phi = (1 + r)^{-5}$. Proceeding backward to year 2000, the maximum total expected net benefits can likewise be computed for each combination of projects. Denoting feasible combinations of projects as d for each initial state d^*:

$$B_{d*,\,2000} = \theta B_{2000}(d^*) + \max_{d*\leq d\leq D} \{\phi B_{d,2005}\} \tag{17.26}$$

where $d^* \leq d \leq D$ limits the design states d to those that can exist given an initial state d^* and a final state D.

In general, the recursive formula for all 5-year periods y and initial states d^* is:

$$B_{d*y} = \theta B_y(d^*) + \max_{d*\leq d\leq D} \{\phi B_{d,y+5}\} \tag{17.27}$$

For the initial year in which the design is known, and for subsequent years (such as 1980 and 2010) for which the screening model has defined the design, no choice of d has to be made. For values of y less than or equal to 1980, the state D refers to that which is desired in 1980 and not 2010. Once the recursive functions are solved for each 5-year period, it is a simple matter to select the optimal schedule of project completion dates over all years. Although admittedly limited, this dynamic procedure is considered adequate for preliminary screening prior to a more detailed simulation analysis.

ANALYSIS PROCEDURES

As stated at the outset, the primary purpose of this study was to examine the problems involved in integrating optimization and simulation methods for both defining and evaluating alternative design and operating rules for large, complex river basin systems. The Delaware River provided realistic and sufficient data for model development and comparison. Throughout this study no attempt was made to find the optimal economic solution. What was sought was a relatively small set of clearly superior management plans, each somewhat flexible to possible changes in future conditions. Both types of models, but especially the simulation model, were employed to examine the effects of changes in both economic and engineering parameters on the design and operation of the entire system.

A river basin is best analyzed as a whole only after several subbasins have been examined separately. Analyses of the major tributaries were made to explore the characteristics of different parts of the basin and to gain an impression of the interdependence of the different subbasins. Also of interest was the reduction in total expected net benefits if constraints were added to provide for more equitable geographic distribution or balance of benefits among subbasins or among the four states (Delaware, New Jersey, New York, and Pennsylvania) included in the basin.

The effect of controlling water quality through regulation and augmentation of the streamflow also was examined. To study the reduction in net benefits that might occur if this method of stream quality control were used, requirements to augment low flows were imposed. No attempt was made to evaluate the benefits or losses resulting from changes in stream quality. Dissolved oxygen, biochemical oxygen demand, acidity, and salinity were the quality parameters considered. Estimates of the streamflows required, in addition to secondary wastewater treatment, to maintain various quality standards, could be based, for example, on work by the Federal Water Pollution Control Administration. Using these minimum required flows, the static screening models provided information on the reduction in the expected annual net efficiency benefits associated with various probabilities of maintaining these minimum flow requirements. Such procedures permit the simultaneous evaluation of water quality and the other beneficial uses, and provide a means of estimating the economic cost associated with any other objective.

RESULTS

One of the difficulties in evaluating any study of a system in which a global optimum is not known is that it is never quite clear how good one's solution is. In this case, however, there is reason to expect that the solutions achieved come fairly close to the global optimum. For one thing, variation in several of the engineering and economic parameters, say over a range of $\pm 10\%$, yielded an insignificant change in net benefits. This implies that the global optimum is probably not significantly better than the best solution obtained by the analysis.

Possibly because of the insensitivity of the net benefits to changes in design near the optimum, the net benefits derived from the screening models were always within 8% of those derived from the simulation models. This result is particularly gratifying, given the simplifying assumptions required in the development of the screening model. Of course, these results are based on only a relatively small number of comparisons. Given more time and money, larger differences might have been found. Nevertheless it appears that, at least for the models and data used in this study, the net benefits derived from the screening analyses of each subbasin as well as for the system as a whole are essentially the same as those derived from the simulation analyses.

Not only do the two models lead to comparable values of the optimum, but, more importantly, they generate comparable values of the decision variables. The results indicate that these two very different models can yield preferred physical designs that are very similar to one another. For one tributary of the Delaware, the Schuylkill River, simulation could not find a better solution than that indicated by the screening model, even after an exhaustive search. This was an exception, however, since for most other subbasins some improvement was obtained through sensitivity analysis using the simulation model. But no "new" projects, that is, projects at zero capacity in the screening solution, were ever found to be preferred by the simulation analysis. Usually, the simulation analysis led to some adjustments in the proposed project capacities or targets.

Again, given more time it is possible that projects not defined by the screening solution could have been found that would increase the total net benefits. No doubt the fact that they were not found in this study has much to do with a fortunate set of assumptions made during the screening analysis.

It is interesting to note that the use of mean monthly flows in a deterministic screening model of the entire Delaware system indicated that any change in the existing system only decreased annual net benefits. This result, clearly inaccurate, motivated the development of the stochastic screening models described above.

The results of the dynamic screening-simulation analysis were equally as interesting as those of the static model. They showed that the demands for recreation, water supply, power, and flood control are sufficiently great in the Delaware basin to make it desirable to construct most of the projects proposed for the year 2010 within the first 20 years of the 40-year planning period. Perhaps if budget constraints had been imposed, the scheduling of projects would have been more evenly spread out over the total 40 years. Of primary interest in any such dynamic analysis, of course, is the first period's decision, since later analyses with improved data may suggest changes in the decision of future periods. Using the dynamic-simulation model to explore the sensitivity of reasonable changes in future economic and technological conditions, no significant improvement in the first scheduling decisions was found.

Once the best solution was obtained from both the static and dynamic analyses, it was compared to solutions suggested by the numerous technical reports that antedated this model-building effort. The best existing plan offered only 73 % of the present value of the expected net benefits obtainable from designs yielded by the model studies. This is an indication of the potential benefits of the use of these types of analyses for assisting, but not replacing, the more traditional economic–engineering planning processes.

Complete records were not kept on the man-hours and money spent in data preparation and model development and testing. Clearly these activities contributed most to the total cost of this analysis. Whatever the cost, it was several orders of magnitude less than the additional net benefits derived from the $(100/73) - 1 = 37\%$ improvement over the best existing estimate of the optimal solution. For the Schuylkill River alone, this represented more than \$1,000,000 in expected annual net benefits. Hopefully, the techniques and learning experiences resulting from this analysis of the Delaware River basin will help reduce the costs of future screening-simulation analyses of other large, complex river basin systems.

CONCLUSIONS

Both static and dynamic optimization models were used to reduce the range of possible alternative designs and operating rules for further more detailed evaluation by static and dynamic simulation techniques. Without the information provided by the screening models, it would have been both impractical and prohibitively expensive to simulate a sufficient number of design and policy alternatives to be able to conclude with reasonable confidence that an optimal or near-optimal set of alternatives had been found. Yet without the ability to simulate the results derived from the solutions of the screening models, there would be little opportunity to test the effect of the many limiting assumptions that often must be made when structuring a mathematically tractable optimization model of a complex river system. Based on the limited results of this Delaware River basin study, the combined screening-simulation method of analysis appears to be both a practicable and an efficient means of defining and evaluating alternative designs and operating rules for large river basin systems.

ACKNOWLEDGMENTS

Portions of this work were completed at Cornell and Harvard universities with partial support from the Office of Water Resources Research, U.S. Department of the Interior. The writers are also indebted to numerous individuals, particularly Robert Dorfman, Myron B Fiering, and Harold A. Thomas, Jr., whose assistance made this study possible.

REFERENCES

1. H. G., Acres, Ltd., 1972. *Water Quality Management Methodology and Its Application to the Saint John River*, Niagara Falls, Ontario.
2. DORFMAN, R., 1965. Formal Models in the Design of Water Resource Systems, *Water Resources Research*, Vol. 1, No. 3, pp. 329–336.
3. FIERING, M. B, 1964. A Multivariate Technique for Synthetic Hydrology, *Journal of the Hydraulics Division, ASCE*, Vol. 90, No. HY5, pp. 43–60.
4. FIERING, M. B, 1967. *Streamflow Synthesis*, Cambridge, Mass.: Harvard University Press.
5. GYSI, M., AND LOUCKS, D. P., 1969. *A Selected Annotated Bibliography in the Analysis of Water Resource Systems*, Ithaca, N.Y.: Cornell University Water Resources and Marine Sciences Center, Publication 25.
6. HUFSCHMIDT, M. M., AND FIERING, M. B, 1966. *Simulation Techniques for the Design of Water-Resource Systems*, Cambridge, Mass.: Harvard University Press.
7. KRISS, C., AND LOUCKS, D. P., 1971. *A Selected Annotated Bibliography in the Analysis of Water Resource Systems*, Ithaca, N.Y.: Cornell University Water Resources and Marine Sciences Center, Publication 35.
8. LOUCKS, D. P., ET AL., 1969. *Stochastic Methods for Analyzing River Basin Systems*, Ithaca, N.Y.: Cornell University Water Resources and Marine Sciences Center, Technical Report 16, Aug.
9. MAASS, A., ET AL., 1962. *Design of Water-Resource Systems*, Cambridge, Mass.: Harvard University Press.
10. MORIN, T. L., AND ESOGBUE, A. M. O., 1971. Some Efficient Dynamic Programming Algorithms for the Optimal Sequencing and Scheduling of Water Supply Projects, *Water Resources Research*, Vol. 7, No. 3, pp. 479–484.
11. RUSSELL, C. S., d'ARGE, G., AND KATES, R. W., 1970. *Drought and Water Supply*, Baltimore: The Johns Hopkins Press.

12. Texas Water Development Board and Water Resources Engineers, Inc., 1970. *Systems Simulation for Management of a Total Water Resource*, Austin, Tex., Report 118.
13. THOMAS, H. A., JR., AND FIERING, M. B, 1962. The Mathematical Synthesis of Streamflow Sequences, in A. Maass et al., *Design of Water Resource Systems*, Cambridge, Mass.: Harvard University Press.
14. U.S. Army Corps of Engineers (U.S. Army District at Philadelphia), 1962. *Delaware River Basin, New York, New Jersey, Pennsylvania and Delaware*, Delaware River Basin Report 1961, House Document 522, 87 Congress, 2nd Session, Aug. 16, Washington, D.C.: U.S. Government Printing Office.
15. U.S. Water Resources Council, 1964. *Evaluation Standards for Primary Outdoor Recreation Benefits, Policies, Standards and Procedures in the Formulation, Evaluation and Review of Plans for Use and Development of Water and Related Land Resources*, Supplement 1, Washington, D.C.: U.S. Government Printing Office.

18

Determination of the Discharge Policy for Existing Reservoir Networks*

Guy Leclerc and David H. Marks

In this chapter optimization models are used to look at the question of operating a small existing reservoir system to meet various conflicting objectives that are not easily quantifiable in commensurate units. The two important aspects of the study to stress are the way that the stochastic nature of the streamflow may be handled, in this case by a technique known as *chance-constrained programming*, and, perhaps even more important, the way that different objectives for the system can substantially change the configuration of the optimal system. Two objectives concerning the way the goals for recreational use in a series of small reservoirs may be measured are considered. One objective is to minimize the total weighted drawdown over time in all the reservoirs; this leads to the least volume of water released for other objectives but impacts in a heavier manner on some reservoirs over others. The other objective, which attempts to define a more equitable distribution of drawdowns, is to minimize the maximum weighted drawdown so as to spread the disadvantages among the various reservoirs. This, however, causes much more water to be released. Both of these considerations are presented in the context of a case study on the Rivière du Nord, in the Province of Québec, Canada.

INTRODUCTION

Economic development of a river basin is, among other factors, governed by the magnitude of the flow that is guaranteed to prospective users. The

*Adapted from *Water Resources Research*, Vol. 9, No. 5, October 1973. pp. 1155–1165.

expected supply offered by the streamflow shows large variations during the year; in Canada, in particular, a significant component of the total annual flow comes from spring snowmelts. The water demands may be spread over the entire year or may, like irrigation and water-based recreation, occur at specific times that generally do not overlap the periods of large streamflows. To match supply with demand, reservoir systems are built and operated.

The economics of a reservoir network is a function not only of construction costs but of the strategy adopted to operate the reservoirs. An efficient operating policy maximizes the net return of the investment. For reservoir design, this implies minimization of reservoir capacity to meet its objectives at a given level of reliability; for existing reservoirs, an efficient policy implies maximization of the magnitude of the streamflow guaranteed at a specified level of reliability.

An operating policy is a set of rules guiding the determination of the discharges from each reservoir. It must be established at the beginning of a season, when the magnitude of the streamflow is still unknown. The operating policy not only must account for the uncertainty in the streamflows but also for conflicting objectives of the network. The risk of making a wrong decision, whose effect may be carried over several seasons, is consequently minimized.

The derivation of a discharge or operating policy is an involved task around which much analysis has been centered. ReVelle et al. have proposed that chance-constrained programming, with a linear operating rule, be used to estimate the optimal capacity of a single multipurpose reservoir (2). The same approach was followed here in the framework of a linear programming screening model. The model was used to derive the operating policy for a network of eight existing reservoirs. This analysis was a component of a broader study undertaken at the Ecole Polytechnique in Montréal, Québec.

PRINCIPAL METHODS OF SOLUTION

Many investigators have addressed the important problem of defining operating policies for reservoirs: Roefs presented an excellent description of the available methods of solution and documented his description with many references (3). Recently, two methods of solution have received much attention: stochastic simulation and mathematical programming. The first method examines the operation of the reservoir network for various operating policies and sets of synthetic inflows. The optimal operating policy is defined as the one that maximizes net benefits. Stochastic simulation preserves the random nature of the inflows and, through sequential computations which represent the evolution of the operations, also preserves the physical sequence of the process. This approach does not guarantee an optimal solution; it merely describes the outcome for a specified configuration. It may be expensive if many discharge policies are investigated and if many reservoir configurations are considered.

The second method formulates the problem of reservoir operation in terms of mathematical optimization techniques, particularly linear and dynamic programming. In this method, the decision variables directly define the best operating policy for a given configuration. In the stochastic simulation method the best policy is chosen among those analyzed; in the mathematical programming method, the policy is directly defined because it is related to the decision variables.

Mathematical programming guarantees an optimal solution to the problem formulated. However, because the formulation approximates the real problem, the optimal solution obtained by mathematical programming is an approximation of the optimal strategy for the real system. A good solution may, however, be derived at a relatively low cost. Moreover, sensitivity of the solution to changing conditions may be analyzed inexpensively.

The method of solution adopted here lies in the second generic group; the optimization technique is linear programming. The problem is formulated around the linear operating rule originally proposed by ReVelle et al. for the estimation of the optimal capacity and operating policy for a single multipurpose reservoir (2). This approach has since been extended to a network of four multipurpose reservoirs by Nayak and Arora, who also analyzed the behavior of the solution under different sets of initial conditions (1).

DESCRIPTION OF THE PROBLEM

The Rivière du Nord, a tributary of the Ottawa River, flows through the resort area north of Montréal, Canada. The city of St.-Jérome, located in the lower section of this area, takes its water from the Rivière du Nord. The streamflow at St.-Jérome, where a minimum flow of 250 cubic feet per second is required, is composed of the natural inflow of the Doncaster River, a tributary of the Rivière du Nord, and the discharges of the reservoir network, for which a discharge policy is sought.

The reservoir network, shown in Figure 18.1, is composed of eight natural lakes controlled by small dams. Their physical characteristics are given in Table 18.1. These reservoirs are used for regulation of the streamflow and for extensive water-based recreation. These two objectives are in conflict because, when water is discharged from a reservoir, surface drawdowns are generated; these make the reservoir less attractive for recreation and reduce the benefits associated with this objective.

It would be desirable to determine the discharge policy for the network

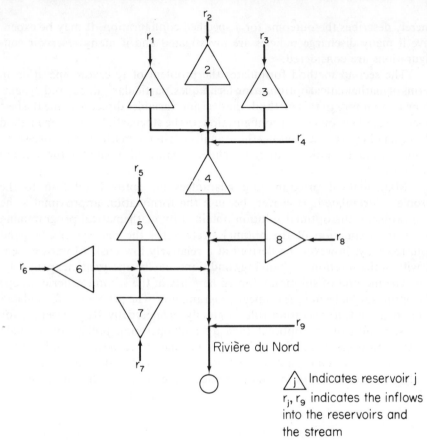

Figure 18.1 Schematic outline of the reservoir network on the Rivière du Nord.

Table 18.1 Physical Characteristics of the Reservoirs

Reservoir Number (j)	Area (ft^2 × 10^6) (a_j)	Storage Capacity (ft^3 × 10^6) Minimum (S_j min)	Storage Capacity (ft^3 × 10^6) Maximum (S_j max)	Maximum Surface Elevation (ft) (H_j max)
1	32.6	0	260.9	8
2	11.2	0	44.6	4
3	29.3	0	322.0	11
4	36.2	0	217.5	6
5	13.9	0	83.6	6
6	53.0	0	211.9	4
7	27.9	0	167.3	6
8	47.3	0	568.7	12

on the basis of the net benefits due to streamflow regulation and to water-based recreation. But the evaluation of these benefits is in itself a major task. It cannot be addressed in the framework of this screening model. Nor, as is often the case, is enough quantitative information available at this time to justify even a limited evaluation of the benefits.

A surrogate approach has consequently been adopted, where the regularization objective is represented by a constraint on the minimum flow that will be allowed at St.-Jérome, and the recreational goals are incorporated in the objective function. This approach allows one to analyze the effect on the recreation objective of different goals taken as different levels of constraints, for regularization of the streamflow. The result of this surrogate approach is to center the problem around the allocation of the surface drawdowns among the reservoirs. Finally, to account for economic differences among the reservoirs, a multiplier, C_{ji}, of the surface drawdowns is introduced.

LINEAR OPERATING RULE

Let the volume of water discharged from reservoir j during season t, $d_{j,t}$, be computed by the linear operating rule:

$$d_{j,t} = S_{j,t-1} - b_{j,i} \qquad (18.1)$$

where $S_{j,t-1}$ is the storage in this reservoir at the end of season $t - 1$, and $b_{j,i}$ is a factor which reflects the stochastic nature of inflows and is set in advance as a guideline for the operators of the reservoir. The purpose of the analysis is to define the optimal set of $b_{j,i}$ for each of the i periods of a year. The $b_{j,i}$ are the decision variables in the analysis. Once they have been set, the operators can determine how much the discharge will be in any season. The discharge in a particular season is a function of the storage at the end of the previous season and of the adjustment factor. The discharge is not constant from year to year for a given season because the storage at the end of the previous season varies from year to year.

The adjustment factors are set for any season i, say any month in a year. The index i can be related to the index t, defining a particular period in a sequence, by:

$$i = \text{remainder of } [t \div 12] \qquad (18.2)$$

That is, for a model which divides the year into months, January ($i = 1$) could correspond to periods $t = 1$, $t = 13$ for the following year, and so on.

The discharges vary inversely with the adjustment factors. In general, a large $b_{j,t}$ implies a small discharge. It should be emphasized that the adjustment factors need not be positive. A negative $b_{j,t}$ implies that a minimum amount of inflow into the reservoir is required to meet a commitment to provide a discharge $d_{j,t}$ since the amount already available, $S_{j,t-1}$, is not sufficient.

The level of a reservoir at the end of any period is defined by the continuity equation:

$$S_{j,t} = S_{j,t-1} - d_{j,t} + r_{j,t} \tag{18.3}$$

where $r_{j,t}$ is the water inflow to reservoir j during season t and the other terms are as previously defined. Upon substitution of the linear rule, Eq. (18.1), the continuity equation can be written as:

$$S_{j,t} = r_{j,t} + b_{j,t} \tag{18.4}$$

Conversely, Eq. (18.1) can be rewritten as:

$$d_{j,t} = b_{j,t-1} - b_{j,t} + r_{j,t-1} \tag{18.5}$$

Equations (18.4) and (18.5) define the storage and the discharge in terms of the decision and exogenous variables.

USE OF THE LINEAR RULE

A small example illustrates the behavior of the linear operating rule. In particular, it shows that this rule accounts for situations in which reservoirs already exist, and in which overflows must consequently be made from time to time. This situation will occur whenever the reservoir capacity is insufficient to store all the water surplus accumulated during a season.

Consider, for example, a reservoir of 200-unit capacity situated in the sample network shown in Figure 18.2. A minimum flow, q_t min, of 100

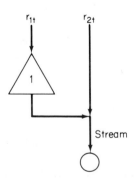

Figure 18.2 Reservoir network for example problem.

units is guaranteed during each season; the objective of the discharge policy is to minimize the surface drawdowns in the reservoir. The data for this hypothetical example are given in Table 18.2.

Table 18.2 Inflows into the Reservoir and Stream for the Hypothetical Example

Season Number, t	Inflows	
	$r_{1,t}$	$r_{2,t}$
0	50	50
1	100	80
2	20	100
3	30	120
4	10	15
5	210	100

For the configuration given, the streamflow will equal the discharge from the reservoir, as given by Eq. (18.5) plus the natural flow into the stream, $r_{2,t}$. This flow must exceed the prescribed minimum, so:

$$(b_{1,t-1} - b_{1,t} + r_{1,t-1}) + r_{2,t} \geq q_t \min \tag{18.6}$$

The adjustment factor for any season can thus be defined in terms of known data and previous factors:

$$b_{1,t} \leq b_{1,t-1} + r_{1,t-1} + r_{2,t} - q_t \min \tag{18.7}$$

For our problem, if we had set the factor for period zero as $b_{1,0} = 50$, we could define:

$$b_{1,1} \leq 80 \tag{18.8}$$

From Eq. (18.4) and from knowledge of the reservoir capacity, we can define the storage level at the end of period 1 as:

$$S_{1,1} = b_{1,1} + 100 \leq 200 \tag{18.9}$$

Since the surface drawdown is minimized when the end storage is maximized, we select $b_{1,1} = 80$. Table 18.3 gives the results of the computations for the remaining seasons. This shows that even in existing reservoir systems when a sequence of higher than normal flows causes considerable discharges, the linear decision rule is still applicable and the adjustment factor guides operation for present and future uses.

Table 18.3 Results of the Hypothetical Example

Season Number i or t	Adjustment Factor $b_{1,i}$	Reservoir Storage Level $S_{1,t}$	Discharge $d_{1,t}$	Stream-flow q_t
1	80	180	20	100
2	180	200	0	100
3	170	200	30	150
4	115	125	85	100
5	−10	200	135	235

FORMULATION OF THE SCREENING MODEL

A linear programming model is formulated to define the optimal set of adjustment factors. The minimum flow objective is described by a minimum flow constraint and the recreation objective is incorporated in the objective function. A surrogate measure, the drawdown, is adopted to describe the recreational losses due to surface drawdowns. A multiplier, C_{ji}, is also introduced to account for economic differences between the reservoirs and for the time of the year when the drawdown occurs.

Constraints

In all, there are five kinds of constraints for this problem. These deal with continuity, minimum flow, the drawdowns, the discharges, and the reservoir capacity. They will be defined in turn.

The continuity constraint for each independent reservoir is defined by Eq. (18.3). For reservoirs which are fed by other reservoirs, however, Eq. (18.3) must be modified to include their discharges. Such is the case for reservoir 4 of the network on the Rivière du Nord, as can be seen from Figure 18.1. For this particular reservoir the continuity equation becomes:

$$S_{4,t} = S_{4,t-1} - d_{4,t} + r_{4,t} + (d_{1,t} + d_{2,t} + d_{3,t}) \qquad (18.10)$$

A similar adjustment would, of course, have to be made in all like situations.

The constraint insuring that the minimum required flow, q_t min, exists in the stream in all periods is simply a generalization of Eq. (18.6) used in the example problem. Specifically, we can write:

$$\sum_j (b_{j,i-1} - b_{j,i}) \geq q_t \min - (r_{9,t} + \sum_j r_{j,t}) \qquad (18.11)$$

where $r_{9,t}$ is the natural inflow of the tributary.

The surface drawdown in any reservoir j at the end of any season t,

$G_{j,t}$, may be defined as the difference between the maximum elevation of the reservoir, H_j max, and the actual elevation. This latter quantity may be approximated as the storage in any reservoir divided by its area. Defining the area of any reservoir as a_j, we can write:

$$G_{j,t} = H_j \max - S_{j,t}(a_j)^{-1} \tag{18.12}$$

This becomes, by simple substitution of the continuity equation defining $S_{j,t}$, Eq. (18.4):

$$G_{j,t} + (a_j)^{-1}b_{j,t}\,h = H_j \max - (a_j)^{-1}r_{j,t} \tag{18.13}$$

for all independent reservoirs. For reservoir 4, which is dependent, we may write:

$$G_{4,t} + (a_j)^{-1}\left[b_{4,t} + \sum_{j=1}^{3}(b_{j,t-1} - b_{j,t})\right] = H_4 \max - (a_j)^{-1}\left[r_{4,t} + \sum_{j=1}^{3}r_{j,t-1}\right] \tag{18.14}$$

Over all, the drawdowns cannot exceed the maximum and minimum elevations of the pool, so that:

$$H_j \min \leq G_{j,t} \leq H_j \max \tag{18.15}$$

The requirement that the discharge is positive is:

$$d_{j,t} = S_{j,t-1} - b_{j,t} \geq 0 \tag{18.16}$$

Since the storage level at any time can be defined in terms of the drawdowns:

$$S_{j,t-1} = S_j \max - (G_{j,t-1})a_j \tag{18.17}$$

the nonnegativity constraint can be conveniently rewritten as:

$$a_j(G_{j,t-1}) + b_{j,t} \leq S_j \max \tag{18.18}$$

Finally, the requirement that the storage capacity of the reservoirs be met is stated as:

$$S_j \min \leq S_{j,t} \leq S_j \max \tag{18.19}$$

Chance-Constrained Programming

Using the constraints given in Eqs. (18.10)–(18.19), it is clear that we can formulate a linear programming problem. Unless we take special steps, however, we will fail to take into account the essentially stochastic nature of a reservoir network. This feature can be accounted for by using what is known as Chance-constrained programming, as described by ReVelle et al. (2).

Chance-constrained programming takes the stochastic nature of a problem into account by the following simple device. Each stochastic variable is replaced by what might be thought of as a certainty equivalent. Specifically we can replace each such variable by the critical magnitude it may attain, in any particular period, with a specified level of risk α. This transformation of a variable { } is usually denoted by crit_α{ }.

For our problem, the inflows $r_{j,t}$ are the stochastic variables. As probability distribution functions are, indeed, known for these variables, we may substitute:

$$\text{crit}_\alpha\{r_{j,t}\} \quad \text{for } r_{j,t} \tag{18.20}$$

$$\text{crit}_\alpha\left\{r_{9,t} + \sum_j r_{j,t-1}\right\} \quad \text{for } \left\{r_{9,t} + \sum_j r_{j,t-1}\right\} \tag{18.21}$$

and:

$$\text{crit}_\alpha\left\{r_{4,t} + \sum_{j=1}^{3} r_{j,t-1}\right\} \quad \text{for } \left\{r_{4,t} + \sum_{j=1}^{3} r_{j,t-1}\right\} \tag{18.22}$$

in the constraints given previously. The constraints are then fully defined for chance-constrained programming.

Objective Functions

Two objective functions were considered for the determination of the discharge policy. The first one minimizes the total summation of the drawdown in the network; it is written:

$$\text{Min } z = \sum_j \sum_i C_{j,i} G_{j,i} \tag{18.23}$$

where $C_{j,i}$ is a weighting function associated with the relative economic cost of drawdowns in each reservoir for each period. The objective function of Eq. (18.23) implies that the losses due to surface drawdowns are cumulative and are a direct function of the drawdown observed at the end of the period. This characteristic is realistic since the level of recreation activities is, indeed, a function of the actual drawdown, and the total annual losses are the summation of the losses generated in the system during each season. Equation (18.23) also implies that the losses are a linear function of the drawdowns. This may not be the case because 1 ft of drawdown may cause a lesser reduction in the recreation activities than 3 or 4 ft. For instance, if a shallow channel links two regions of a lake, 1 ft of drawdown may not hamper communication between the two regions, whereas 3 or 4 ft may not leave enough water for motorboats to navigate from one region of the lake to another. To account for this nonlinearity, a piecewise linearization may be made which does not cause any difficulties in terms of global optimality because the damage function would appear to be convex.

The first objective function is known as the minimum drawdown solution; it merely seeks to minimize drawdowns, irrespective of where the water comes from. Such an approach could, in fact, be highly inequitable. It could empty one reservoir completely, ruining all recreational possibilities for some, while not taking any water from other reservoirs. A more equitable solution may be one where all reservoirs contribute to the streamflow and absorb a given share of the losses due to surface drawdowns.

To analyze the influence of such social constraints, a second objective function is also considered. Specifically, we formulate a minimax objective:

$$\text{Min}[z = \text{Max } C_{j,i} \, G_{j,i}] \quad \text{all } i, j \tag{18.24}$$

which will lead us to minimizing the largest weighted drawdown in the network. Equation (18.24) is nonlinear but is easily linearized by the introduction of the set of constraints:

$$z \geq C_{j,i} \, G_{j,i} \tag{18.25}$$

which transforms the objective function of Eq. (18.23) simply to that minimizing z.

ANALYSIS OF THE RESULTS

The chance-constrained linear programming model was run for the Rivière du Nord. The solution was obtained using MPS, the standard IBM mathematical programming system, on an IBM 370/155 digital computer. The solution for the minimum drawdown objective function (206 rows and 399 variables) took 0.73 minute of central processor time, and the minimax solution (301 rows, 494 variables) took 1.22 cpu minutes. The approach is, therefore, quick and computationally efficient.

The physical characteristics of the reservoirs were taken as given in Table 18.1. The critical levels of the streamflows which were used are given in Table 18.4. The reliability level for these critical flows was arbitrarily chosen such that $\text{crit}_\alpha\{\ \}$ corresponds to a magnitude equaled or exceeded 21 times in a 23-year historical record. The minimum flow requirement for the river was taken as equal to 250 cfs throughout the year. Finally, no preferences were assumed between the reservoirs or the seasons, so all $C_{j,i} = 1$.

Minimum Drawdown Objective

The optimal solution for the minimum drawdown case has a magnitude of 65.3 ft. Table 18.5 presents the $b_{j,t}$ obtained for the optimal solution. Graphical descriptions of the behavior of the drawdown in typical reservoirs are shown in Figures 18.3 and 18.4.

Table 18.4 Critical Flows Used in the Analysis of the Rivière du Nord

	Critical Inflows (cfs)									
	$r_{j,t}$ for Reservoir j, Eq. (18.20)							At point 9, $r_{9,t}$ Eq. (18.21)	At dependent reservoir 4, $r_{4,t}$, Eq. (18.22)
Month	1	2	3	5	6	7	8		
J	0	0	5.0	2.0	0.7	1.4	0.4	269	28.0
F	3.3	0.1	0	4.8	2.1	0	0.7	231	20.2
M	2	0	4.7	3.8	8.1	0	3.3	296	20.4
A	21.8	33.6	91.7	24.4	62.9	47.5	46.8	1674	48.6
M	6.2	7.6	20.1	5.9	10.3	8.1	17.7	1325	181.1
J	0.9	4.8	10.4	1.2	0	0	0	476	45.1
J	0	1.5	2.4	0.3	0	0	0	239	22.6
A	0	0.1	0	0.3	0	0	0	163	9.9
S	0	2.3	3.2	0	0	0	0	179	9.0
O	1.0	5.5	10.4	1.0	0	3.1	0	181	12.3
N	5.0	3.2	14.8	4.0	5.3	1.7	6.7	287	30.2
D	5.5	2.7	8.4	3.7	3.6	3.0	6.3	358	36.9

Table 18.5 Adjustment Factors as Guidelines for Setting Discharges $(b_{j,i})$: Total Minimum Drawdown Solution

	Reservoir							
Month	1	2	3	4	5	6	7	8
J	247	45	309	170	79	19	164	433
F	247	44	310	120	71	21	167	435
M	256	45	310	173	74	26	167	486
A	204	−44	107	−253	20	46	42	446
M	242	12	242	−28	68	185	146	522
J	259	32	295	185	80	212	167	569
J	261	41	316	190	83	143	167	569
A	261	44	322	201	83	0	167	461
S	261	39	314	41	84	0	167	449
O	217	30	282	−34	81	0	147	449
N	220	29	261	83	73	0	155	449
D	233	37	300	184	74	14	159	467

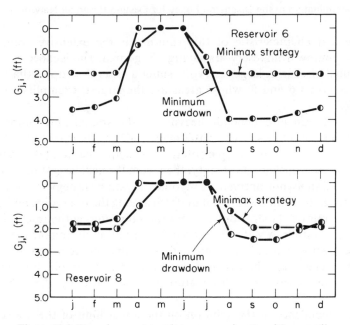

Figure 18.3 Drawdowns at two large reservoirs, 6 and 8, according to minimum total drawdowns for the system and minimax strategies, to minimize maximum drawdown, for twelve months.

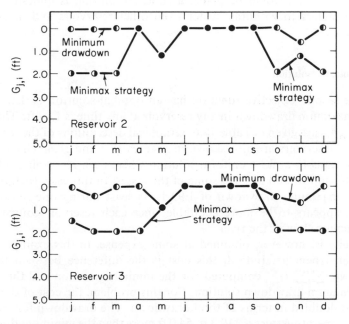

Figure 18.4 Drawdowns at two small reservoirs, 2 and 3, according to minimum total drawdowns for the system and minimax strategies, to minimize maximum drawdown, for twelve months.

The first characteristic of this solution is the extensive contribution to the streamflow from reservoirs having a large area. This is expected because, for an equal discharge, large reservoirs suffer a lesser drawdown than smaller ones. Reservoirs 6 and 8, whose areas are the largest, exemplify this characteristic.

The solution also depends heavily on the spatial and temporal distribution on the inflows, and on the magnitude of the maximum allowable drawdown. The obvious myopic solution, which would first take water from the largest reservoir until it is empty, then from the next largest reservoir, and so on, is definitely not optimal. A myopic operating strategy is not followed because the natural inflows are not uniform among the reservoirs, and because the upper limit on the drawdown is not the same among the reservoirs. It is preferable to withdraw water from a small reservoir if, during the next period, the drawdown is reduced or eliminated. This amounts to accepting a short-term loss higher than the loss incurred at another reservoir, but this minimizes the long-term loss. This feature is exemplified by the behavior of reservoir 2.

The dependence of the solution on the upper limit of the drawdown is shown by the behavior of reservoir 8. This is the largest reservoir in terms of volume, the second largest in terms of surface area, and has the largest maximum allowable drawdown. If it were empty, a large contribution to the objective function would be made, a contribution that is undesirable. The solution prefers to withdraw water from other reservoirs and in doing so minimizes z.

Minimax Solution

The second objective function has an optimal solution of 1.99 ft, that is, the maximum drawdown in any reservoir at any time is 1.99 ft. The corresponding $b_{j,i}$ are given in Table 18.6. Graphical descriptions of the drawdown of typical reservoirs are also shown in Figures 18.3 and 18.4. This objective function leads to a more uniform solution where each reservoir contributes to the streamflow and absorbs some of the losses. In this case, each reservoir shows a maximum drawdown of 1.99 ft, at least during one season. This solution appears to be more equitable since each reservoir contributes to the minimum flow of the river.

Equity is, however, obtained at some expense, in this example, of the recreational benefits. Indeed, this cost is the difference between the total drawdown, $\sum_j \sum_i G_{j,i}$, computed for the minimax solution and the optimal total minimum drawdown solution. For this problem, the cost of the system is almost doubled in terms of total drawdown. The drawdown for the more equitable operating rules is 119.3 ft, 54.0 ft more than the minimum drawdown obtained using the alternative objective function.

Table 18.6 Adjustment Factors as Guidelines for Setting Discharges ($b_{j,i}$): Minimax Solution

Month	Reservoir							
	1	2	3	4	5	6	7	8
J	206	22	264	97	51	105	108	473
F	188	22	264	74	43	101	112	473
M	191	22	251	83	46	85	112	466
A	196	−44	49	−174	1	19	21	399
M	242	12	242	38	65	185	146	522
J	259	32	295	185	80	212	167	569
J	261	41	310	176	83	212	112	569
A	261	44	322	129	55	107	112	509
S	203	39	314	50	56	107	112	475
O	193	8	236	−4	53	107	104	475
N	196	22	225	72	56	93	107	457
D	209	15	242	96	46	107	104	458

DISCUSSION

The approach followed in this study has several advantages. It gives a quick solution to a complex problem and allows the analysis of its behavior under different sets of conditions, easily and inexpensively.

Chance-constrained programming, although it accounts for the randomness of the natural inflows, does not define the magnitude by which the system fails; this may be a serious problem because a small failure may be relatively unimportant, whereas a large failure may have long-term effects. This difficulty can be overcome with a simulation study of the operating rules selected by the optimization models performed, for instance, with the historical data.

Another difficulty arises when chance-constrained programming is used. In this case study, a large fraction of the total annual runoff is generated during the snowmelt season; this season is short and may occur during March, April, or May. When the crit$_\alpha\{$ $\}$ is selected for March, its value would likely correspond to a year when the snowmelt occurred in April and May. The same behavior is also expected for April and for May, resulting in total natural inflows that are significantly smaller than those observed even in a dry year.

A similar difficulty arises among the reservoirs. For a given season the critical inflow to reservoir 1 may come from year 1, whereas the critical inflow to reservoir 2 may come from year 10, and so on. Chance-constrained programming available in 1973 therefore assumes that the inflows are thus serially and spatially uncorrelated. This deficiency is not severe in this study because the reservoirs are small and are always full at the end of the snowmelt season. It may, however, prove to be a serious drawback to the screening model for use in the analysis or design of large reservoirs. One possible solution to this difficulty is the determination of a critical year whose characteristics, yet to be defined, would be related to the level of risk. Alternatively, programming techniques may eventually be developed to account for correlated inflows.

This linear programming analysis is, however, very quick and efficient. To balance its deficiencies, which imply that its solutions do not represent the absolute optimum because of its inability to model every aspect of the physical system, it has the strong advantage of rapidly providing insights into the structure of the problem and also of defining regions of optimal configuration which may subsequently be investigated in greater detail. As has been seen, it can do this even for complex, stochastic problems, provided suitable modifications can be made. In short, this linear programming approach is a good screening model.

ACKNOWLEDGMENTS

This study was made possible through collaboration with École Polytechnique, Montréal, Canada. The authors are indebted to the research group studying the Rivière du Nord, particularly A. Leclerc, J. M. Mejia, J. Rousselle, and P. Egly.

REFERENCES

1. NAYAK, S. C., AND ARORA, S. R., 1971. Optimal Capacities for a Multireservoir System Using the Linear Decision Rule, *Water Resources Research*, Vol. 7, No. 3, pp. 485–498.
2. ReVELLE, C., JOERES, E., AND KIRBY, W., 1969. The Linear Decision Rule in Reservoir Management and Design, 1. Development of the Stochastic Model, *Water Resources Research*, Vol. 5, No. 4, pp. 767–777.
3. ROEFS, T. G., 1968. *Reservoir Management: The State-of-the-Art*, Wheaton, Md.: IBM Washington Scientific Center, July.

19

Unreliability in Railroad Network Operations*

JOSEPH F. FOLK AND JOSEPH M. SUSSMAN

This chapter presents simulation as a stand-alone technique for the analysis of complex systems that defy more formal mathematical analysis. The case study under consideration, that of operation of a railroad network, reinforces the point that optimization in its broadest interpretation includes the enumeration of a few feasible configurations through trial and error in complex systems as well as mathematical manipulation of simple models in more tractable problems.

THE PROBLEM

Although railroads handle over 40% of all U.S. freight ton-miles, the industry is plagued with a low rate of return on capital and a dwindling market share. The majority of the traffic still handled by railroads is low-value, low-revenue raw materials; high-value, high-revenue merchandise is increasingly handled by truck.

*Extensively adapted from *Models for Investigating the Unreliability of Freight Shipments by Rail* by J. F. Folk, Ph.D. Thesis, MIT Department of Civil Engineering, Cambridge, Mass., 1972. A version of this paper was also presented at the Fall Meeting of the Operations Research Society of America, Atlantic City, N.J., 1972.

Whatever other reasons there may be, the market split between rail and truck must be largely explained by examining the total costs of transportation, in money and time, to the shipper (6,9). The monetary costs consist of haulage and other direct charges, which are generally lower for rail than for truck, and losses on merchandise in transit, which occur at comparable rates for both modes. The relevant time costs concern the average transit time and its reliability. Transit time, although generally longer for rail, is often not the primary factor in a shipper's choice of mode. Attention thus focuses on the unreliability of transit time, which is known to be a serious problem in rail service (7).

Transit unreliability can be illustrated by considering Figure 19.1, which is derived from railroad operating data. Although the mean transit time for this origin–destination pair is 6.5 days, the variation about this mean is obviously substantial. For this analysis, transit time unreliability is equated with this variation. Alternative measures of reliability are discussed by Martland (7).

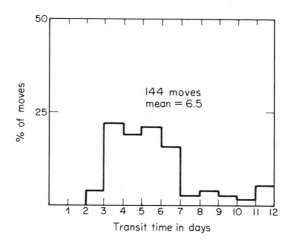

Figure 19.1 Transit time in days for a given origin-destination pair.

Poor reliability results in higher shipper and consignee costs. For example, if one considers the inventory system, unreliability clearly translates into needs for larger safety stock in the warehouse of the consignee if stockouts are to be avoided (7). Hence it is reasonable to hypothesize that unreliability is a determinant of the shipper's choice of transportation, an explanation for declining market share of the railroads. An understanding of the causes of the unreliability of transit time in railroads is thus an essential precondition for improving the situation.

RAILROAD OPERATIONS

A typical railroad shipment consists of three major elements, as illustrated by Figure 19.2. First, a shipper orders an empty car, and then waits
for it to be delivered before loading it and releasing it to the railroad. Second,
there is the local pickup and delivery service which handles the car through
one or more local yards. Third, the loaded freight car travels on one or more
mainline trains. Unreliability is associated with each of these elements.

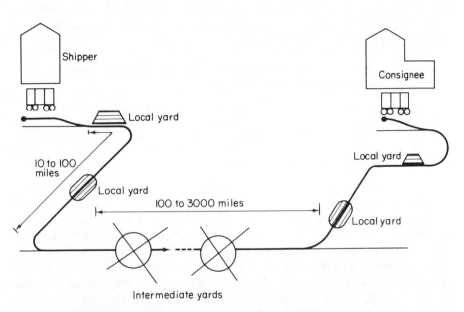

Figure 19.2 A typical journey of a loaded freight car from shipper
to consignee.

Based upon current understanding of the operation of the railroads,
it was deemed most effective to focus on an examination of the unreliability
of the mainline move. This was taken to include any switching at origin and
intermediate yards. For the purpose of this analysis, therefore, railroad transit
time was defined as the interval between the arrival of the loaded car at the
originating yard until its departure from the terminating yard.

Let us, therefore, look more closely at the movement of cars along the
mainline. After a loaded car arrives at the local originating yard, it is switched
onto a mainline train. This mainline train may carry the car directly to the
local terminating yard in the destination city. More typically, especially
for long hauls, the car is switched on to some other mainline train at an inter-

mediate yard. This may happen several times to a single car. The total car move is thus made up of line haul segments and yard segments, wherein a car must make a connection between its inbound and outbound trains.

The yard segments are of particular interest because their switching process is often unreliable. Each time a car is switched, the potential for a missed connection exists. Missed connections are critical in that they lead to car delays in the order of 12 to 24 hours, the time until the next appropriate outbound train; large variations in transit time; and hence unreliable performance. Table 19.1 shows the magnitude of transit time delay and transit time variance as a function of the probability of a missed connection at a yard. These probabilities are quite realistic in the light of the analysis of railroad operating data performed (8).

Table 19.1 Average and Standard Deviation of Delay Time as a Function of the Probability of a Missed Connection

Probability of Missing Connection	Average Delay (hr)	Standard Deviation of Delay (hr)
0.1	2.4	7.2
0.2	4.8	9.6
0.3	7.2	11.0

A car may miss its outbound connection for a variety of reasons: the outbound train may simply not run due to a lack of power, crew, traffic, or other causes. Or it may not be able to accommodate a car if there is excess traffic which exceeds the train's length or weight restrictions. This may happen particularly if a cancellation the day before has backed up traffic. The car may also miss its connection because it was misclassified or broke down. Finally, a car may miss a connection because the inbound train carrying it may arrive behind schedule. This could be prevented, of course, by holding the outbound train to allow the car to make the connection despite the lateness of arrival, but this may well lead to further problems. Specifically, Belovarac and Kneafsey have shown that the primary cause of late arrivals at a yard is late departure from the preceding yard (2). Hence, holding trains to allow particular connections to be made may well lead to inbound lateness at succeeding yards and cascading possibilities of other missed connections throughout the network.

All in all, the question of yard performance and missed connections is complex. The various design components and operating policies of the rail network interact heavily to affect performance. Even though complex, the system can be usefully simplified for the purposes of analyzing various operating policies available to the railroads. Because this type of stochastic network problem is not readily amenable to analytic procedures, especially because of the discontinuities in the traffic, of the kind imposed by limitations on the length of the train, we developed a model for simulation of the network. It was purposely simple so that it could study a particular issue, network unreliability. A variety of other more complex network models are discussed by Folk (3).

RAILROAD NETWORK MODEL

Model of the Network and Yards

The network configuration of the model is shown in Figure 19.3. Yard B receives trains from nodes A_1 to A_j, and sends this traffic to other yards, including yard D. Yard D receives trains from yard B and from nodes C_1 to C_k, and sends this traffic to E_1 through E_n and beyond. The model considers flows in one direction only and thus could not be used to study problems such as the cycling of motive power or crews. The primary objective of this modeling effort, however, is on the reliability of car movements from origin to destination. Assuming that locomotives and crews are available for any outbound train, a model with one-directional flow does not appear to leave out any important features.

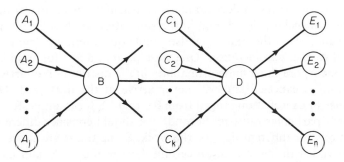

Figure 19.3 The basic network represented in the railroad network model.

The simulation of classification yards in the network is extremely simple. A railroad classification yard, in the real world, consists basically of three different sets of tracks: receiving tracks, where inbound trains arrive and are inspected prior to classification; classification tracks, where cars are sorted depending on the destination of the car; and departure tracks, where several strings of cars from the classification tracks are formed into an outbound train. Based on observations by the authors at several different classification yards, it appears that the amount of time a car spends in the receiving area is very difficult to model since it can be affected by train priority and commodities; yard congestion; lateness of arrival; time until the appropriate outbound train is scheduled to depart; and the amount of traffic already prepared for these outbound trains. So, for this analysis ,which was concerned mainly with the network rather than the yard, attention was focused on the classification yard.

In the simplified form of the network model, the "arrival" time of a block of traffic at a yard was taken as the time that it appears on the classification tracks. As a further simplification, it was assumed that if a train from A_i "arrives" at B at time t, then all the cars on this train arrive on the classification tracks in yard B at time t. This arrival time t is sampled from a distribution with specified mean and variance, where the mean of this distribution is chosen to be close to the time that these cars are scheduled to arrive on the classification tracks. The scheduled time available for connection for a car at a yard, to be consistent with the above, was taken as the time between this car's scheduled arrival on the classification tracks and the scheduled departure of its outbound train.

Traffic Generation in the Model

Substantial data from several railroads were analyzed on train histories, train arrival times at yards, and daily variations in train length (5). This analysis was necessary to develop realistic distributions of arrival times and traffic levels which could be used in the simulation analysis.

The analysis of the train data also points out several modeling concepts that need careful consideration. One should, first, allow for train cancellations by incorporating a finite probability in the distribution of arrival times that a train is canceled and will never arrive at the next yard. Observed frequencies of cancellation ranged from 0.0 to 0.12, for example, for different trains on a particular railroad. Second, one should consider different traffic distributions for different days of the week. Some trains are canceled more often, or run with fewer cars, on certain days of the week, and most local trains do not run on Sundays. Third, one should attempt to have the distributions of the traffic reflect system-wide yard policies. For instance, if trains are dispatched using a policy of never holding an outbound train, one might expect that train arrivals at succeeding nodes in the network would be more on time.

For each day in the simulated network, the model samples a distribution of the traffic levels, measured in the number of cars, and arrival times at yards for this traffic. The traffic leaving yards B and D is considered "dependent" traffic, since its amount and departure times depend on traffic from previous yards as well as the yardmaster's dispatching policy on outbound trains at these yards. Traffic from the A nodes and C nodes can be taken to be "independent traffic," both in terms of the traffic levels and the arrival times at B and D, respectively.

After sampling for traffic levels and arrival times, the model notes the total amount of traffic for each outbound train at each yard, and the arrival times of these blocks of traffic. A specified yard policy on forming outbound trains, called a "dispatching" policy of the model, is used to determine if and when the outbound train will depart, based on traffic ready to go at a certain time. This policy is applied and trains dispatched. Statistics on the performance of the network for the different policies can then be collected.

Yard Policies on Making Up Outbound Trains

The model examines the effect of alternative policies for holding and canceling trains, based on traffic levels and arrival times of this traffic, at each intermediate yard such as B and D. In effect, the decision at B of whether or not to hold the outbound train for D depends on the traffic from nodes A_1 to A_j and the arrival times, either known or anticipated, of this traffic at yard B. Many policies are available, and it is not intuitively clear which might be best.

An example of a possible "dispatching" policy for a yard is shown in Figure 19.4. The dispatching rule is to assemble and send out the train when the traffic available exceeds the policy line at any time. For this example, if a train of 100 cars were available at the scheduled departure time of 10 A.M.,

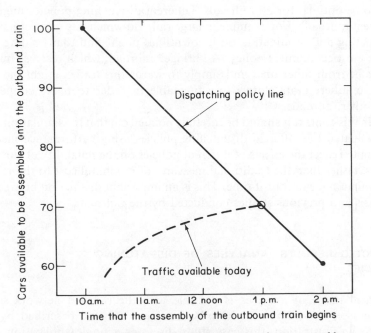

Figure 19.4 A possible dispatching policy at a yard as suggested by: Beckmann, McGuire and Winsten (1).

the train should leave right away. If, on the other hand, today's accumulation of traffic were as shown, the train should be made up at 1 P.M. with only 70 cars. Finally, if less than 60 cars were available by 2 P.M., the train should be canceled. The policy says, in effect, that we should hold trains for up to 4 hours in order to send out at least a 60-car train. The kind of policy shown in Figure 19.4 is a compromise between the more rigid policies of always leaving on schedule regardless of the amount of traffic ready to go, which could be represented by a vertical line through 10 A.M., and of never leaving until at least 100 cars are ready, which would be represented by a horizontal line through 100 cars. It should also be pointed out that a dispatching policy might also specify the maximum length of a train.

The straight-line form of dispatching policy has been suggested, by Beckmann et al. (1), as representative of actual railroad departures, and our investigations with railroads reinforce this impression. Important differences do, however, appear to exist between railroads. One railroad we examined appears to follow a policy of adhering fairly strictly to a schedule with the option of running advance sections if there is sufficient traffic ready to move ahead of schedule. Another railroad we looked at favors holding traffic to build up long trains. Officials from all railroads agreed that there was considerable uncertainty about which policies were optimal and that more research was needed to evaluate effects of these quite different policies.

It should not be thought, however, that the same dispatching policy would be optimal for all railroads. Different dispatching policies might suit different railroads. For example, a large railroad which has most of its traffic originating and terminating on its own lines might find that holding trains is the most economical policy. A "bridge" railroad, which receives much of its traffic from other lines and simply forwards this traffic, might, however, have to adhere more closely to a timetable in order to make connections with other railroads.

In this context it should be carefully pointed out that the simulation analysis evaluated the different dispatching policies using performance measures chosen to reflect the effects of different policies on the reliability of car movements, rather than the traditional measures of minimum cost to the railroad or minimum mean transit time. This is an important distinction between this analysis and previous studies conducted by the railroads.

SELECTED RESULTS: ANALYSES OF DISPATCHING POLICIES

A broad range of analyses is possible with the railroad network simulation model, and many runs have been carried out and described by Folk (4). As an illustration, this case study discusses a single application of the model: the examination of the effect of different dispatching policies on the reliability of the transit time of cars moving through the network, assuming different time distributions of the arrival times of incoming traffic.

Description of Analyses

Eight different dispatching policies for yards B and D were considered. These policies are shown graphically in Figure 19.5. They cover a range from a maximum holding time of 12 hours and a minimum train of 70 cars (policy 1), to a maximum hold of only $\frac{1}{2}$ hour and no minimum-train-length restriction (policy 8).

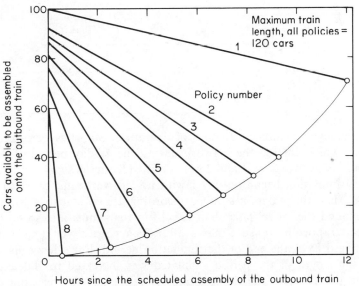

Figure 19.5 Dispatching policies examined.

These eight policies are tested under three sets, or "cases," of arrival-time distributions of the "independent" traffic from the A and C nodes (Table 19.2). The mean of the arrival times has in all cases been measured relative to the scheduled departure times of the outbound trains (Table 19.3). For case II, the mean arrival time of the input traffic is the same as for case I, but the standard deviation has been reduced from 4.0 to 2.0 hours. For case III, the standard deviation of the arrival-time distribution is 2.0 hours, but now the mean arrival time is 2 hours earlier than in cases I and II. Also, for case III, the mean of the distribution of the train's delay over the link B-D was reduced 2 hours, so all traffic was scheduled to arrive at yards B and D 2 hours earlier than for cases I and II.

Table 19.2 Characteristics of the Distribution of the Arrival Times of the Traffic
from the *A* and *C* Nodes at the Classification Tracks at Yards B and D

| | Characteristic of the Distribution | | |
Case	Mean (hr)	Standard Deviation (hr)	Type
I	−2.0	4.0	Normal
II	−2.0	2.0	Normal
III	−4.0	2.0	Normal

Table 19.3 Scheduled Departure Times of Trains from Yards

| Trip | | Scheduled |
From	To	Time (hr)
B	D	0
D	E_1	0
	E_2	4
	E_3	8

   The analyses simulated 21 days of operation. Statistics were recorded
for traffic which entered the network during the 16-day period of days 3
through 18 inclusive. Days 1 and 2 were omitted from the summary statistics
since the simulation began on day 1 with no leftover traffic at either yard
B or D. Also, the program stopped recording traffic statistics for traffic
which entered the model later than day 18 in the simulation, so no traffic
would be "trapped in transit"; that is, all traffic recorded would have moved
through yard D by the end of the simulation period. For a few runs of the
model, the length of the period simulated was increased to 100 days, to
minimize the effect of the random-number sequences for selecting traffic
levels and arrival times. Strictly speaking, of course, the model simulates
units of time that could be half a day as well as anything else besides a day.
   The specific network simulated for these analyses includes three *A* nodes,
two *C* nodes, and three *E* nodes (Figure 19.6). Each link in the network is
scheduled to have one train per day. An average of 30 cars per day go from
A_j to B, 90 cars per day on all other links. As there are three E_n destinations
for each of the three A_j and two C_k nodes, there are 15 distinct groups or
blocks of traffic on the network. The A_j to E_n blocks are of an average daily
length of 10 cars, and the C_k to E_n blocks are 30 car lengths on average. The
analysis simulated the movement of an average of 270 cars per day, that is,
of 4320 cars for the 16-day period.

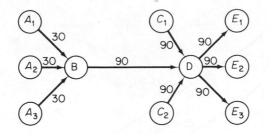

Figure 19.6 Specific railroad network analyzed and its flows in cars per day.

Yard Statistics

The simulation model can also generate statistics on the activities at the intermediate yards, B and D. One interesting statistic is the number of trains canceled under each of the policies (Table 19.4). Since the analysis assumed that crews and locomotives were always available, the only reason for a cancellation would be insufficient railroad cars to constitute the minimal required length of the train. As we would expect, the number of cancellations decreases as the minimum allowable length of the train decreases for higher policy numbers.

The average lateness of the trains in leaving the yards is also shown in Table 19.4. As would be expected, the average lateness is smaller for policies which adhered closer to schedule, and as the distribution of arrival time becomes tighter.

Table 19.4 Statistics on the Trains Made Up at Yards B and D

Polioy Number	Minimum Train Longth (cars)	Maximum Wait (hr)	Cancellations (%)		Average Lateness (hr)	
			Case I	Case III	Case I	Case III
1	70	12	15.6	10.9	1.9	0.3
2	40	9.2	7.8	6.2	1.6	0.3
4	24	7	3.1	1.6	1.0	0.2
6	8	4	0	0	0.5	0.1
8	1	0.5	0	0	0.2	0.1

Reliability Performance

A first measure of the performance of the network is the weighted mean standard deviation (WMSD) of the time the traffic leaves yard D for each of

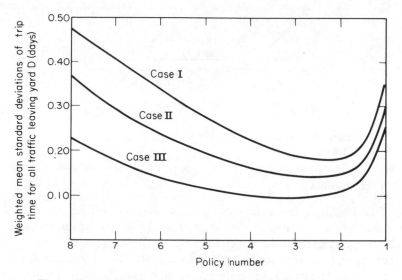

Figure 19.7 Weighted mean standard deviations in trip time vs. dispatching policy.

the traffic blocks on the network. Since there is one train per day scheduled over each link, a car leaving yard D can be described as 0 days late, 1 day late, 2 days late, and so on, depending on how many connections the car missed at either B or D. To calculate the WMSD for each run of the simulation model, one estimates the standard deviation for the travel times through the network of each of the 15 traffic blocks leaving yard D, and then weights the standard deviation for each block by the number of cars in the block. To check the results for different random number sequences used in the simulation, the effect of each policy was simulated several times, and an average WMSD was obtained.

The WMSDs for the eight different policies are shown in Figure 19.7 for each of the three cases of arrival-time distributions. As can be seen, policies 2, 3, and 4, which are fairly long-hold policies, resulted in the lowest weighted mean standard deviations for these cases. They also had the lowest mean lateness for all cars leaving yard D. As the dispatching policy shifts toward a no-hold policy, that is, as one moves toward policy 8, the difference between the WMSDs of the three different cases increases. Since we have assumed that a train's arrival time is the time it arrives on the classification tracks, one would indeed expect that the network performance of cars with "tight" connections would be sensitive to the arrival-time distribution of incoming traffic.

Missed Connections

Another measure of network performance is the percent of "unsuccessful" connections of all traffic. For our network, a car from an A node has two scheduled connections to make in the network, one at yard B and one at yard D, whereas a car from yard C has only one. Thus, for this performance measure, the traffic from the A nodes, since it is counted twice, weights equally with the traffic from the C nodes, which is twice as large.

The percent of unsuccessful scheduled connections from the different policies under each case of arrival-time parameters is shown in Figure 19.8. The results are similar to those obtained by considering the WMSD; policies 2, 3, and 4 giving the lowest percentages of unsuccessful connections. Also, as the arrival-time distribution gets better, that is, as one moves from case I to case III, the percentage of unsuccessful connections decreases for each policy.

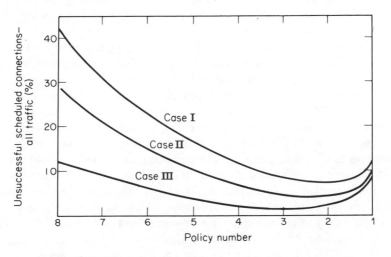

Figure 19.8 Percent unsuccessful scheduled connections vs. dispatching policy.

Interpretation of the Results

Several results appear to be somewhat counterintuitive to what might have been expected. The policy of holding trains for substantial amounts of time, policy 1 for example, did not actually result in large average departure delays of the trains made up under these policies, contrary to what one might have expected. In fact, seldom was any train held more than 4 hours. These rather short departure delays can be attributed both to the distributions of arrival time used in the analysis and to the presence of cars which miss their connections. The probability of holding a train for a long period of time, which, in effect, is the probability that several trains simultaneously are very late, is quite small for the particular distributions of arrival time used in the analysis. Moreover, if there is significant leftover traffic from the previous day at any yard, the outbound train for the following day might not be canceled, even if one or two of the inbound trains today were very late. This is because the model assumes that if a car misses its scheduled outbound train one day it is available for an on-time departure the next. Situations with high numbers of missed connections led to train departures closer to the scheduled departure time than the policy might suggest.

Another interesting result, somewhat counterintuitive, concerns the policies which had very short minimum trains, specifically, policies 5 through 8. Although their minimum train lengths were 20 cars or less, their actual lengths were seldom less than 50 cars. Again, this is a result both of the arrival-time distributions and the amount of leftover traffic. For policy 8, the strictest "no-hold" policy, 40 or 50 cars were often left over on any day, which ensures that the next day's train is at least 40 or 50 cars long.

Other results were more or less expected. The policy of never holding an outbound train, policy 8, was not a good policy for cars with "tight" connections and large standard deviations in arrival times. As the mean arrival time became earlier and its standard deviation was decreased, case III, there was a sharp drop in the number of missed connections under policy 8. Conversely, for the policies where the connections were not so tight, the differences in the number of missed connections for the different distributions of arrival times were much smaller. The performance of the long-hold policy, policy 1, was not significantly affected by differences in the distributions of arrival times used in the analysis. Most of the missed connections under policy 1 were the result of train cancellations, which occurred when less than 70 cars were ready to go within the holding-time limit of 12 hours. Since the analyses only examined differences in the distributions of the arrival time and not of the traffic level, and since the holding-time limit was long enough to allow even the very late arrivals to make the outbound train, there was little difference among the three cases of arrival times under policy 1. In short, although significant differences were noted in the number of missed connections for the different policies, these differences became much less substantial as one "improved" the arrival-time distributions.

Alternative Traffic Generation Processes

Before using the results of the analysis in forming operating policies, one should carefully examine the assumptions embedded in the model. In this case, one should particularly consider the effect of the procedure used to generate traffic from the A_j and C_k originating modes. The results discussed so far were obtained on the assumption that the distribution of the traffic levels and arrival times for the traffic from A_j and C_k was independent of the dispatching policies in effect at the intermediate yards, B and D. One could, alternatively, choose the traffic from A_j and C_k according to "reflective" distributions which reflect the type of dispatching policies in effect at the intermediate yards.

Analyses were performed using "reflective" data. Compared to the results using "nonreflective" data which we have already discussed, the reliability of the transit time changed very little for the hold policies, and improved slightly for the no-hold policies.

CONCLUSIONS

The analyses of train dispatching policies and arrival-time distributions have important implications on real-world operating strategies. For example, the results indicate the benefits, in terms of reliability, of improving arrival-time distributions of incoming traffic at yards. Also, the results show the different consequences, in terms of missed connections and train cancellations, of following too severe a dispatching policy; that is, never waiting for late traffic, or always waiting for late traffic.

As mentioned previously, the results presented here show only a small fraction of the many different applications of the model. They do, however, demonstrate the feasibility of using simulation to test rather complicated operating policies for a railroad network.

ACKNOWLEDGMENTS

This work was carried out under the auspices of the Transportation Systems Division of the Civil Engineering Department at MIT, and was performed within the context of the Railroad Service Reliability Project with the support of the Federal Railroad Administration of the U.S. Department of Transportation. Special thanks are due to A. S. Lang and to project monitors John Williams and John West for providing valuable assistance with respect to the overall goals and strategy of this research.

REFERENCES

1. BECKMANN, M., McGUIRE, C. B., AND WINSTEN, C. B., 1955. *Studies in the Economics of Transportation*, New Haven, Conn.: Yale University Press.
2. BELOVARAC, K. J., AND KNEAFSEY, J. T., 1972. *Determinants of Line Haul Reliability, Studies in Railroad Operations and Economics*, Vol. 3, Cambridge, Mass.: Department of Civil Engineering, MIT, Research Report R72-38.

3. FOLK, J. F., 1972. *A Brief Review of Various Network Models, Studies in Railroad Operations and Economics*, Vol. 7, Cambridge, Mass.: Department of Civil Engineering, MIT, Research Report R72–42.
4. FOLK, J. F., 1972. *Models for Investigating Rail Trip Time Reliability, Studies in Railroad Operations and Economics*, Vol. 5, Cambridge, Mass.: Department of Civil Engineering, MIT, Research Report R72–40.
5. FOLK, J. F., 1972. *Models for Investigating the Unreliability of Freight Shipments by Rail*, Ph.D. Thesis, Department of Civil Engineering, MIT, Cambridge, Mass, June.
6. KOLSEN, H. M., 1968. *The Economics and Control of Road-Rail Competition*, Sydney, Australia: Sydney University Press.
7. MARTLAND, C. D., 1972. *Rail Trip Time Reliability; Evaluation of Performance Measures and Analysis of Trip Time Data, Studies in Railroad Operations and Economics*, Vol. 2, Cambridge, Mass.: Department of Civil Engineering, MIT, Research Report R72–37.
8. REID, R. M., ET AL., 1972. *The Impact of Classification Yard Performance on Rail Trip Time Reliability, Studies in Railroad Operations and Economics*, Vol. 4, Cambridge, Mass.: Department of Civil Engineering, MIT, Research Report R72–39.
9. SILLCOX, H., 1971. Stand-by and Second Best, paper presented before the Graduate School of Business Administration, Harvard University, Boston.

PART IV

Evaluation

20

Introduction

Since around 1965, the techniques and even the conceptual basis for evaluation of public projects have been undergoing fundamental change. The procedures that are now being developed by research and are being adopted by industry and government, are significantly different from the benefit–cost analyses that have been standard in the United States. The cases in this section have been selected to illustrate the potential and drawbacks of these new approaches. We have also attempted to abstract from their experience to suggest a desirable process of evaluations for the future.

The new procedures reflect three streams of development which coincidentally converge at present. First, there is the theory of multiobjective evaluations. As of mid 1973, this procedure was under consideration for official adoption by the U.S. Water Resources Council, an organization which sets the pattern for the U.S. Army Corps of Engineers, the Bureau of Reclamation, and other developmental agencies in the United States. This approach may be said to derive from the realization, discussed in detail subsequently, that, since it is not possible to define a single overall measure of value to society, it is preferable to consider the different impacts of any system separately. The use of multiobjective evaluation is illustrated by the cases in Chapters 21 and 22.

Decision analysis is the second major development. This is a procedure which, first of all, systematically incorporates explicit notions of the nature of risk into the evaluation. Specifically, it uses precise assumptions about the probability distribution functions instead of, as has been usual, an arbitrary risk premium. Most importantly, however, the theory of decision analysis has led to explicit ways of assessing multidimensional utility functions. Although there is so far hardly any experience in the use of these analytic functions and it is not clear that they work, they represent the kind of information about preferences that is essential to a real multiobjective analysis. Chapters 23 and 24 illustrate the use of decision analysis in systems planning.

Finally, there has been a growing realization among systems planners that evaluation is not only a process, but a process in which a formal index of evaluation may not play a dominant role. Also, many are convinced, on what appears to be reasonable evidence, that it is even inappropriate if not wrong to think that anyone should use an analytic index to impose a solution in what is really a political process. Furthermore, since there may not even be a single decision maker, it may not even be possible to impose an analytic solution. Chapters 25 and 26 illustrate this point of view. As a conclusion, Chapter 27 reviews the recent changes in values concerning public projects in the United States, and provides a descriptive model of how society exercises control over new systems.

TYPOLOGY OF EVALUATION TECHNIQUES

To place the recent developments into context, it is useful to classify evaluation techniques according to the kinds of issues they take into account. One possible typology is illustrated in Table 20.1. This table distinguishes

Table 20.1 Typology of Evaluation

Type	Non-linear	Includes Risk	Multi-dimensional	Many Decision Makers	Issues
I	No	No	No	No	Discount rate
II	Yes	No	No	No	Nonlinear value
III	Yes	Yes	No	No	Utility function
IV	Yes	Yes	Yes	No	Multiattribute
V	Yes	Yes	Yes	Yes	Welfare economics

Attributes of Evaluation spans the Non-linear, Includes Risk, Multi-dimensional, and Many Decision Makers columns.

five possible cases of increasing complexity. Each higher-order type incorporates a new element, such as risk or nonlinear utility, which was not included in the lower-order case. Conversely, each lower-order type is included, as a simplified case, in the higher-order types. Given the attributes of the evaluation which are singled out for consideration, it is evidently possible to think up other cases that would represent different combinations of the attributes from the ones illustrated in Table 20.1. The particular types shown, however, are especially interesting because they appear to represent the actual chronological development of increasingly sophisticated techniques of evaluation. Following what thus appears to be the sequence of developments, each type of evaluation is described in turn below.

Type I: Standard Benefit–Cost Analysis

This is the simplest case. It represents the standard benefit–cost evaluation procedures in which future costs and benefits are discounted to some common point in time, and compared. The doctrine of this, the traditional approach of engineering practice, is embodied in what is known as engineering economy.

As indicated in Table 20.1, the standard benefit–cost procedure assumes:

1. That the value, V, of a benefit, B, or cost, C, increases linearly with the amount of benefits or costs at any time. This implies that a project with 10 times the amount of any benefit, say, acres of wilderness preserved, is 10 times as valuable in that regard.

2. That notions of probability need not be incorporated explicitly and that it is appropriate to use expected values. This is reasonable as long as values are, in fact, linear in the amount of benefits.

3. That there is but one dimension of benefits and costs or, more precisely, that all the several dimensions can be justifiably collapsed and measured in one dimension. Typically, money is taken to be the measure of all things. If it is not feasible or practical to quantify a benefit or cost, such as an aesthetic one, it does not even get considered.

4. That there is but one decision maker or, more precisely again, that all parties to the decision are agreed upon a single criterion of evaluation, typically that of maximizing profit. This assumption is reasonable so long as all groups accept that it is meaningful to measure all benefits and costs, such as loss of life and acres of wilderness, on a common basis and with the same weights on each kind of benefit and cost.

Each and all of the above assumptions may be acceptable in a wide range of circumstances. They are almost certainly justified for relatively small projects such as, for example, the choice of whether to build a bridge out of steel or reinforced concrete at a given site. So this discussion by no means intends to suggest that designers ought to disregard the lessons of engineering economy. These techniques ought to be used whenever the assumptions on which they are based are valid. For the more complex situations, the other types of evaluation should be used.

The principal issue concerning the use of a type I evaluation is, once its assumptions are accepted, the choice of the discount rate. The discount rate, typically given in terms of percent per year, is the measure by which it is possible to compare benefits and costs that occur at different points in time. From approximately 1950 to 1973, the general practice in the U.S. government has been to assume a discount rate of between 3 and 5%. Exceptions to this rule do exist; some agencies have used discount rates of up to 10%, and many assume that it does not exist or is zero. As part of the general revision of evaluation procedures, to be described in more detail below, this rate is scheduled to be raised to 7%, which is much closer to the internationally accepted figures of between 8 and 12%.

The effect of these higher, more accurate, discount rates is to discourage capital-intensive projects such as large dams, and to encourage the imaginative use of operational strategies for developing resources, as by the use of irrigation projects. In particular, for instance, the higher discount rates encourage planners to solve problems caused by peak loads on a system by relatively expensive but short-term methods rather than by building capacity which will remain idle most of the time.

Type II: Consumers' Surplus

Next in our typology we come to the evaluation procedures which recognize the nonlinearity of the values in terms of benefits and costs. This real value of any benefit is known as its utility, and the utility function describes the real value of the benefits. The nonlinearity of the utility function, which contradicts the first assumption on which the standard benefit-cost procedures are based, is a pervasive phenomenon. To cite a homey example: one's utility for the first plate of food is fairly high when one is hungry, but the utility for successive plates diminishes as one becomes sated until, finally, it may even become negative. Similarly, a traveler may be extremely anxious to save 10 minutes on his way to the airport (so he may catch his plane) and relatively unwilling to spend much money to save much more time. As a general rule, both individuals and the public have nonlinear value functions and, specifically, a diminishing marginal utility for benefits.

When a person does, in fact, have a diminishing marginal utility for benefits, it can be presumed that the value of what he receives is greater than the cost. Indeed, as suggested by Figure 20.1, he would be likely to demand more of a good until its utility, at the margin, was equal to its cost. This would occur at Q^* in the figure. Since the person's marginal utility is diminishing as the quantity of a good increases, which implies a demand curve sloping toward the southeast, it follows that his utility or value for less than Q^* of a good is greater than its price. The difference between the ultility and the price, summed over all quantities used, is known as the consumers' surplus.

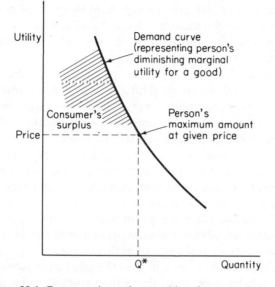

Figure 20.1 Consumers' surplus resulting from nonlinear value function with diminishing marginal utility.

The type II evaluation procedures attempt to incorporate the consumers' surplus into the measurement of benefits. Basically, this approach recognizes that benefits often have a real value much greater than their price. It then calculates benefit–cost ratios using these higher values. The approach can certainly represent an advance over the standard benefit–cost analysis, and has been used as such to a considerable degree in England, especially since the early 1960s. An obstacle to using this approach lies in the difficulty of estimating the demand functions for any good. (See the discussion in Part II, especially Chapter 5 and case studies of Chapters 10 and 11).

Type III: Decision Analysis

During the last few years, it has become possible to construct analytic representations of a person's, specifically a decision maker's, utility with regard to risk. This has led to the development of decision analysis as a formal tool for evaluation. This approach to evaluation, type III, has been increasingly used since its introduction into practice in the middle and late 1960s.

The essential element of decision analysis, from the point of view of this discussion, is that it embodies procedures to quantify any individual's own utility over risk. Like the utility functions used in the type II evaluations, which concern an individual's utility over quantity, these utility functions are usually nonlinear. Unlike the utility functions over quantity, however, the utility functions over risk are not expressed in terms of common units, such as money, which different groups might be willing to pay for any speci-fied number of goods. The utility functions over risk are, rather, expressed in relative units valid for an individual and not comparable to those of others. (Formally, the utility functions are expressed on a special form of cardinal scale, the ordered metric, and are valid for any positive linear transformation.) The procedures for assessing utility over risk have been extensively validated: they not only lead to consistent descriptions of preference structures, but also accurately predict the choices that will be made.

The process of decision analysis is conceptually simple. First, all possible sequences of decisions and their consequences are laid out. Since there can be several choices at any stage, and since each choice may branch into several consequences, it is common to speak of this representation as the decision tree. Second, all possible outcomes are indicated together with the a priori probability of occurrence. Next, the utility function of the decision maker is assessed, and the utility or real value of each outcome is calculated. Finally, the optimal choice at each stage, and thus the optimal sequence of choices, is calculated on the basis of maximizing the expected value of the utility. This process has been shown to work quite successfully in practice.

To date, most uses of decision analysis have been focused on the situa-tion where there is but one dimension to the set of outcomes, for example, when all consequences can be measured in monetary terms as profits or losses. Such situations most often occur in business and, indeed, decision-analytic procedures for evaluation are being rapidly adopted by modern management.

Type IV: Multiattribute Analysis

In a great number of situations, it is extremely difficult to visualize how some of the consequences of any choice should be compared. What, for example, would be the most satisfactory way to compare the several dimen-sions of alternative plans for a new highway: cost, number of lives lost in accidents, loss of environmental quality? There is no doubt that we can, as individuals, compare such projects because man always has made choices between alternatives which had such diverse consequences. The issue is that the obvious analytic procedures for comparing these different consequences do not seem satisfactory. Specifically, many people feel unhappy with the usual procedures that assume that it is reasonable to add up the separate values a group might have for different items. Such a procedure presumes that one's feelings about, or preferences over, different attributes are independent.

This is often contrary to our personal experience: our desire for one object does, in general, depend upon our level of satisfaction on other dimensions. A person may not care much about stereophonic sound when he is poor and hungry, for example, but care very much once he has a job and money.

Multiattribute evaluations, type IV, attempt to account for the non-linear, nonadditive nature of any individual or group's utility function. As indicated in the suggestions for further reading which follow, recent developments demonstrate that it is now possible to assess an individual's real preferences over several attributes. This can be done by making minimal assumptions about the nature of the utility functions. Once the multiattribute utility function is encoded, it can be used in the evaluation process just like a single-dimensional utility function. The multiattribute evaluations simply use multiattribute utility in the decision analysis. These procedures are just beginning in the early 1970s to be applied to practical problems.

Decision analysis, either single or multiattribute, is a powerful extension of the traditional procedures insofar as it explicitly and systematically accounts for risk. Because it inherently focuses on the preferences of an individual, it is best suited to situations in which an individual does make the choices and is, in fact, the decision maker. By extrapolation, it can also be used when the individuals are representative of a larger, homogeneous, and like-minded group which will make the choices. Such situations are most often found in the private sector, especially in business. For those groups it may not be too much to suppose that their members are in general agreement on the objectives and their relative desirability; if they were not, they might leave and join other groups.

Conversely, decision analysis is not directly suitable for situations in which individuals, and the like-minded groups they may represent, are in conflict. Such situations are most likely to occur in the public sector, where many different interests must jointly agree upon a common objective. Decision-analytic procedures can then be used to explore the desires of the several groups, but it is probably inappropriate to use them to determine what decision ought to be made; after all, whose preferences would one use? For public decisions, it might then be best to think of using the techniques of decision analysis in what might be better called a preference analysis of each of several groups concerned with an issue.

Type V: Multiobjective Evaluation and Negotiation

The techniques of multidimensional evaluation are the most recent set of evaluation procedures to enter the process of implementation. These procedures attempt to lay out explicitly the preferences of the different groups concerned with a project for the sets of possible consequences. In this way, they intend to allow the analyst to estimate what choices are preferable to the several groups, and how differences might be resolved. The existing procedures do not, however, describe any particular procedure for achieving a compromise.

As currently defined by the governmental agencies that intend to use multiobjective evaluation, the procedure involves two analytic functions. First, the maximum levels of attainment along the several dimensions or attributes of the consequences are calculated. This defines what is known as either the production possibility frontier or, in the language of multiobjective evaluation, the transformation curve. An example of how this is done is given in Chapter 21. Second, the analyst is supposed to describe the indifference, that is, the isoutility curves of the several parties concerned with a project. These two functions are then combined in the multiobjective evaluation. This procedure, as illustrated in Chapter 22, does not embody a formal analytic method for defining which choices are best and should be made. Multiobjective analysis, in what is a radical departure from previous evaluation procedures, does not attempt to impose or prescribe a technocratic solution on public choice. Multiobjective evaluation recognizes that questions of public choice are ultimately ethical questions which, in a representative society, may most appropriately be left to the political process.

It is important to note that the existing procedures of multiobjective evaluation do not propose clear, analytic methods for determining the preferences of any group. In fact, the existing literature indicates that the practitioners of multiobjective evaluation are essentially unaware of the important developments that have occurred in the multiattribute assessment of utility functions. This may be due to the fact that multiobjective evaluation has been developed largely by economists, who had no particular reason to be aware of the significant advances being made in psychophysics and operations research. The result, in any case, is that current versions of multiobjective analysis could be significantly improved by incorporating the procedures for assessing multiattribute utility functions. This development is certain to occur soon, since the need is obvious and the procedures for accomplishing it have already been validated.

A comment on the typology of evaluation is in order here. Since the type IV evaluation procedures incorporate the advanced procedures for assessing preferences or utility functions, which current multiobjective or type V evaluations do not, they are in some sense, at least, more sophisticated than the latter. It might seem that the sequence between types IV and V is out of order. Our preference for maintaining the sequence proposed in Table 20.1 is based on the belief that the consideration of many decision makers encompasses the type IV evaluation, which focuses on a single decision maker. We recognize, however, that the current versions of multiobjective evaluation do not meet our definitions. But we have every confidence that, since techniques for assessing multiattribute utility functions are available, current versions of multiobjective evaluation will evolve rapidly toward our definition.

That multiobjective evaluation does not define the best alternative, but rather leaves this selection to judgment, is not a defect. It is by now a well-demonstrated proposition of welfare economics that it is impossible either to compare the utility of different groups on an absolute scale or, consequently, to define a single overall utility function valid for all groups. Since it is not possible to define what this social welfare function should be, neither is it possible to define what the overall social optimum should be. At most, one might identify what society—that is, its component interest groups—would agree to do. Multiobjective analysis recognizes this real limitation to analytic knowledge and wisely does not attempt to impose a choice upon the diverging interest groups concerned with a project.

Multiobjective evaluation can, however, be improved. Not only should it incorporate multiattribute utility functions, as indicated previously, but it should also probably attempt to adopt relevant principles of game theory so that we can describe what choices society will make. This is different, of course, both conceptually and in fact, from what choices they perhaps ought to take. The ability to define what choices the public would prefer would enable systems planners to spend their detailed efforts on designs which will have the greatest probability of acceptance. One only needs to consider the enormous amount of energy that has been wasted in recent years on urban and regional plans which have been rejected by the public to appreciate the desirability of being able to predict the outcomes of the eventual negotiation process that takes place over a large-scale systems design. Equally, of course, it would be desirable to define the possible results of different strategies for negotiation. As there are not, as yet, any effective ways to predict the outcome of negotiation in practice, it is only possible to speculate on how this might be used. We believe that this may eventually prove to be one of the most interesting areas of research on evaluation.

ADDITIONAL READING

A comprehensive introduction to the several types of evaluation discussed above is provided by

DE NEUFVILLE, R., AND STAFFORD, J. H., 1971. *Systems Analysis for Engineers and Managers*, New York: McGraw-Hill Book Company, especially Chapters 6, 7, 8, 10, and 11.

With regard to type I evaluation procedures, two recent editions of standard books on the practice of engineering economy are those of

DeGarmo, E., 1967. *Engineering Economy*, New York: The Macmillan Company, 4th ed.

Grant, E., and Ireson, W., 1964. *Principles of Engineering Economy*, New York: The Ronald Press, 4th ed.

And the issue of the choice of discount rate is succinctly presented by

Baumol, W. J., 1968. On the Social Rate of Discount, *American Economic Review*, Vol. 58, No. 4, pp. 788–802, Sept.

A less formal, and perhaps more readable, version of this was given as

Baumol, W. J., 1969. On the Appropriate Discount Rate for the Evaluation of Public Projects, in H. H. Hinrichs and G. M. Taylor, eds., *Program Budgeting and Benefit–Cost Analysis*, Pacific Palisades, Calif.: Goodyear Publishing Co., Inc.

Since much of the use of type II evaluations has been confined to government, especially that of Great Britain, not too many detailed accounts of its use are accessible. One which should be fairly easy to obtain, however, is

Beesley, M. E., 1973. *Urban Transport: Studies in Economic Policy*, London: Butterworths.

A comprehensive review of the traditional evaluation procedures appears in

Prest, A. R., and Turvey, R., 1966. Cost–Benefit Analysis: A Survey, in *Surveys of Economic Theory*, Vol. III, New York: St. Martin's Press.

A readable presentation of decision analysis and of the techniques for assessing single dimensional utility functions is to be found in

Raiffa, H., 1968. *Decision Analysis: Introductory Lectures on Choices Under Uncertainty*, Reading Mass.: Addison-Wesley Publishing Company, Inc.

The procedures for assessing multiattribute utility functions, needed for a type IV, as well as a type V, evaluation, are presented in

Keeney, R. L., 1971. Utility Independence and Preferences for Multiattributed Consequences, *Operations Research*, Vol. 19, No. 4, pp. 875–893, July-Aug.

With regard to multiobjective evaluation, type V, the most cogent descriptions of the theory and proposed practice have been presented under the aegis of the U.S. Water Resources Council. As of this writing, the best available reference is probably

U.S. Water Resources Council, 1971. Proposed Principles and Standards for Planning Water and Related Land Resources, Vol. 36, *Federal Register*, 24144–24194, Washington, D.C., Dec.

It is to be expected, however, that more refined versions of this statement will be available in the years ahead.

Finally, there are essentially no descriptions of the practical procedures for using a type V or negotiated kind of evaluation of public projects. A rich theoretical literature is available covering both detailed points and overall concepts. One interesting reference that may be useful as an introduction is

HOWARD, N., 1971. *Paradoxes of Rationality: Theory of Metagames and Political Behavior*, Cambridge, Mass.: MIT Press.

One attempt to suggest the implications of game theoretic approaches to planning problems is given by

DE NEUFVILLE, R., AND KEENEY, R. L., 1973. Multiattribute Preference Analysis for Transportation Systems Evaluation, *Transportation Research*, Vol. 7, No. 1, pp. 63–76, March.

SUGGESTED PROCEDURES FOR EVALUATION

Based upon our understanding of the state of the procedures available for evaluation of systems, we suggest a three-stage process:

1. To represent the full set of alternatives open to the planner, together with their possible consequences.
2. To explore the preferences of the different groups that are associated with a choice of projects.
3. To suggest, within the limits of possibility, the possible outcomes of the interaction of the technical opportunities and of the values of society.

Each of these steps may be either quite complex or quite simple. This would depend on the nature of the problem and of the assumptions, as to the number of attributes or objectives or as to risk and linearity of the utility function, that may be suitable. We would propose that, before selecting a particular type of evaluation, the analyst assess which assumptions it might be reasonable to make. Then he can determine how detailed he must be in each step of the problem.

CASE STUDIES

The case studies on evaluation were selected to illustrate the problems and possibilities of the new approaches for system evaluation: single and multiattribute decision analysis, and multiobjective evaluation. Also, a case is included to illustrate and emphasize that decision making is often not a single act but a process. Finally, the section concludes with a proposal for what an evaluation process might look like in practice.

The first case in Chapter 21 describes in detail two procedures for evaluating the transformation curve. The example that is worked out is actually a small scale version of a real project carried out at MIT for the Government of Argentina. The authors were responsible for executing the actual multiobjective analysis for that project. Unfortunately, it was not possible to release the detailed results of that project because of contractual obligations with Argentina. The methodological results are, however, essentially the same.

The case study in Chapter 22 illustrates how transformation curves, such as the ones generated in the previous case, can be combined with indifference curves to suggest regions of optimal design. The case also indicates the difficulties this method, as now defined, has in estimating utility or preference functions, and in describing possible compromise situations. Current procedures lead to a solution only in the simple case, where all parties fortuitously agree on a common plan.

The analysis in Chapter 23 introduces the procedures of decision analysis. Here, a one-dimensional approach is used with a continuous definition of the possible consequences of any choice. The problem is that of designing fill for a swampy area. The outcomes are defined by the settlements that take place, and these are determined by what the compressibility index turns out to be. In this example the emphasis is on the use of the entire procedure, and a quite simple utility function is used.

The study in Chapter 24, on the other hand, focuses on the use of a multiattribute assessment of the utility function. This case illustrates the first time a multiattribute utility function was assessed for the purpose of evaluating a possible public project. This study is interesting in that it illustrates both the power of the analysis and the relative ease with which multiattribute utility functions can be assessed, and it suggests the limitations of decision analysis for defining the best outcome for a situation in which there are several conflicting interest groups. The complementarity of multiattribute preference analysis and multiobjective evaluation then becomes apparent.

The description of IBM's decision to proceed with the development of the System 360 series of computers, a set of decisions which eventually involved $5 billion, shows how decisions about systems actually unfold. The point of this case, Chapter 25, is not methodological like the others. It is simply to suggest that, frequently, the evaluation of a plan for a system neither depends upon an analysis nor culminates in a single decision. Rather, the evaluation and selection is a continuous process of negotiation, in which analysis may play an important role but one that is supportive rather than determining.

The process of evaluation proposed in Chapter 26 suggests one way in which the evaluation process might best be conducted. Manheim's proposals essentially reject almost any algorithm quantifying and comparing benefits and costs. This attitude recognizes that often, especially with regard to highway projects in the United States but also elsewhere, the benefit–cost procedures have been used to try to impose a technocratic solution where, because of the nonlinearity of the utility functions and their differences between groups, no technocratic solution was appropriate. Manheim suggests instead a range of procedures, labeled community interaction, for inquiring after social values and for encouraging detailed negotiations between interest groups. His procedures involve extensive hearings, working group meetings, and general education of the public, and he uses the results to draw out an understanding of the feelings of society toward major choices.

Manheim's proposed process is actually similar to the procedures we have suggested which would combine the capabilities of multiattribute preference analysis and multiobjective evaluation. The principal differences lie in the degree of interaction with the community that the planner is advised to maintain. Manheim's proposals are time-consuming and expensive, and are designed to overcome the limits of less involving procedures which may obtain responses only from a small group of persons or only about a few alternatives and attributes. Whether these limits are serious, and whether it is worth while to spend the extra effort over simpler, more analytic procedures is an empirical question.

The trends we see suggest a confluence of all three approaches discussed in this part: decision analysis, multiobjective evaluation, and community participation. We see a movement toward a more formal multiobjective evaluation which incorporates the assessment of multiobjective utility functions and an explicit analysis of risk. We also see a concurrent movement toward a greater involvement of the community in delineating objectives and preferences by a variety of modes. We anticipate, in short, a grand synthesis of the approaches into a body of procedures that will enable the planner or designer to analyze the options available, and to present relevant information about the choices open to the interested groups, and to suggest areas of possible compromise on a choice.

In addition to the influence of analytical procedures for choosing preferred public projects, there is also, in the United States, a significant change in the values that are preferred by American society. There appears to be a pervasive shift toward environmental objectives and, consequently, away from purely economic efficiency. These changes, and the associated actual developments in decision-making processes are reviewed in Chapter 27. The chapter and the section then concludes with a suggested descriptive model of the interactive, dynamic process of how society controls decisions about new projects.

21

Multiobjective Analysis in Water Resource Planning*

JARED L. COHON AND DAVID H. MARKS

This paper discusses an application of multiobjective theory to the analysis of development alternatives for a large-scale river basin. The problem of choosing how to allocate a scarce public good, such as water, to many different uses, and the complexity of water resource systems make the analysis of such investment alternatives quite difficult. These difficulties have given rise to the application of systems analysis techniques to river basin planning problems. However, the applications to date, most of which are based on the single objective of maximization of economic efficiency, fail to capture the multiobjective nature of water resource planning problems. Such important considerations as regional development and environmental quality represent other planning objectives that should be included in design models. The purpose of this chapter is to demonstrate how many planning objectives can be integrated into optimization models.

Using techniques such as mathematical programming and simulation, it is possible to formulate the problem of finding good development alternatives in planning a water resource system with many objectives. A typical river system represents a complex stochastic physical system embedded in an economic, social, and institutional framework which is difficult to model. Decision variables in the system range from deciding whether water resource investments should take place, to questions of where and when facilities such as reservoirs, diversions, power generation equipment, and irrigation sites

*Adapted from Water Resources Research, Vol. 9, No. 4, August 1973, pp. 333–340.

should be operated over time. The concept of trying to model some of these interrelated decisions analytically in the context of the physical and non-physical environment has been classified by Dorfman as a screening model (5). Such a model would be an optimization problem of the form:

Maximize net benefits = total benefits − costs

subject to continuity constraints on water flow

technological constraints on the alternatives

policy constraints

The authors who have presented models of this nature—Blanchard (2), Loucks (9), Maass et al. (10), Poblete and McLaughlin (11), Rogers (12), Smith (14), and Wallace (15)—have generally used a linear programming format. With one exception (12), all the objective functions used were uni-dimensional in that they considered only the economic efficiency objective: maximization of net national income. As mentioned, however, most public investment problems are multiobjective in nature and it is important that screening models reflect this.

The main requirements for the consideration of multiobjective water resource problems in screening models are the theory of multiobjective analysis and a solution technique for a linear programming problem with a multiple- or vector-valued objective function.

MULTIOBJECTIVE ANALYSIS

The theory of multiobjective analysis is presented in more detail in the paper by Major appearing as Chapter 22. For our purposes a hypothetical example is discussed to highlight the intuitive content of the theory and to provide motivation for multiobjective linear programming.

Consider a project, such as has been proposed in the State of Alaska, to build a pipeline for the transportation of crude oil from the frozen, oil-rich northern shore to the relatively warmer south shore. The oil deposits are huge and, if exploited, represent a substantial increase to national income. However, environmentalists claim that the construction of the pipeline to pump oil at 140°F would result in the melting of the frozen tundra. The pipeline would, thus, threaten wildlife by obstructing annual migrations.

Two planning objectives relevant to this problem are maximization of net national income benefits and the preservation of environmental quality. The former may be measured in dollars and the latter, say, by the number of animals of certain species existing in Alaska. Suppose that the animal population before construction of the pipeline is estimated at 100,000 head.

Suppose also that the oil companies have several alternatives in building the pipeline, for example the pipeline may be buried or placed above ground at various depths or elevations. Displaying the alternatives to show their contribution to each objective would give rise to an illustration such as that in Figure 21.1. In this case net national income benefits are measured in dollars

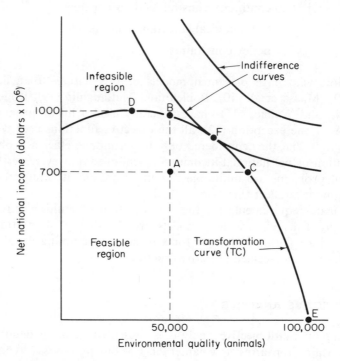

Figure 21.1 Graphical representation of a multiobjective analysis for the hypothetical example.

on the vertical axis, while the horizontal axis measures environmental quality in number of animals of selected species living in Alaska (the more animals, the more environmental quality). Each point in the space represents a pipeline alternative. For example, point A represents an alternative which contributes $700,000,000 to national income and preserves 50,000 animals. The curve labeled TC is the transformation curve of net benefits and represents the boundary of the set of feasible alternatives. That is, a point outside the transformation curve (to the northeast) is physically infeasible. All points inside the transformation curve (between the origin and TC) are feasible, but not as good as at least one point on the boundary. For example, point A is feasible but is worse than point B because national income benefits can be increased by moving from A to B without decreasing environmental quality as measured by the number of animals preserved. Similarly, point C is also better than point A by a similar argument. Note, however, that nothing can be said, a priori, about the relative advantage of B to C or vice versa until we are agreed upon the tradeoffs we are prepared to make between the different objectives.

The selection of the optimal alternative for the problem depicted in Figure 21.1 would be relatively straightforward if society's preferences or, alternatively, the tradeoffs society would accept as represented by the social indifference curves were known with certainty. As known from basic theory [see de Neufville and Stafford (4), for example], the optimal alternative would be a point on the highest indifference curve, or the one farthest to the northeast, which still gives a feasible solution. In Figure 21.1 the optimum point would be F, at which the indifference curve and transformation curve are tangent. The negative of the slope of the line of tangency at F, in this case

$$\frac{1,500,000,000}{150,000} - 10,000 \text{ \$/head}$$

indicates, for this hypothetical indifference curve, society's relative marginal social values for, or tradeoffs between, national income benefits and environmental quality at this point. In this case, society would have indicated that it is willing to give up one head of wildlife if it gets \$10,000 in national income in return. Or, similarly, it would have been willing to give up \$10,000 of national income to preserve one head of wildlife.

Unfortunately, preference information is usually sketchy and is almost always highly uncertain. However, even without indifference curves, the analytical approach represented in Figure 21.1 can still be of great value in the decision-making process. Rather than attempting to indicate point F as the optimal alternative we, as analysts, can provide the decision makers with the transformation curve and all the useful sensitivity and tradeoff information that it embodies. This is precisely the role of the multiobjective linear programming screening model. The theoretical characterization of such a model and two solution techniques are discussed below.

MULTIOBJECTIVE LINEAR PROGRAMMING

Multiobjective linear programming differs from linear programming because it must deal with more than one objective function. A basic linear programming problem is of the form:

$$\text{Max } Z = \bar{c}^T \bar{x}$$
$$\text{subject to } \bar{A}\bar{x} \leq \bar{b} \tag{21.1}$$
$$\bar{x} \geq 0$$

where Z is a scalar; $\bar{c}$ is an $n \times 1$ vector, and c_j, an element of $\bar{c}$, is in \$/unit of x_j; $\bar{x}$ is an $n \times 1$ vector of decision variables; $\bar{A}$ is an $m \times n$ matrix; $\bar{b}$ is an $m \times 1$ vector and b_i is in units of resource; and all quantities except the x_j and resultant Z are known. Naturally, in this and the subsequent theoretical discussions, we could equally well speak of a minimization instead of a maximization problem. A general form of the multiobjective problem with s objectives is, however:

$$\text{Max } \bar{Z} = \bar{C}\bar{x}$$
$$\text{subject to } \bar{A}\bar{x} \leq \bar{b} \qquad (21.2)$$
$$\bar{x} \geq 0$$

where all parameters are as before except that $\bar{Z}$ is an $s \times 1$ column vector and $\bar{C}$ is an $s \times n$ matrix. It is clear from an examination of Eqs. (21.1) and (21.2) that the two problems are identical except for their objective functions.

While the normal linear programming problem is easily solved using well-known techniques such as the simplex method, the multiobjective problem is not easily solved because $\bar{Z}$, which is a vector, cannot be directly maximized or minimized. Referring again to the oil pipeline example, a normal linear programming problem results if one of the two objectives is ignored. For example, we could obtain:

Max net national income benefits

subject to technological constraints

other constraints

and an optimum can be found (point D in Figure 21.1). When considering both national income and wildlife population, however, little can be said about the optimum when looking at the physical system without preference information. We can only say that this optimum lies somewhere on the transformation curve of net benefit. Mathematically, the problem is:

Max net national income benefits
net environmental quality

subject to technological constraints
other constraints

A problem such as this with two objectives is written mathematically as:

$$\text{Max } \bar{Z} = [Z_1, Z_2]^T = \bar{C}\bar{x}$$
$$\text{subject to } \bar{A}\bar{x} \leq \bar{b} \qquad (21.3)$$
$$\bar{x} \geq 0$$

where $\bar{C}$ is now a $2 \times n$ matrix and $\bar{x}$, $\bar{A}$, and $\bar{b}$ are defined as before. The constraint set defined by $\bar{A}\bar{x} \leq \bar{b}$ on n-space maps into the feasible region defined by $\bar{Z}$ on 2-space as shown in Figure 21.1. The boundary of the feasible region is the transformation curve, TC.

In single objective linear programming, the optimality of a solution is uniquely defined. $\bar{x}^* \in X$ is optimal when maximizing if:

$$Z(\bar{x}^*) \geq Z(\bar{x}) \qquad \text{for all } \bar{x} \in X, \ \bar{x}^* \in X \tag{21.4}$$

where X is the set of all feasible solutions, that is, solutions which satisfy the constraint set, and is defined as:

$$X = \{\bar{x} \mid \bar{A}\bar{x} \leq \bar{b}, \ \bar{x} \geq 0\} \tag{21.5}$$

In the multiobjective problem, the concept of optimality is replaced by the notion of "noninferiority." A solution, $\bar{x}_1 \in X$, is inferior if there is some solution $\bar{w} \in X$ for which:

$$\bar{Z}(\bar{w}) \geq \bar{Z}(\bar{x}_1) \tag{21.6}$$

$$\begin{aligned} \text{that is, } Z_1(\bar{w}) &\geq Z_1(\bar{x}_1) \\ \text{and } Z_2(\bar{w}) &\geq Z_2(\bar{x}_1) \end{aligned} \tag{21.7}$$

where at least one of the expressions in Eq. (21.7) must be satisfied as a strict inequality. Similarly, a solution $\bar{x}_1^*$ is said to be noninferior if there is no $\bar{w} \in X$ such that:

$$\bar{Z}(\bar{w}) > \bar{Z}(\bar{x}_1^*) \tag{21.8}$$

Whereas the solution of a single objective linear programming problem is the optimal solution, the solution of multiobjective linear programming is the definition of the set of noninferior solutions, otherwise known as the noninferior set.

The set of noninferior solutions will always lie on the boundary of the feasible region. As shown in Figure 21.1, any interior point, such as point A, of the feasible region will be inferior to at least one boundary point, such as points B and C. However, all the transformation curve need not be in the noninferior set. In Figure 21.1, the portion of the transformation curve between the vertical axis and point D is inferior to point D. The reader can easily prove this to himself by applying Eq. (21.7) to any point on the segment.

The solution of a multiobjective linear programming problem, that is, the generation of the noninferior set, proceeds by first transforming the vector-valued objective function into a scalar-valued function which allows solution by conventional methods. The solution of the transformed problem gives a point in the noninferior set. The parameters used in the transformation are then varied systematically to yield a number of points which satisfactorily represent the noninferior set. Unfortunately, there is no general definitive work in this area of mathematical programming, owing to the newness and limited number of applications of the technique. An excellent, but very theoretical, presentation of multiobjective linear programming is given in Beeson (1). The other references cited here are mainly presentations of specific solution techniques with the exception of Kapur (8).

Two approaches to the transformation of the objective function into a scalar quantity are various weighting techniques and the constraint method. Weighting techniques as described by Zadeh (16), Savir (13), Geoffrion (7), and Kapur (8) transform the two-dimensional problem of Eq. (21.3) into:

$$\text{Max} \sum \lambda_i Z_i = Z_2 + \lambda_1 Z_1$$
$$\text{subject to } \bar{x} \in X \tag{21.9}$$

in which the objective function is now a scalar quantity. One of the λ_i is selected to be equal to unity, specifying objective i as the numeraire. All other objectives are weighted by selected λs in terms of the numeraire. In the pipeline example, suppose that national income is selected as the numeraire, such that $\lambda_{NI} = 1$. λ_{EQ}, the weight for environmental quality, would then have the units (\$ of national income)/(head of animals).

If the noninferior set is strictly convex, it is easily generated by a weighting technique. Successive solutions of the transformed linear programming problem of Eq. (21.9), with systematically varied values of the λ_i weights, traces out the noninferior set. This is easily carried out using the parametric programming features available in most linear programming computer packages.

When the noninferior set is not convex, the weighting technique fails to generate the entire set. Figure 21.2 shows a transformation curve, all of which defines the noninferior set, but which is not convex. The procedure is started with $\lambda_1 = 0$, $\lambda_2 = 1$, which yields the solution at point A. The value of λ_1 is increased up to $\hat{\lambda}_1$. The solutions obtained lie on the segment AB of the transformation curve. When $\lambda_1 = \hat{\lambda}_1$, however, there are three possible solutions, none of which are better than the others. In practice, the output from the linear programming solution package, for example IBM's MPS (Mathematical Programming System), would probably not provide enough information to determine points B, C, and D. It is expected that only one of the three points would be available. Although the output from MPS does indicate which decision variables would enter the basis for alternative optimal solutions, it is doubtful that one could select the appropriate combinations of the variables to find these solutions. This is especially true for large-scale problems when there are many alternative optima.

The procedure continues by increasing λ_1 to values greater than $\hat{\lambda}_1$, which gives solutions on the segment DE of the noninferior set. Note that the solutions "skip" from the segment AB to the segment DE, excluding the

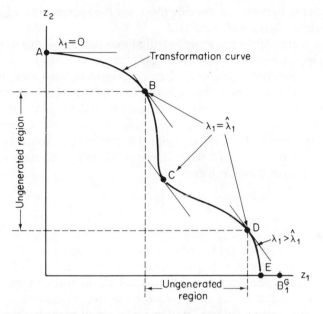

Figure 21.2 Weighting and constraint methods with a nonconvex noninferior set.

segment BD, which is in the noninferior set. Thus the weighting techniques fail to generate the entire noninferior set when it is nonconvex.

An approach that will generate the whole noninferior set with any arbitrary shape is the constraint method described by Facet (6). The original problem of Eq. (21.3) is transformed into:

$$\text{Max } Z_2$$
$$\text{subject to } \bar{x} \in X \qquad\qquad (21.10)$$
$$Z_1 \geq B_1$$

The objective function in Eq. (21.10) has been made scalar by including objective Z_1 as a constraint in the problem. The algorithm for generating the noninferior set is straightforward. B_1 is set at zero or at some predetermined lower bound and then increased incrementally until the solution becomes infeasible. At every value of B the problem is solved, yielding a point in the noninferior set. For example, in Figure 21.2, B_1 is varied from zero to B_1^G, at which point there is no feasible solution to Eq. (21.10).

A shadow price, λ_1, is associated with the constraint $Z_1 \geq B_1$ in Eq. (21.10). At every solution point of Eq. (21.10), λ_1 is the value of the tradeoff of objective Z_2 for Z_1. Furthermore, the value of λ_1 found from a solution to Eq. (21.10) on the segments AB and DE would yield the same solution if it were used in Eq. (21.9). The nonconvexity of the noninferior set destroys this relationship between the two problems on the segment BD of the transformation curve shown in Figure 21.2.

The generality of the constraint method makes it preferable to weighting schemes. Operational considerations also favor the constraint method. Specifically, parametric variation of the constraint is more straightforward than changing the weights in the objective function. It should be mentioned that when there are many objectives, say more than three, both the weighting and constraint methods may be computationally infeasible. Other methods which directly search the feasible region for noninferior solutions are necessary for such problems. To date, however, there are no such techniques applicable for large-scale problems.

THE EXAMPLE PROBLEM

The problem under consideration is the development of the water resources of a hypothetical river basin, shown schematically in Figure 21.3. The river either is the boundary of or flows through four regions. Many development alternatives are proposed for the river, including the building of reservoirs for storage, the import of water from out of the basin, the export to agricultural uses from which there will be no return flow, and the construction of power generation and irrigation sites. An important allocation problem exists because each region has plans for the river which, by themselves, could exhaust the available water resources.

The water available in the area under study is both limited in total amount and unevenly distributed throughout the year. The flow in the river results almost entirely from snowmelt in the mountains and, therefore, exhibits distinct seasonal fluctuations in discharge.

Multiobjective Formulation

An optimization model was first formulated with the main decision variables being the questions of whether and to what size such development alternatives should be built. It was of the form:

$$\text{Maximize } Z(\bar{x}) = \text{net national income benefits}$$
$$\text{subject to } \bar{x} \in X \tag{21.11}$$

where $\bar{x}$ contains all decision variables under consideration and X is defined by the continuity, technological, and policy constraints specified for the physical system. Thus the purpose of the optimization is to find that particular combination of decision variables that is feasible, where $\bar{x} \in X$ implies that no specified constraints are violated, and is the best solution in terms of

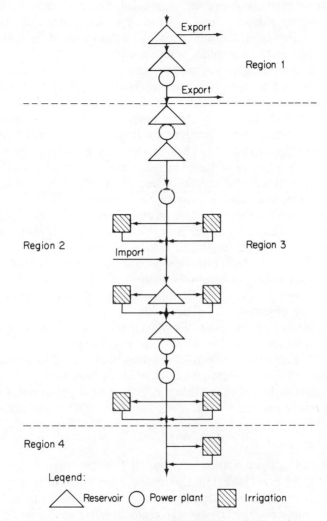

Figure 21.3 The river basin and development alternatives.

the stated objective. A complete description of the exact form of this model, developed for the Rio Colorado region of Argentina, can be found in Cohon (3). The simplified model used here has approximately 240 constraints and 600 variables. Separable, piecewise linear approximations were made for nonlinear costs, benefits, and technical relationships.

In addition to the objective of development, represented by the maximization of net national income benefits, the regional governments are principally concerned with the allocation of the water among the four regions. This distributional objective was taken as the second planning criterion for the example analysis. It should be mentioned, however, that there may be other national development objectives such as balance of payments and economic stabilization. If these objectives or others proved to be important, they could and should be included in the model.

The other major objective for the planning of the river basin, the allocation of the water among the four regions, is difficult to resolve because the regions differ greatly in present levels of development and wealth. Regions 1 and 4 are relatively rich, with highly developed infrastructures for irrigation and hydroelectric energy production. Regions 2 and 3 are barren, with very few people living in them. Thus, if the only planning objective were economic efficiency, most of the available water would be allocated to regions 1 and 4, which could use the water more efficiently, owing to their existing infrastructures and proximity to existing markets. The national government recognizes this but wants to develop regions 2 and 3 to integrate demographic and economic patterns in the country. Furthermore, the main decision-making body is a committee with equal representation from each of the concerned regions. The committee would not accept an allocation that did not provide for development of regions 2 and 3.

Regional concern over the distribution of the river's water can be represented by the objective of an equitable regional water allocation. The obvious problems in measuring this objective can be avoided in the multiobjective model since there is no requirement that the objectives be commensurable. The allocation objective can, thus, be measured in arbitrary units.

The regional water-allocation objective for the case study was represented as the minimization of the absolute deviation of regional water use from average regional water use. For p regions this can be mathematically written as

$$\text{Min WALL} = \sum |WU_i - WUA| \qquad (21.12)$$

in which WALL is the deviation in usage, WU_i is the water used by region i, and WUA is the average regional water use, all in cubic meters per second. This characterization implies that equity is achieved by an equal allocation. This is really unnecessary. The average regional water use could be replaced with no complications by some predetermined allocation. Deviations from this allocation would then quantify the objective. The allocation could also be in terms of regional income from water-related activities rather than in terms of water received alone, with little change in formulation.

The expression in Eq. (21.12) is an absolute value which is nonlinear and cannot be included directly in the linear programming model. A transformation that will enable us to include Eq. (21.12) in the linear program is:

$$\left. \begin{array}{l} \text{Min WALL} = \sum (G_i + T_i) \\ \text{subject to} \quad WU_i - WUA = G_i - T_i \\ \qquad G_i, \ T_i, \ WU_i, \ WUA \geq 0 \end{array} \right\} i = 1, \dots, p \qquad (21.13)$$

in which G_i and T_i are the positive and negative deviation of WU_i from WUA, respectively. Only G_i or T_i, but not both, can be nonzero for each of the p constraints, as can be seen by inspection of the form of the constraints and the objective function. For a given deviation, $G_i - T_i$, the sum $G_i + T_i$ is minimized when G_i or $T_i = 0$.

Screening Model

The screening model for this problem has to include a representation both of the physical system and the national income objective, given by Eq. (21.11), and of the water allocation objective expressed in Eq. (21.13). To use the constraint method of solution, which is the preferred approach, it is necessary that one of the two objectives be placed in the constraint set. Putting Eqs. (21.11) and (21.13) together in the appropriate manner by including the national income objective as a constraint gives:

$$\text{Min WALL} = \sum (G_i + T_i)$$
$$\text{subject to } \quad WU_i - WUA = G_i - T_i, \quad i = 1, \ldots, p$$
$$\text{and with } \bar{x} \in X \tag{21.14}$$
$$Z(\bar{x}) \geq B_j$$
$$\text{all variables nonnegative}$$

in which B_j is an arbitrarily defined constant.

The complete screening model is, in fact, slightly more detailed than Eq. (21.14). A constraint that defines the average regional water use in terms of the regional water uses, and p constraints which define regional water use in terms of consumptive water use variables in $\bar{x}$, must be included. The final form of the screening model is:

$$\text{Min WALL} = \sum (G_i + T_i)$$
$$\left.\begin{array}{l} \text{subject to } WU_i - WUA = G_i - T_i \\[2mm] \quad WUA = \dfrac{1}{p}\sum_i WU_i \\[2mm] \quad WU_i = \sum_t \sum_{s \in R_i} (E_{st} + EX_{st}) \end{array}\right\} i = 1, \ldots, p \tag{21.15}$$
$$\bar{x} \in X$$
$$Z(\bar{x}) \geq B_j$$
$$\text{all variables nonnegative}$$

in which t is the number of seasons, R_i is the set of sites in region i, and E_{st} and EX_{st} are the average diversion for irrigation and average interbasin export at site s during season t in cubic meters per second.

Solution and Results

Solutions to the screening model, Eq. (21.15), define points on the transformation curve between the national income objective and the regional water allocation objective, that is, the set of noninferior solutions. In the interest of computational efficiency, an iterative solution strategy was developed. In step 1, the single objective problem of maximizing net national income benefits, Eq. (21.11), was solved. The resulting value, $Z_{max} = B_{max}$, gave an upper bound on B_j because Eq. (21.15) is infeasible for $B_j > B_{max}$. In each of the subsequent steps, several solutions were obtained by parametrically varying B_j from an initial value to B_{max} in prespecified increments. After each step, an inferior range of solutions was defined and used to decrease the range of B_j to be considered at successive investigations of the noninferior set.

Solutions were obtained by using MPS on an IBM 370/155 computer. The approximate cost of each step is shown in Table 21.1. It is interesting to note the economies of scale of parametric programming. Whereas the cost of one solution was $9 (step 1), the cost of six parametric solutions in steps 3 and 4 was $23 and $25, respectively. Although the number of solutions increased sixfold, the total step solution cost less than three times as much.

Table 21.1 Solution Procedure and Computation Costs

Step No.	Number of Solutions	Computation Cost ($)
1	1	9
2	5	20
3	6	23
4	6	25
Total	18	77

Each of the noninferior solutions is listed in Table 21.2. The letter assigned to each of the solutions correspond to the points shown in Figure 21.4. The second and third columns of Table 21.2 show the values of the two objectives at each solution. The last column shows the value of the shadow price or dual variable associated with the constraint:

$$Z(\bar{x}) \geq B_j \qquad (21.16)$$

in the model. The shadow price at any point shows, at the margin, the increase in the deviation from the equal allocation which would result from a unit increase in B_j. The slope of the transformation curve at the corresponding

Table 21.2 Noninferior Design Solutions

Point on Transformation Curve (Fig. 21.4)	Net National Income Benefits, $Z(\bar{x})$ (pesos × 10^{12})	Deviation from Equal Allocation, WALL (m^3/sec)	Shadow Price [(m^3/sec)/ (pesos × 10^{12})]
A	1.8	0	0
B	1.9	41.0	450
C	2.0	102.1	640
D	2.05	136.6	690
E	2.06	143.6	700
F	2.07	150.6	700
G	2.08	157.7	720
H	2.09	168.0	1,140
I	2.10	397.7	47,870
J	$Z_{\max} = 2.10005$	436.0	—

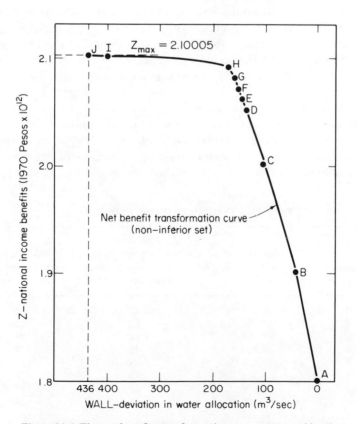

Figure 21.4 The net benefit transformation curve generated by the constraint method.

point is, because of the way the axes of Figure 21.4 are given, the negative of the reciprocal of the shadow price.

There are several noteworthy features about the transformation curve plotted in Figure 21.4. First, it will be noticed that the value of WALL decreases on its axis from the origin. It was drawn in this manner, because the regions prefer smaller deviations in water allocation to larger ones. Second, the national income axis starts at a value of 1.8×10^{12} pesos. For values less than 1.8×10^{12}, WALL = 0, and these points, although feasible, are inferior. The third interesting aspect of the curve is that it stops abruptly at point J where WALL = 436 and $Z = Z_{max}$. Points to the left of J with WALL > 436 must have values of $Z \leq Z_{max}$ for feasibility, and therefore they are inferior.

The Screening Model as a Decision-Making Tool

It should be clear from the level of detail and size to which the screening model is limited that such models are only part of the decision-making process. As a result of the many simplifying assumptions that must be made in formulating and solving the screening model, its results should be further analyzed by more detailed and realistic simulation models. Nevertheless, it is of interest to inspect the transformation curve of Figure 21.4 to try to gain insight into the design problem.

Each of the points on the transformation curve imply different design alternatives, some of which are shown in Table 21.3. As we move from point A to J, national income increases while the deviation in regional water allocation increases because relatively more water is given to regions 1 and 4

Table 21.3 Features of Different Designs

Design Solution	Water Export to Region 1 (m³/sec)	Capacity of an Irrigation Project in Regions 2 and 3 (hectares × 10³)	Water Use in Region 4 (m³/sec)
A	84	70	122
B	91	63	127
C	98	52	124
D	103	46	128
E	104	45	127
F	105	44	125
G	106	42	123
H	107	23	126
I	107	15	166
J	107	3	173

while less is being given to regions 2 and 3. This happens because relatively high national income benefits are possible in regions 1 and 4. Furthermore, the changing water allocation directly affects the capacities of the facilities to be built at different sites. This is indicated most dramatically for a selected irrigation project in regions 2 and 3. This irrigation area, which is in the relatively low national income benefit regions, shrinks from a size of 70,000 hectares for design A to 3000 hectares for design J. Similar changes in capacities occur at the other sites.

Given the transformation curve in Figure 21.4 and the information of Tables 21.2 and 21.3, an optimal plan could be selected by the decision-making process. In general, little can be said about the optimal alternative without information about preferences or observed decisions. In this case, however, one would expect that the decision-making process would select a plan in the vicinity of point G. Using a geometrical argument, it could be the optimum for a typical indifference curve like the one in Figure 21.1. From a tradeoff analysis, one would not expect a point on the segment JH to be selected in Figure 21.4 because relatively little national income is gained when moving from H to J, but the water allocation becomes significantly uneven. Similarly, when moving from F to A the water allocation is not greatly improved, but a rather large amount of national income is sacrificed.

Whatever point on the curve is ultimately selected in the decision-making process, the contribution of the transformation curve is clear. If the planning objectives are selected so that they consider all the issues of the problem, the transformation curve will clearly embody all the important tradeoff information required for a rational decision. Of course, this is only true when the many limiting physical and economic assumptions in the screening model can be accepted or considered in more detailed simulation models.

SUMMARY

The analyses of public investment problems are typically concerned with the selection of projects from among a large number of alternatives so as to maximize social welfare. The study of water resource development problems are of this type, with the added difficulty that the development alternatives are interrelated through a complex physical and economic system. These complexities require the application of systems analysis techniques to such problems.

Traditionally, however, analyses that did include the formulation of optimization and simulation models were based on guidelines which simplistically equated the maximization of social welfare with the maximization of economic efficiency. Thus previous applications have failed to capture the multiobjective nature of public investment problems. In water resources applications, planning objectives such as environmental quality and regional development were generally ignored or taken as secondary considerations.

The motivation for including many objectives in planning models is provided by recent applications of the theory of multiobjective analysis. The purpose of this study was to demonstrate how multiobjectives could be included in optimization models and to discuss possible solution techniques. Furthermore, the application of multiobjective analysis to decision making was briefly discussed.

ACKNOWLEDGMENT

We wish to acknowledge the Secretariat for Water Resources of the Republic of Argentina for their support of the project from which this work was derived.

REFERENCES

1. BEESON, R. M., 1971. *Optimization with Respect to Multiple Criteria*, Ph.D. Thesis, University of Southern California, Los Angeles, Calif.
2. BLANCHARD, B., 1964. *A First Trial in the Optimal Design and Operation of a Water Resource System*, S.M. Thesis, MIT, Cambridge, Mass.
3. COHON, J. L., 1972. *Multiple Objective Screening of Water Resource Investment Alternatives*, S.M. Thesis, MIT, Cambridge, Mass.
4. DE NEUFVILLE, R., AND STAFFORD, J. H., 1971. *Systems Analysis for Engineers and Managers*, New York: McGraw-Hill Book Company.
5. DORFMAN, R., 1965. Formal Models in the Design of Water Resource Systems, *Water Resources Research*, Vol. 1, No. 3, pp. 329–336.
6. FACET, T. B., 1970. A Solution to the Multiple Objective Linear Programming Problem, Ralph M. Par, MIT, Cambridge, Mass., Dec.
7. GEOFFRION, A. M., 1967. Solving Bicriterion Mathematical Programs, *Operations Research*, Vol. 15, pp. 39–54.
8. KAPUR, K. C., 1970. Mathematical Methods of Optimization for Multi-Objective Transportation Systems, *Socio-Economic Planning Sciences*, Vol. 4, pp. 451–467.
9. LOUCKS, D. P., 1969. *Stochastic Methods for Analyzing River Basin Systems*, Ithaca, N.Y.: Cornell University Water Resources and Marine Sciences Center, Technical Report 16, Aug.
10. MAASS, A., ET AL., 1962. *Design of Water-Resource Systems*, Cambridge, Mass.: Harvard University Press.
11. POBLETE, J. A., AND McLAUGHLIN, R. T., 1970. *Time Periods and Parameters in Mathematical Programming for River Basins*, Technical Report 128, Ralph M. Parsons Laboratory for Water Resources and Hydrodynamics, MIT, Cambridge, Mass., Sept.

12. ROGERS, P., 1967. *A Systems Analysis of the Lower Ganges—Brahmaputra Basin*, International Symposium on Floods and Their Computation, Leningrad, U.S.S.R., Aug.

13. SAVIR, D., 1966. *Multi-Objective Linear Programming*, Report ORC 66–21, Operations Research Center, University of California, Berkeley, Calif., Nov.

14. SMITH, D. V., 1970. Stochastic Irrigation Planning Models, Center for Population Studies, Harvard University, Cambridge, Mass., Jan.

15. WALLACE, J. R., 1966. *Linear-Programming Analysis of River-Basin Development*, Sc.D. Thesis, MIT, Cambridge, Mass.

16. ZADEH, L. A., 1963. Optimality and Non-Scalar-Valued Performance Criteria, *IEEE Transactions on Automatic Control*, Vol. AC-8, pp. 59–60.

22

Multiobjective Redesign of the Big Walnut Project*

DAVID C. MAJOR

This study deals with the application of multiobjective analysis to the redesign of a water resource project, the proposed Big Walnut dam and reservoir in Indiana, which had been previously designed by conventional techniques. The technique of benefit–cost analysis has been highly developed in water resource applications in the United States but, until recently, the benefits and costs so measured have been limited to a single objective, the increase of national income. Increments to national income are usually measured by noting the economic savings of preventing flood damage, estimating navigational savings, and the like. This emphasis on the national income (or "efficiency") objective shows little concern for a range of other social and income distribution objectives relevant to water resource development and thus centers the evaluation process on a myopic view of the project. Multiobjective analysis is a generalization of traditional efficiency-oriented benefit–cost analysis. It permits recognition of the many objectives: national income, regional income, environmental quality, defense, and others relevant to water resource development, and requires that these objectives be taken explicitly into account in program design. The scholarly foundations for the implementation of multiobjective analysis were laid down in the principal publication

*Extensively adapted from Multiple Objective Redesign of the Big Walnut Project by David C. Major and Associates, Water Resources Council–MIT Cooperative Agreement WRC-69-3 and Multiobjective Resources Planning presented at the Fourth Australasian Conference on Hydraulics and Fluid Mechanics, Monash University, Melbourne, Australia, Nov.–Dec. 1971 by David C. Major and W. T. O'Brien.

322

of the Harvard Water Program, *Design of Water-Resource Systems* (2), and in Marglin's *Public Investment Criteria* (3).

A special task force of the Water Resources Council, a federal interagency group, proposed in late 1969 a set of uniform criteria for bringing multiobjective analysis into the everyday evaluation process of federal water agencies (8). To test these proposed new criteria, a series of 19 case studies was carried out by agency and university teams on project designs proposed by one or another federal agency (10). The Big Walnut was one of these projects, and the MIT group was chosen to study it. The purpose of these studies was to see if the project analysis could be reformulated in multiobjective terms and how this would affect project design.

The four investment objectives proposed by the Council in the standards on the basis of which the tests were conducted were national economic development, regional economic development, environmental quality, and social well-being. The first objective is the traditional criterion of water system design; the other three are new as explicit design criteria, and are themselves multidimensional. The reformulation of the Big Walnut project described in this chapter deals primarily with the first and third objectives: increasing national income and improving environmental quality. For the original versions of the Council criteria and later proposed versions, the reader is referred to its reports (8, 9, 11).

BIG WALNUT: THE RIVER, THE ECOLOGICAL AREA, AND THE RECOMMENDED PROJECT

Big Walnut Creek flows through west-central Indiana for about 55 miles before merging into the Eel River. It lies in the Wabash River basin with its waters feeding eventually into the Wabash and then the Ohio Rivers. The region traversed by the creek is chiefly agricultural, with small towns and settlements, the largest of which is Greencastle, with a population approaching 10,000. The nearest large cities are Indianapolis, some 35 miles to the east, and Terre Haute, an equal distance to the west. The general topography of the region consists of flat or gently rolling agricultural land, interspersed with small woodland acreages. There are, however, landscape features in the region that stand in contrast to the general topography. One such area lies on Big Walnut Creek, north of the Highway 36 bridge. Here the creek winds through a narrow, steeply sloped region of ecological significance.

The ecological region is an area of small bluffs, bounded by two well-preserved wooden covered bridges. The area of particular significance encompasses about 350 acres (although more acreage would be used to buffer the area in a preservation plan) and is at present largely owned by two individuals. The area is not widely visited, a circumstance that has helped to maintain its charm and integrity. Scientific interest in the area stems from the fact that the area nurtures a relatively "mature" ecological system, a glacial relict biota that is the product of the retreating Wisconsin ice sheets. The area essentially reproduces a Canadian "north woods" environment, which has enabled the continued existence of such plants as the Canadian yew, eastern hemlock, and other plants that were isolated as the ice receded to the north. The result is an area that is unique within the region, and reproduced again only hundreds of miles to the north. Added to this relict community is one of the largest blue heron rookeries in the state as well as some of the last virgin forest stand in Indiana, including the state's largest hemlock, and what is possibly the largest living sugar maple in the world. It is this area, along with some other areas along the river that are not of special ecological significance, that would be flooded by the project that was originally recommended by the Corps of Engineers for the Big Walnut.

The project recommended by the Corps is, except for its relationship to the ecological area, a straightforward Corps project, formulated according to normal efficiency-oriented federal water resources investment criteria (4, 5, 6). The plan for the project was developed in the course of a multipurpose basin planning effort for the Wabash basin, including water supply, flood control, and recreation. As originally proposed by the Corps, the Big Walnut dam would have been constructed of rolled earth, and would be about 130 feet high and 3 miles long. The drainage area served by the project is 197 square miles in extent; the river flow at the dam site has a mean annual value of about 250 cubic feet per second, a relatively small stream.

The water supply of the dam would go primarily to the city of Indianapolis, whose water is supplied by the Indianapolis Water Company (IWCO), a private utility. The water supply storage space in the project would be purchased by the State of Indiana according to state policy and then sold to IWCO. Although there is no formal contract for this to be done, it is thought that all alternative sites have higher costs than the Big Walnut so that it will be in the interest of IWCO to purchase water from this source. In any case, Indiana officials generally regard themselves as in a "water-short" area, a fact reflected in the state's policy of developing all sites for "maximum capacity" and buying the allocated storage for water supply even without formal contracts for resale.

The flood control storage planned as part of the Big Walnut project would protect Greencastle and locations farther downstream. The recreational opportunities afforded by the dam would cater to visitors from the surrounding area and, presumably, from farther away. There are other reservoirs in the area for recreation so that, however relevant the market demands for recreation on the Big Walnut, the project would not provide unique recreational opportunities.

CONTROVERSY OVER THE BIG WALNUT

Although not perhaps an area of national environmental quality significance, the Big Walnut ecological area is of both scientific and aesthetic interest to persons and groups in Indiana and bordering states (1). A visit to the
area is enough to convince one of its attractiveness, particularly in the context
of the relatively flat Indiana landscape. The area has been known to specialists
for some time, but not to the general public.

Those who were interested in the area were taken by surprise when the
Corps announced its proposed project. A controversy ensued whose outlines
are by now familiar in the United States: vigorous conservationist opposition,
in this case led by the Izaak Walton League, to a proposed project of a federal
agency. One result of this controversy was a decision by the Corps to consider
various alternatives to the Big Walnut project. An advisory committee was
set up to assist them in this consideration. The committee recommended that
the ecological area be preserved by construction of a bypass around it from an
upper storage reservoir to a lower one. This bypass would have been both
expensive and, according to the Corps, of uncertain effect, and so this recommendation was rejected. The Corps then recommended its original plan once
again, with the addition of certain facilities to permit use of the limited natural
area that would remain after the completion of the reservoir. In publishing
this modified plan, the Corps also published the results of its original site-
selection work on the Big Walnut in order to make the various alternatives
available to interested parties (7).

The Corps' revised plan did not mollify the opposition since, although
there were added facilities, there was no change in the level of encroachment
on the ecological area. In the fall of 1969, Congress, recognizing the nature of
the controversy, provided $35,000 for additional study of the ecological prob-
lems that would be posed by the implementation of the reservoir.

PROJECT REDESIGN: MULTIPLE OBJECTIVES

The formulation of the project in terms of relevant multiobjectives requires (1) the choice of relevant objectives, (2) the analysis of preferences
of different interest groups, and (3) the development of the transformation
curve that shows the frontier of tradeoffs between objectives. These are discussed in order.

Objectives for the Big Walnut

The sources of information for the study's choice of objectives were
(1) the written material on the Big Walnut available from the Corps and

other participants in the controversy, (2) correspondence with the participants, and (3) field trips to the Big Walnut site and the Louisville District of the Corps.

On the basis of these sources of information, it was decided to analyze the Big Walnut project in terms of two principal objectives, that of increasing the national income (efficiency) and of preserving all or part of the ecological area. These correspond to the then principal objectives of the Corps and the conservationists who participated in the discussion about the Big Walnut. The Corps' analysis also considered alternatives to meet fixed water quality standards, which amounts to a separate objective.

Five additional objectives relevant to the decision became clear from a careful review of Corps' documents and decisions. These were found by analysis of the original Big Walnut report with its appendix containing 24 alternatives including the original project, and discussions with Corps personnel. These secondary objectives were:

1. Minimizing the number of families relocated.
2. Avoiding mudflats.
3. Meeting water supply projections apart from the efficiency benefits of doing so.
4. Avoiding the inconveniences and hazards of flooding that are not measurable as efficiency losses.
5. Avoiding relatively high project expenditures.

None of these additional objectives was dealt with systematically by the Corps, that is, by the setting up of a system of benefits and costs toward an objective and the orderly accounting of these benefits and costs for several alternative project designs. Also, no explicit mention of a regional objective for the Big Walnut was found in the documents.

The graphical presentation of the analysis of the Big Walnut in terms of two objectives implies two axes, one for net national income benefits and the other for net acres of the ecological area preserved. Both axes should refer to appropriately discounted net benefits. In this case, discounted net national income benefits were stated in average annual equivalents to make the analysis compatible with that of the Corps. In addition, both axes should refer to the appropriate scope of analysis. The national income axis should, for example, refer not only to net benefits from different alternative designs of the Big Walnut project, but also to the incremental net benefits obtained by the shift of certain demands for system outputs from the Big Walnut to other projects or to other sites on the Big Walnut itself, given different levels of the environmental quality constraint on the Big Walnut. In other words, for this problem it would be desirable to assume the entire Wabash system to have been optimized and then to consider the tradeoffs between environmental quality and national income at each level of preservation of the Big Walnut ecological area.

Because of the limitations of time and money for the case study, the national income analysis was not extended in this way. Instead, work was done with the benefits of alternative designs of the Big Walnut alone, an approach that presents the decision framework as for an isolated project. Total net national income benefits for the entire system would, then, actually be somewhat greater in each case than those calculated.

It might also be argued that the environmental quality axis should refer only to acres of the Big Walnut ecological area since this area is, strictly speaking, unique. However, it would appear that when multiobjective methods are applied nationally, there will be some necessity to group environmental areas by approximately equal quality. A further complication lies in the problem of treating the duration of flooding appropriately. In the analysis, "normal flood pool" was converted to "acres preserved." However, even an occasional extreme flood might damage the ecological area from a scientific point of view. To deal with this problem a simulation model of the river together with expert ecological advice could be used.

Preferences of Different Groups

We focus on the preferences of the parties involved in the planning process, the Corps of Engineers and the conservationists, led in this instance by the Izaak Walton League. Looking first at the Corps' preferences, it is clear from the Corps' discussion of alternative projects that their planners stuck close to their mandate to maximize national efficiency benefits. However, it is also evident in the shape of the revised plan, which has certain added facilities intended to enhance the environmental quality aspects at a small cost in efficiency, that the Corps tends to choose what appears to be the efficiency maximizing plan but has a willingness to move somewhat from this plan to accommodate serious nonefficiency objectives. In the revised plan, for example, the Corps recommended additions with a cost of $0.5 million. This is essentially the behavior that one would expect from the traditions of the standard single objective cost-benefit analysis as modified by government regulations (4, 5, 6); in this respect, the Corps is doing what it should do as an executive agency.

Such behavior, on the two-dimensional preference surface, yields indifference curves resembling those in Figure 22.1. These curves represent the Corps' willingness to compromise with environmental quality advocates to a certain extent in terms of losses in national income. However, these curves do not slope sharply, a reflection of the fact that the Corps rejected the expensive diversion alternative proposed by the advisory committee. The Corps' indifference curves are not defined for our problem beyond the point

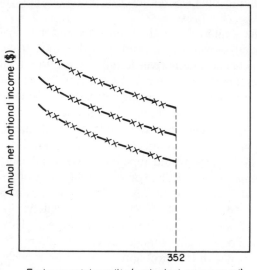

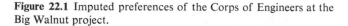

Environmental quality (ecological acres saved)

Figure 22.1 Imputed preferences of the Corps of Engineers at the Big Walnut project.

where the entire ecological area (352 acres) is preserved, since beyond this point only the efficiency objective remains relevant.

From the Corps' discussion of the alternatives, we were drawn to additional information about the Corps' preference toward objectives other than the two with which this study is principally concerned. It appeared obvious from the alternatives that a different site at Reelsville was superior to the Big Walnut site, since it has approximately equal national income benefits and does not impinge at all on the ecological area. Nevertheless, inquiry established that the national income benefits for Reelsville are not as great as reported since the recreation benefits for the site are less than the number furnished by the Bureau of Outdoor Recreation, primarily because of a trestle that would cross the reservoir. In addition, net benefits would be reduced by the addition of pumping costs higher than those for the Big Walnut. The Corps also rejected Reelsville for reasons having to do with objectives other than efficiency. First, we were told that a greater number of families would have to be relocated for the Reelsville site than for the Big Walnut. Second, the mudflats that would be generated by Reelsville were regarded as distinctly negative from an environmental standpoint. Third, the greater initial outlay for Reelsville did not appeal to the Corps' planners, who apparently operate under an implicit budget constraint. All these pieces of information relating to preferences could be analyzed in a longer study.

Indifference curves for the conservationists have quite a different shape than those for the Corps. Figure 22.2 shows one set of curves, but it is well to remember that these are aggregations composed of preferences of individuals whose curves have varying shapes, even if all individuals are classed in a group, such as "conservationists" in this analysis. From the standpoint of ecological study, for example, the value of the area is its mature integrity, which might decrease radically with only a small amount of flooding. On the other hand, the area would still be extremely attractive for hiking and nature watching, even if part of the area were flooded from time to time. Thus a set of conservationists' preference curves embodies different types of individual preferences.

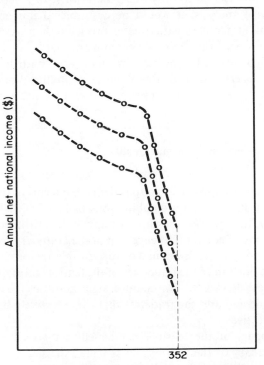

Environmental quality (ecological acres saved)

Figure 22.2 Imputed preferences for conservationists at Big Walnut project.

In general, we found that there is a large element of concern for the integrity of the area and that much of the satisfaction from visiting the area would be destroyed for many if the area were impinged upon, even slightly. Thus we can expect that conservationists' preferences will have shapes that reflect the special value placed on the ecological area. The steeply sloping portions in the center of the figure reflect the fact that a very large national income gain would have to be realized for the conservationists to assent to a small amount of encroachment on the area. Once this gain was realized, however, and the area's integrity had been lost, only small additional amounts of national income gain would be required to compensate for the loss of the remainder of the area. This is reflected in the slightly sloping portions of the curves to the left. To the right of the point at which the entire area is preserved, the conservationists' curves are, like those of the Corps, not defined, because when the entire area is preserved, only the national income objective remains relevant.

The curves we have drawn to express the preferences of the Corps and of the conservationists do so only in a general way. In a detailed application of multiobjective analysis, it would be desirable to implement procedures for determining preferences with greater precision than was required for this analysis. [Such procedures are now becoming available, as shown by the case study on Mexico City in Chapter 24, in which a multiattribute preference or utility function is estimated for the Ministry of Public Works. These procedures have yet, however, to be implemented in a formal economic analysis. —Eds.]

National Income-Environmental Quality
Transformation Curve

The optimum design for a group is defined in terms of the preferences of that group and the maximum combinations of benefits that can be achieved. Specifically, the optimum for any group will be at the point of tangency of their best curve of equal preference and of the boundary to the set of feasible benefits, as long as both define convex feasible regions over the domain of interest. The portion of the boundary of the feasible set sloping from northwest to southeast defines the dominant designs and thus the alternatives that should be considered and the tradeoffs that can effectively be made between the several objectives.

The boundary of the feasible net benefit set goes by several names. Here, the boundary of the feasible set is called the net benefit transformation curve. (The production possibility frontier is an analogous concept.) In this case we seek the transformation curve which shows the maximum combinations of net income and ecological acres saved. This will provide us with a means to define the preferred designs for the Big Walnut of both the Corps and the conservationists.

An estimate was made of the optimal efficiency allocation of each level of storage at the Big Walnut site to each of the system outputs: flood control, recreation, and water supply and water quality control. This was done by making assumptions about the effects on benefits and costs of varying seasonal storage capacity, and about the complementarity, or lack of it, between the use of storage for one purpose and storage for another. Flood control benefits versus storage were taken as calculated by the Corps according to their normal procedures. Recreation benefits versus storage were reestimated as a function of reservoir surface area, using a technique that allocated space to the least water-acreage-intensive recreation activities first. This method provides maximum recreation benefits under the Corps' assumption that all recreation activities are valued at $1 per person per day. The benefits calculated for recreation here and in the Corps' report are only for water-based recreation. Benefits from water-related activities (camping, hiking, etc.) should be calculated in a fuller study.

Water supply and water quality control benefits for different sizes of the reservoir were estimated as follows. First, a relation between storage and safe yield for water supply alternates was developed from monthly streamflow data for 1951–1960. Then a curve of annual cost versus yield was developed by taking the Corps' value for 78 million gallons per day, the projected water supply from the Big Walnut, and assuming a factor to account for economies of scale. These two curves together yield an annual cost versus storage curve that was taken as the benefit curve for water supply and water quality flows. This procedure is the same as the Corps' procedure and thus does not reflect an adjustment for deviations from the alternative cost measure of benefit that is probably required for the Big Walnut. In addition, this curve does not include pumping costs, an important factor. (The effects of omitting this factor may not affect net benefits as much as total outlay figures and benefit–cost ratios because of the use of an alternative cost measure of benefits.)

The cost curve for the Big Walnut was derived from information in the Corps' report of supplemental site studies (7). This curve, when subtracted from the overall benefit curve derived from the separate benefit curves described above, provides a net benefit curve that was plotted against flood pool

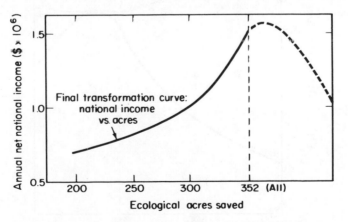

Figure 22.3 Transformation curve: national income versus acres preserved.

elevation. Flood pool elevation was then converted to acres of the ecological area preserved, by an analysis of the topographical maps of the project, and a final transformation curve derived in terms of net benefits toward the two objectives (Figure 22.3). As explained in connection with the discussion of the preference curves, the final transformation curve does not actually extend beyond the maximum number of acres saved (352) since no design can save more than this number of acres.

The results of considering this transformation curve together with both the preferences of the Corps (Figure 22.1) and the preferences of the conservationists (Figure 22.2) are that both groups would, in all probability, select the project marked A in Figure 22.4 maximizing national income benefits, since this project is outside the defined area of tradeoff and would completely preserve the environmental quality area. In this case the multiobjective problem of the Big Walnut would be easily solved.

The transformation curve, however, should be taken only as illustrative,

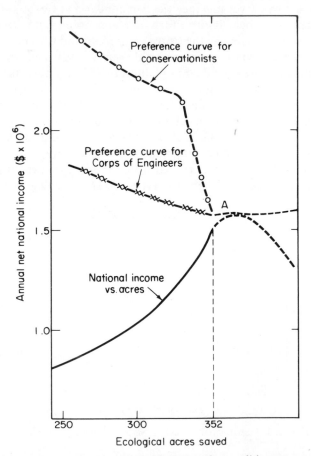

Figure 22.4 Optimal project at Big Walnut for conditions assumed.

for several reasons. First, a more thorough study of all information for estimating both efficiency and environmental benefits and costs should be made for the Big Walnut, probably using a simulation model to capture the highly stochastic nature of the system. In addition, of course, benefits and costs for national income could be discounted at rates of interest both higher and lower than the interest rate of 3 1/8 % used by the Corps in accordance with federal instructions at the time of planning. A discount rate for acres should also be selected and applied if the acres are to become available in future years rather than in the present. As an example of the possible effects of a more thorough study, recreational benefits could be considered in a more formal manner than described above. We now think that our method overemphasized recreation benefits at small storage levels. If this is so, the transformation curve shifts in such a way that the national income maximum tends to enter the area in which the entire ecological area is not preserved.

Second, the transformation curve drawn is for the Big Walnut when it is considered as an isolated project. As indicated previously, the more appropriate curve is the one for the ecological area versus the national income benefits generated within the system as a whole with each level of the environmental quality constraint.

Third, the two objectives considered here are, we think, the two most important objectives in the Big Walnut decision problem. However, to the extent that they are not the only objectives, the two-dimensional curve does not present the entire decision problem.

Fourth, additional investment criteria are relevant in the Big Walnut problem, in particular, scheduling criteria, an interest rate for environmental quality benefits, and the calculation of off site environmental quality costs.

Resolution of Disagreement

We cannot generally suppose that a project can be designed without conflict between objectives. To illustrate this general case, we indicate what would happen should the maximum national income point on the transformation curve fall within the limits of the ecological area, as in Figure 22.5. It is this situation in which tradeoffs between objectives can and must be calculated. First we estimate multidimensional benefit–cost ratios from the point of view of the major groups involved. The figures presented below are examples of the analysis, not final estimates of the ratios involved in the Big Walnut project.

Given the Corps' preference curve as illustrated, the optimal project for the Corps is the project yielding the net benefit combination at B. The value in dollars placed by the Corps on preserving the marginal acre of the ecological area is the negative of the slope of the tangent at B, which could be estimated, from the detailed information available, at $1700 per acre, (0.0017×10^6),

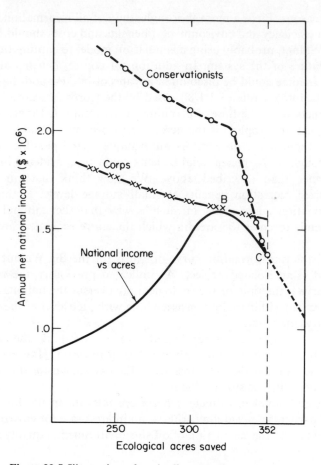

Figure 22.5 Illustration of tradeoffs at Big Walnut, assuming that the maximum national income point falls within the limits of the ecological area.

in average annual net national income benefits. The annual average gross benefit and costs were approximated from working papers as $2.1 and $0.5 million, respectively. From Figure 22.5 it can be estimated that 320 acres would be preserved, and it is assumed that no acres are foregone elsewhere in the system because of the interindustry process used to build the optimal project.

The multiobjective benefit–cost ratio for this project, B, assuming that these values are correct, is then, figuring in millions of dollars:

$$\left(\frac{B}{C}\right)_B = \frac{2.1 + 0.0017(320)}{0.5 + 0.0017(0)} = \frac{2.6}{0.5} = 5.2 \tag{22.1}$$

based upon the preferences of the Corps.

Given the conservationists' preference curves, the optimal design is represented by C in Figure 22.5, which indicates that the conservationists would prefer to preserve the entire area. The tangent to the appropriate preference curve near the limit of the area suggests that they would be willing to give up as much as $24,000 in annual net national income benefits to preserve the last acre of the area. In fact, they would have to give up only $15,000 in average annual net national income benefits to preserve the last acre (the slope of the transformation curve near the limit of the area). The benefit–cost ratio for their optimal project, C, using the latter value as a weight, is:

$$\left(\frac{B}{C}\right)_c = \frac{1.7 + 0.015(352)}{0.4 + 0.015(0)} = \frac{7.0}{0.4} = 17.5 \qquad (22.2)$$

For this ratio, as for the ratio presented above, the values for efficiency gross benefits and costs are approximations developed from the seminar working papers on which the transformation curve is based. The number of acres preserved is taken from Figure 22.5 and the number of acres lost through the interindustry process is assumed to be zero.

Note that the two ratios just presented cannot be compared, since different weights are used. The weights are imputed monetary values for nonmonetary objectives, taken at the margin. Because of the differences between groups and individuals interested in a multiobjective decision problem it is unlikely, in general, that they will agree in the first instance upon the relative values to be used in any possible benefit–cost ratio calculation.

However, it is possible to suppose that in the political process a compromise between the projects they individually prefer is possible. Each group might sacrifice some of its benefits for the purpose of achieving a compromise while still perceiving that the scheme is desirable overall. It is not possible to identify unambiguous theoretical or practical grounds for determining where this compromise should fall, as between B and C. Nor was it possible in this case for us to forecast where it will fall. It seems that the issue would, ultimately, have to be resolved through some kind of negotiation process which would depend upon the institutional constraints and the cooperative nature of the participants.

This brings us to a recognition of the fact that, since we have no analytic way to put together a social welfare function, we rely on the political process to read social preferences and to reflect them in project and program design. In this process we can almost always expect conflict at various stages, as in the original discussions on the Big Walnut dam; certain compromises and agreements, as at the present stage of the Big Walnut controversy; and perhaps further disagreements at later stages in the decision process. The aim of multiobjective analysis of investment problems is to contribute to the proper working of the political process by making clear to decision makers the multidimensional consequences of alternative programs and projects.

STATUS OF THE BIG WALNUT PROJECT IN 1973

The status of the Big Walnut project as of 1973 was as follows. The Louisville District of the Corps of Engineers is planning to propose a reservoir on a new site about 3 1/2 miles downstream from the controversial site that would have resulted in the inundation of the ecological area.

The new site provides a fascinatingly direct tradeoff, since all system outputs remain approximately the same as for the old site, the number of homes taken is approximately the same (72 versus 69), the ecological area is preserved, and the cost of construction for the given level of outputs is increased. (The Corps does not yet know by exactly how much; one off-the-cuff estimate was about several million dollars.) Thus the tradeoff involved is exactly between increased cost and the ecological area.

The choice of the new site seems to have been made in an open and thorough way. Corps' planners walked around the projected seasonal and flood pool levels with a local professor of biology who has been much concerned with preserving the ecological area. His opinion was obtained on the possible damage that might be done to the area from the proposed new seasonal and flood pool levels; he thinks that the danger is slight and that the proposed new site will not hurt the area.

In fact, he and other conservationists are aware that the proposed new site, if approved, might be the only feasible way of preserving the area, since it is close enough to allow the Corps to propose the inclusion of the ecological area in the total project area as a natural preserve. If this is not done, it may not be possible to protect the privately held area against development.

ACKNOWLEDGMENTS

The case study described here was executed in cooperation with a group of students at MIT: Carlos Bravo, Jared Cohon, Walter Grayman, Brendan Harley, Dennis Lai, Carl Magnell, Edward McBean, and Dr. W. T. O'Brien. We are also grateful for the financial support of the U.S. Water Resources Council, which made this work possible.

REFERENCES

1. LINDSEY, A. A., SCHMELZ, D. V., AND NICHOLS, S. A., 1969. *Natural Areas in Indiana and Their Preservation*, Lafayette, Ind.: Purdue University, Department of Biological Sciences, p. 269ff.
2. MAASS, A., ET AL., 1962. *Design of Water-Resource Systems*, Cambridge, Mass.: Harvard University Press.
3. MARGLIN, S. A., 1967. *Public Investment Criteria*, Cambridge, Mass.: MIT Press.

4. U.S. Bureau of the Budget, 1952. *Budget Circular A-47*, Washington, D.C.: U.S. Government Printing Office.

5. U.S. Federal Inter-Agency River Basin Committee, Subcommittee on Benefits and Costs, 1950 (rev. 1958). Report to the Federal Inter-Agency Committee: Proposed Practices for Economic Analysis of River Basin Projects, *The Green Book*, Washington, D.C.: U.S. Government Printing Office.

6. U.S. President's Water Resources Council, 1962. *Policies, Standards and Procedures in the Formulation, Evaluation and Review of Plans for Use and Development of Water and Related Land Resources*, Washington, D.C. Available as U.S. Senate Document 87–97, Aug.

7. U.S. Senate Document 90–96, 1968. Interim Report 3, *Wabash River Basin Comprehensive Study, Indiana, Illinois, and Ohio*, 2 vols., Washington, D.C.: U.S. Government Printing Office. (Certain appendices are available as working documents only and are not included in Senate Document 90–96).

8. U.S. Water Resources Council, 1969. Report of the Special Task Force: *Proposed Standards for the Evaluation of Water and Related Land Resources*, Washington, D.C., June.

9. U.S. Water Resources Council, 1970. Report of the Special Task Force: *Principles for Planning Water and Land Resources*, Washington, D.C., July.

10. U.S. Water Resources Council, 1970. Report of the Special Task Force: *A Summary Analysis of Nineteen Tests of Proposed Evaluation Procedures on Selected Water and Land Resource Projects*, Washington, D.C., July.

11. U.S. Water Resources Council, 1971. Proposed Principles and Standards for Planning Water and Related Land Resources, Vol. 36, *Federal Register* 24144–24194, Washington, D.C., Dec.

23

Design of Fill for San Francisco Bay*

JOSEPH I. FOLAYEN, KAARE HÖEG, AND JACK R. BENJAMIN

Soil engineers frequently feel a general lack of confidence in their predictions of soil behavior and, consequently, of the exact performance of the facilities they design. The uncertanties encountered in the design of earthworks are found partly in the determination of the properties of a soil and of the loads that will be applied, and partly in the phenomenological model and theory employed in the analysis. In only a few cases has the application of decision analysis, and even of probability theory, to the planning and design of earth structures and foundations been an important contribution of the analysis.

THE PROBLEM

The case study involves the reclamation of former marshland in the San Francisco Bay. The development required the placement of 5 ft of fill over an area of about 1200 acres. The fill covered a 30- to 50-ft-thick, apparently uniform deposit of Bay mud, which is a very soft, compressible organic clay. Because of the requirements of the development planned for the reclaimed area, it was considered important to achieve the grades planned with a tolerance of ±0.1 ft.

*Extensively adapted from Decision Theory Applied to Settlement Predictions, published in the American Society of Civil Engineers *Journal of Soil Mechanics and Foundations Division*, Vol. 96, July 1970, pp. 1127–1141; and Joseph I. Folayen, *Reliability of Predicted Soil Settlement*, Department of Civil Engineering, Stanford University, Stanford, Calif., Research Report 108, June 1969.

The ultimate settlement of the fill and the mud below it, ρ, could be computed from the one-dimensional equation:

$$\rho = (H)(m_c) \log \frac{p_0 + \Delta p}{p_0} \qquad (23.1)$$

in which H is the thickness of the compressible layer, in this case the Bay mud; $m_c = C_c/(1 + e_0)$ is a constant in which C_c is the compression index and e_0 is the initial void ratio, both properties of the fill soil; p_0 is the effective vertical stress existing before any load is added; and Δp is the vertical stress added by the fill over the large area. The terms H and p_0 were taken as known by virtue of field measurements, but m_c was considered a random variable.

Two essential design issues were important. First, it was necessary to determine what thickness of fill it was necessary to place to achieve the desired grades and elevations. This required, using a standard decision-analytic procedure, an estimate of the distribution of m_c and a utility function describing the designer's (or his client's) utility over risk. The second task was to determine how much extra information it would be worthwhile to gather about the Bay mud before committing oneself to a design for the fill.

PROBABILITY DISTRIBUTIONS

Two probability distributions were obtained to show different aspects of the problem and a third derived to integrate the information of the first two. A subjective prior distribution was obtained from the judgment of soil engineers with different degrees of experience in this area, and a sampling distribution based upon field measurements was obtained. From these a posterior distribution combining the preceding two was established.

Prior Probability Distribution

Thirteen experienced foundation engineers in the San Francisco area were interviewed with regard to their subjective feelings about the compressibility of Bay mud. Strictly on the basis of the interviews and prior to obtaining actual laboratory test data, subjective probability distributions for the mean of m_c were established (Figure 23.1). A t-statistic distribution was used because it is the joint distribution of the mean and the variance of a normal distribution where both parameters are unknown. Other distributions were also studied, however. Table 23.1 and Figure 23.1 present a summary of four typical responses of experienced engineers to the interview.

The prior sample sizes, n_1, were obtained by asking the engineers how much their feelings were worth to them. This number is presumed to correlate to the number of actual successful laboratory tests they had earlier performed

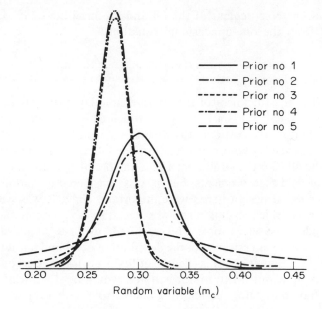

Figure 23.1 Prior probability distributions.

on this material. The fifth prior listed in Table 23.1 is based solely on a general soil mechanics literature survey. It represents the situation when professional experience is lacking and is consequently a weak prior.

Table 23.1 Prior Distributions on the Mean of the Critical Variable, m_c

Parameter	Prior 1	Prior 2	Prior 3	Prior 4	Prior 5
Range of m_c	0.10–0.40	0.10–0.40	0.10–0.40	0.10–0.50	0.10–0.50
Mean, m_1	0.30	0.275	0.275	0.30	0.30
Variance, s_1^2	0.91×10^{-3}	0.23×10^{-3}	0.23×10^{-3}	0.91×10^{-3}	0.81×10^{-2}
Sample size, n_1	82	42	12	4	2
Degrees of freedom, v_1	81	41	11	3	1
Education	BS (1954)	MS (1954)	BS(1952)	MS (1966)	MS (1967)
Years of experience	13	15	17	3	Student

Sampling Probability Distribution

A subdivision of the reclamation project development consisting of 350 acres (equivalent to a square about 4000 ft on a side) was considered. Forty-two sample borings were made approximately 500 ft on a center, with

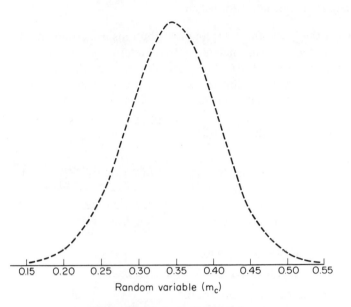

Figure 23.2 Sample distribution.

a penetration or vane shear test halfway in between to check consistency. One-dimensional consolidation tests from 27 different borings were performed.

To test the uniformity of m_c with depth, the samples were divided into two groups. One set of 17 samples from 12 different borings was taken from a depth of 15 ft beneath the generally flat ground surface, and the second set from 15 borings consisted of 22 samples from all depths except 15 ft down to 48 ft. Based upon the analysis, the hypothesis of uniformity was accepted at a 1% level of significance (2).

The data were therefore mixed and all 39 samples used to determine a t distribution on the random variable (Figure 23.2). The range of the sample distribution was found to be 0.21 to 0.45, the mean $m_2 = 0.34$, and the variance $s_2^2 = 0.37 \times 10^{-2}$.

A comparison with the prior distributions shows that the engineers interviewed believed the variance to be much smaller than what the actual sampling data revealed for this case. [Excessively narrow subjective probability distributions are an unfortunate, but regular, feature. Repeated classroom experiments, for example, show that the real answer to almanac questions, such as the number of Robinsons in the Boston telephone directory, lie outside the student's 95% confidence intervals on the distributions about half the time.—Eds.]

Posterior Probability Distribution

When the prior and sampling distributions are conjugate distributions, one can derive the parameters of the posterior distributions quite easily from Bayes' theorem (1, 3). Conjugate distributions may be defined as those distributions that vary directly with functions of the sufficient statistics, "summaries of data that are sufficient to define the generating process." For example, if the prior as well as the sampling distributions are of the t type, the posterior distribution is defined by:

$$f(\mu \mid m, k, v) = \frac{v^{1/2v}}{\beta(\frac{1}{2}, \frac{1}{2}v)}[v + k(\mu - m)^2]^{-1/2(v+1)} \sqrt{k} \qquad (23.2)$$

This equation gives the distributions of the true mean given parameters of the observed system where β is the beta function and:

μ = true mean (unknown)

n_1, n_2 = size of samples 1 and 2

m_1, m_2 = mean of samples 1 and 2

s_1^2, s_2^2 = variance of samples 1 and 2

v_1, v_2 = degrees of freedom in samples 1 and 2

n = pooled sample size = $n_1 + n_2$

m = pooled mean = $(n_1 m_1 + n_2 m_2)/n$

s^2 = sample variance = $[(v_1 s_1^2 + n_1 m_1^2) + (v_2 s_2^2 + n_2 m_2^2) - nm^2]/v = 1/k$

v = degrees of freedom = $v_1 + v_2 - 1$

Parameters with subscript 1 refer to the prior distribution; subscript 2 refers to sampling distribution.

The five resulting posterior probability distributions, derived from Eq. (23.2) and from the prior distributions, are shown in Figure 23.3.

UTILITY FUNCTIONS

The cost of compacted earth fill in place was $1.75 per cubic yard. With 4 ft of fill on top of 40 ft of Bay mud, the estimated ultimate settlement was approximately 34 in. If the actual settlements turned out to be 6 in. (i.e., 17%) less than predicted, which is altogether conceivable, 6 in. of unnecessary fill would have been placed at a cost about $500,000 to the developer. On the other hand, if the settlement was 6 in. more than the preliminary estimate, the planned finished grades would not be maintained, causing loss in scenic

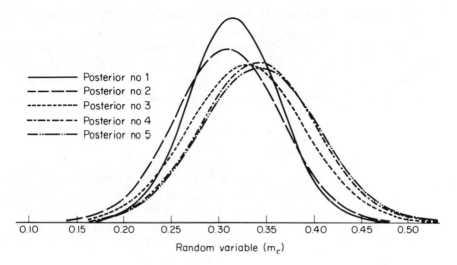

Random variable (m_c)

Figure 23.3 Posterior distribution.

value and some expensive readjustments of utility lines. It is clear that a great deal may be at stake.

To obtain a sense of what utility function might be appropriate for this analysis, the panel of experts were interviewed in some detail. Their responses could be classified into two types of preferences. The first type is the case where the engineer prefers settlement to fall within a certain tolerance of the predicted. Anything outside that range is bad. Although there is some loss within the range of the tolerance, the loss is relatively negligible. The second type of preference is the case where, although the engineer prefers actual settlement to fall within a certain tolerance limit of the predicted, the client prefers actual settlement being "reasonably" smaller than predicted. The reverse may be true in another case, but both cannot be true if the engineer and the client were to maintain their preferences. Two utility functions, a quadratic utility function and an exponential utility function, were fitted to the expressed preferences.

Quadratic Utility Function

The quadratic utility function used was of the form:

$$U(a, m_c) = K(a - m_c)^2 = K(\epsilon)^2 \qquad (23.3)$$

where a is the design value taken for m_c, m_c the actual value, ϵ the error, and K a severity constant. Only one observation of preferences is needed to estimate K and thus the entire quadratic utility function. For illustration, an error of 0.05 in m_c does lead to an error of about 6 in. and a cost of about $500,000. The severity constant then equals 2×10^8.

Since the quadratic utility function is symmetric, it implies that the designer should choose his design value for m_c equal to the mean of the probability distribution, if it is also symmetric. This follows from the equivalent statement that he feels it is equally costly to underestimate as to overestimate settlements.

Exponential Utility Function

Fitting an exponential utility function to the case where the preference is for actual settlement to be less than predicted gives an asymmetric overall utility function. As expected, the consequence is that the designer's optimal choice of design value for m_c is not equal to the mean value of the probability distribution on m_c.

Reliability Estimates

In engineering, reliability is generally defined as unity minus the probability of failure. However, in many problems what constitutes failure is not at all clearly defined. For example, how is failure defined for situations where settlements are a concern? Because economic importance is most often attached to the definition of failure, it is considered essential that reliability statements must reflect the underlying utility function. If the utility function is, say, exponential, whereby penalty is very severe but the returns for success are relatively small, the reliability statement must reflect that situation.

As an example, conventional reliability R_c is unity minus the probability that actual ultimate settlement will exceed 1.2 times the predicted value. Furthermore, corresponding to the quadratic utility function, a quadratic reliability R_q is defined as unity minus the probability that actual settlement will fall outside range $\pm 20\%$ of the predicted. Because it is assumed that error in settlement is due to error in the random variable m_c, the following definitions evolve:

$$R_c = 1 - P(m_c > 1.2a)$$
$$R_q = 1 - P(m_c < 0.8a^* \text{ and } m_c > 1.2a^*) \tag{23.4}$$

in which a* is the optimal design value, which with a quadratic utility function coincides with the mean value of the posterior distribution; and a is a design value which is not necessarily optimal.

The reliability of the predicted settlement may be computed, assuming that the engineer acts as he should, in accordance with his updated posterior probability distribution. For the experienced engineer 1 (see Table 23.1),

$R_c = 0.90$ and $R_q = 0.80$; for engineer 5, who relied solely on a literature survey, the corresponding reliabilities are 0.86 and 0.70. These numbers can be obtained from the cumulative distribution functions presented in Figure 23.4. For both engineers the probability of their prediction being within $\pm 20\%$ of the actual settlement is estimated to be encouragingly high.

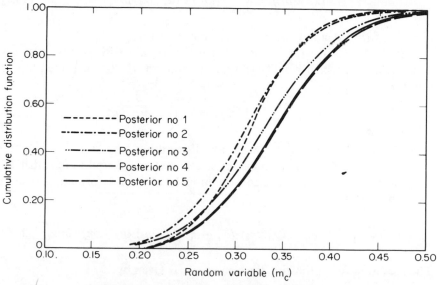

Figure 23.4 Cumulative distribution function.

An interesting comparison is made by computing an upper and lower limit for each engineer so that the actual settlement will fall within these limits with a quadratic reliability of 0.99. For the experienced engineer the upper limit is found using $m_c = 0.35$; the lower limit, by using $m_c = 0.24$. For engineer 5 the corresponding limits are $m_c = 0.41$ and $m_c = 0.23$. As may be observed, the latter set of values represents a much wider range of uncertainty.

Because there are some elements of subjectivity in fitting probability models to a given set of data, different probability distributions were used in addition to the t distribution. As anticipated, there were no substantial changes in the computed reliabilities when the normal, the gamma, and the beta distributions were used successively (2).

OPTIMAL NUMBER OF SAMPLES

It does not pay to increase the sampling cost by an amount greater than the expected terminal gain. On the basis of the prior probability distribution and utility function, the engineer can estimate the optimal sample size.

Assume that the acquisition of no soil samples is being contemplated and suppose a_0^* to be the optimal act based on the prior probabilty distribution on the random variable, that is, using no samples at all. The expected terminal loss is defined by:

$$L(a_0^*, m_c) = \int_{-\infty}^{\infty} f_1(m_c) \, U(a_0^*, m_c) \, dm_c \qquad (23.5)$$

in which $f_1(m_c)$ denotes the prior distribution on m_c.

Similarly assume that an experimental program with sample size n is being proposed which would yield an optimal act a_n^*. The expected loss is then:

$$L(a_n^*, m_c) = \int_{-\infty}^{\infty} f_2(m_c) \, U(a_n^*, m_c) \, dm_c \qquad (23.6)$$

in which $f_2(m_c)$ denotes the posterior distribution.

The value of taking a sample of size n, $v(n)$, is referred to in statistical decision theory as the expected conditional value of information due to sampling and is given by:

$$v(n) = \int_{-\infty}^{\infty} f_1(m_c) \, U(a_0^*, m_c) \, dm_c - \int_{-\infty}^{\infty} f_2(m_c) \, U(a_n^*, m_c) \, dm_c \qquad (23.7)$$

The condition for optimizing the sample size is given by:

$$v(n) = \alpha n \qquad (23.8)$$

in which α is the unit cost of sampling plus testing.

In general, $v(n)$ can be expressed as a function of the severity constant and the expected mean of the variance of the random variable. The optimal sample size n^* is found by differentiating Eq. (23.7) with respect to n (details in reference 2). For the quadratic utility function used, Eq. (23.3):

$$n^* = \sqrt{\frac{K}{k} \frac{s_1^2 v_1}{(v_1 - 2)}} \qquad (23.9)$$

This equation gives an upper bound for the optimal sample size.

It was found that for engineers using priors 1, 2, and 3 it will not pay to take any soil samples (disregarding the advantage of having actual test data for a potential subsequent lawsuit). The analysis shows that engineer 4 should test a number of samples less than 49. For engineer 5 the optimal sample size defined by Eq. (23.9) is undefined because $(v_1 - 2)$ is zero with his prior.

A simplified optimization procedure was developed in an attempt to shed some light on the rigorous preposterior analysis previously described and to arrive at more explicit answers for engineers 4 and 5. The actual situation was that 39 one-dimensional compression tests had been run on samples of the Bay mud. From these, sample sizes of 5, 10, 15, 20, 25, 30, 35, and 39 were chosen in a random fashion but in sequence without replacing the already selected samples. The resulting sampling distributions (Figure 23.5) were combined with the five priors to obtain a set of posterior distributions. Then the expected net gain was plotted against sample size (Figure 23.6). As before, it may be concluded that engineers 1, 2, and 3 needed no tests, while engineers 4 and 5 should test approximately 10 and 15 samples, respectively.

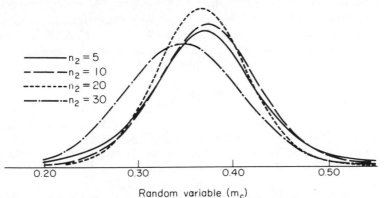

Figure 23.5 Sample distributions for different sample sizes.

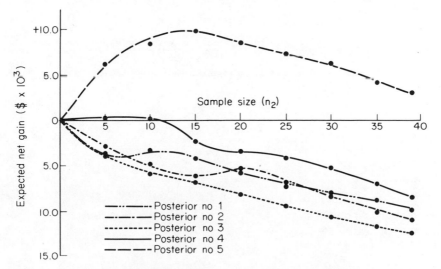

Figure 23.6 Expected net gain versus sample size.

An engineer's decision to invest large sums of money in one-dimensional compression tests on San Francisco Bay mud solely to determine the compressibility does not seem to be the most logical and beneficial. No matter what the number of samples obtained, the inherent variability in the soil compressibility cannot be decreased. It should be recognized that there is an optimal sample size for a given situation depending on many factors, one of which is the engineer's previous experience and judgment.

ACKNOWLEDGMENTS

The work was in part sponsored by a grant from the National Science Foundation, whose support is gratefully acknowledged. The writers appreciate the cooperation of Dames and Moore, Consulting Engineers in the Applied Earth Sciences, and comments and suggestions by Haresh C. Shah. The computations were performed at the Stanford University Computation Center.

REFERENCES

1. BENJAMIN, J. R., AND CORNELL, C. A., 1970. *Probability, Statistics, and Decisions for Civil Engineers*, New York: McGraw-Hill Book Company.
2. FOLAYAN, J. I., 1969. *Reliability of Predicted Soil Settlement*, Technical Report 108, Stanford, Calif.: Department of Civil Engineering, Stanford University, June.
3. RAIFFA, H., AND SCHLAIFER, R., 1961. *Applied Statistical Decision Theory*, Boston, Mass.: Graduate School of Business Administration, Harvard University.

24

Use of Decision Analysis in Airport Development for Mexico City*

RICHARD DE NEUFVILLE AND RALPH L. KEENEY

This study describes the application of decision analysis to a part of a large-scale decision process: selecting a strategy for developing the major airport facilities of the Mexico City metropolitan area. Decision analysis as a formal analytic procedure [see Howard (2) and Raiffa (8)] is designed to aid a decision maker in choosing a good course of action by providing a rationale and methodological procedure for that choice. It is characterized by structuring possible alternatives, creating decision trees which show possible outcomes over time, and the soliciting of judgmental probabilities and utilities on the chance and desirability of certain occurrences. It is particularly pertinent in a case such as this, which involves conflicting objectives, uncertainties about future events, and a long-range planning horizon. The study first describes the technical and nontechnical environment in which the analysis was carried out and the orientation of the analysts to the decision-making process as a whole.

Then objectives and alternatives are structured to identify effective strategies which can be analyzed both in a static and a dynamic sense, where static implies setting all future decisions now and dynamic implies allowing some policy decisions to be made in the future. All these analyses were put together to allow the evaluation of airport development strategies by the central decision maker, the Government of Mexico. Finally, the place and effect of this work within the decision process is discussed.

*Adapted from Chapter 23 in *Analysis of Public Systems*, A. W. Drake, R.L. Keeney, and P. M. Morse, eds., MIT Press, Cambridge, Mass., 1972.

THE PROCESS

Rapid growth in the volume of air travelers, combined with increasingly difficult operating conditions at the existing airport facilities, compelled the Mexican Government to address the question: How should the airport facilities of Mexico City be developed to assure adequate service for the region during the period from now to the year 2000? This was the essential question put to the study team.

As with practically all decision processes about the development of public facilities, even long before formal objectives are sought and defined, the overall question of what to do can be redefined into questions of where, how, and when to implement the services to better focus the problem. Specifically, one needs to be concerned about the following decision variables:

1. The location and/or the configuration of the elements of the system.
2. The operational policy which defines how the services are to be performed and where they will be located.
3. The timing of the several stages in the development.

In this case, because of severe environmental constraints, there are only two different sites adequate for a large, international airport in the Mexico City metropolitan area. The kinds of configurations possible at either site, with respect to the runways for example, were not significantly different in this particular problem. Many different forms of operating the airports, with substantial differences in the quality of service provided, are possible, however. In particular, it is necessary to decide what kinds of aircraft activity (international, domestic, military, or general) should be operating at each of the two sites.

The question of timing is very important, since failure to act at a given time may preclude future options. For example, land available now may not be available in the future when one might want to develop it. On the other hand, premature action can significantly increase total costs to the nation. From the viewpoint of the particular situation involved here, the timing issue and operational policies were the most important aspects of this airport problem.

THE SITUATION

Physical

The existing airport is located about 5 miles east of the central part of Mexico City, on the edge of the remains of Lake Texcoco, a shallow, marshy body of water. The other site is about 25 miles north of the city in an un-

Figure 24.1 General geography of the Mexico City metropolitan area.

developed farming area, near the village of Zumpango. The relative location of the two feasible sites is indicated in Figure 24.1.

Mexico City itself is situated at an altitude of about 7400 ft above sea level. The mountains remain very high in all directions except the northeast. In this direction the range lowers to around 10,000 ft but is still some 3000 ft above Mexico City. Essentially all the flights entering or leaving the Mexico

City area fly over the lower mountains to the northeast, although some do proceed through a smaller and higher pass to the south.

The maneuverability of the aircraft at this high altitude is impaired, especially in hot climates. This requires that the flight patterns be broader than usual and prevents aircraft from safely threading their way through mountainous regions. Thus there are considerable restrictions on the usable airspace around Mexico City. This constraint, which principally affects the capacity of the Texcoco site, is serious since Mexico City handled over 2 million passengers in 1971 and ranks among the busiest airports on the continent.

Access to the airport by ground transportation appears to be reasonable for both sites. The Texcoco site is near the main peripheral highway which can distribute traffic around the suburbs. It is not, however, particularly well connected to the center of the city, to which one has to proceed through congested city streets. The Zumpango site has the clear disadvantage of being farther away, but it can be linked directly to the tourist and business areas via an existing north–south expressway.

The location of Mexico City on a former lake bed makes construction especially expensive at Texcoco. Heavy facilities such as runways not only sink rapidly, but at different rates in different locations, depending on their loads. The existing runways at Texcoco have to be leveled and resurfaced every 2 years. These repairs closed down half the airport for 4 months when they were done in 1971. Because the Zumpango site is on higher and firmer ground, it is not expected to have the same kind of difficulties.

When the Texcoco airport was organized in the 1930s, it was out in the country. But the population of the metropolitan area has grown at the rate of about 5% per year, passing from about 5 million in 1960 to about 8 million in 1971. During this time Texcoco has been surrounded on three sides by mixed residential and commercial sections. This has created problems of noise, social disruption, and safety: a large school is, for example, located under the flightpath some 500 ft from the end of a major runway.

Should a major accident occur on landing or takeoff toward the city it would almost certainly cause many casualties. Using current world-wide accident rates, such an accident could be expected once every 10 years or so (3). Since the approach pattern passes directly over the central parts of the city, noise levels of 90 CNR or more, which is very high, now affect many thousands of people. (The composite noise rating, CNR, is one of many standard indices of noise, which combines decibel level and frequency of occurrence.) These noise levels are bound to persist until at least 1985, until the "quiet" engines now being introduced gradually replaced the more noisy engines. Finally, the construction of any new runways at Texcoco would either require particularly expensive construction in the shallow lake bed or the displacement of up to 200,000 people. A compensating advantage for the Texcoco site is that major facilities already exist.

Institutional

Political power in Mexico tends to be concentrated in the federal government. Ultimately, any decision about a new airport will be ratified by the President. The debate about this decision has been carried on by three major governmental bodies:

1. The Ministry of Public Works, i.e., the Secretaria de Obras Publicas, or SOP.
2. The Ministry of Communication and Transport, i.e., the Secretaria de Comunicaciones y Transportes, or SCT.
3. The Secretaria de la Presidencia, a body with functions similar to those of the Office of Management and Budget in the United States.

PREVIOUS STUDIES

Both SOP and SCT had commissioned large-scale studies of the airport problem within the past few years. The SOP study (10,12), done for its Department of Airports between 1965 and 1967, had recommended that a new airport be built at Zumpango and that all commercial flights be shifted to this facility. The master plan then proposed was not adopted at that time.

The study commissioned by SCT in 1970 (4) resulted in a master plan for expanding the airport at Texcoco by adding new runway and terminal facilities. Interestingly, this report assumed that aircraft could take off away from the city toward the east, and land coming to the city from the east in opposing streams of traffic aimed at adjacent parallel runways. Although this proposal "solves" the noise and capacity problems, its implications for safety are extremely serious at any significant level of traffic, and are unlikely to be acceptable for the expected volumes. This report assumed that "quiet" engines would completely eliminate any noise problems outside the airport boundaries by 1990.

The Government of Mexico did, however, wish to resolve the issue. In early 1971 the new administration committed itself to a restudy. As stated by the President in his State of the Union Message of September 1, 1971, "Construction of a new international airport in the metropolitan area (of Mexico City) is also under study at this time." The study referred to is the one presented here.

EVOLUTION OF THE ANALYSIS

By the time we became involved in the study, the SOP had been formulating the project for a few months. Recognizing that the problem involved multiple conflicting objectives, many uncertainties, and impacts over a long time horizon, SOP had committed themselves to using computers in the analysis process without a formal idea of what analytic techniques would be used. The problem was: How should the alternatives be evaluated? Because of the conflicting plans recommended by the previous reports, one of the first goals was to evaluate various master plans for Zumpango and Texcoco. To do this, a formal decision analysis was carried out; subjective probabilities and utilities were encoded and decision trees were analyzed. The result of this, the static analysis, was a set of optimal time-staged master plans for airport development.

As our work with the SOP progressed, the original problem began to be "solved," but a more basic problem had been uncovered. Namely, we recognized that an overriding issue was: What is the best action to be taken now? Specifically, what is the best decision for the Government of Mexico to take at this time for the development of airport facilities for the capital?

The analysis for this problem clearly had to recognize the opportunities to adjust policy in future years, depending on subsequent events. It was thus a "dynamic" analysis. The original, "static," analysis, where implementation over time was allowed but only fixed strategies were permitted, became a component part of it. As the static analysis was executed first, its presentation precedes that of the dynamic analysis. Hindsight on this project suggests, however, that it might be preferable in future analyses to reverse this process, since the dynamic analysis appears quite useful in defining optimal classes of alternatives which should be subjected to a detailed formal analysis.

Throughout, this study measured and incorporated the best professional judgments of the SOP Director of Airports and his staff in the attempt to analyze the problem from the point of view of the Government of Mexico. As it may be expected that impartial experts might disagree both with the overall structure of the problem as on the details of the analysis, the framework used provides a systematic procedure for examining differences in opinion. This capability was especially important since the analysis was to provide the basis for an orderly advocacy proceeding. Indeed, once the SOP was convinced of what was the best strategy for the government, they had to convince others of this fact. It was also necessary to explain the analysis and its implications to the other major agencies involved, the SCT and the Presidencia. With enough time for a critical review of the study and a serious probing of the differences of opinion between interested parties, the quality of the final decision should be improved.

THE STATIC MODEL

The alternatives for the master plan analysis, directed toward finding the best developmental strategies for the next 30 years, were defined as the sequence of decisions that could be taken at three specified decision points,

1975, 1985, and 1995. No provision was made at this stage for adapting the strategies to account for possible future events. The results of this phase indicated general types of developmental plans which seem attractive.

Objectives and Measures of Effectiveness

The overall objective of the static analysis was to determine how to provide adequate air service for Mexico City for the remainder of the century. To evaluate different alternatives, one needs to specify some measures of effectiveness useful for explicitly considering the effects of each alternative on the various important groups that will be impacted. In this case, the various groups are the government as operator, the users of the air facilities, and the nonusers.

The objective for almost every system design has several dimensions. To get any realistic feeling for the impact of the various alternatives it is therefore necessary to define relevant subobjectives for the problem. The partial objectives chosen for this study were:

1. Minimize total construction and maintenance costs.
2. Provide adequate capacity to meet the air traffic demands.
3. Minimize the access time to the airport.
4. Maximize the safety of the system.
5. Minimize social disruption caused by the provision of new airport facilities.
6. Minimize the effects of noise pollution due to air traffic.

Although there is obviously much overlap, the first two objectives account for the government's stake as operator, the third for the users', and the last three for that of the nonusers.

The measures of effectiveness, X_i, associated with each of these six objectives are reasonably standard in airport planning. They were taken as:

$X_1 \equiv$ total cost, millions of pesos

$X_2 \equiv$ practical hourly capacity in terms of the number of aircraft operations per hour

$X_3 \equiv$ average access time in minutes, weighted by the number of travelers from each zone of Mexico City

$X_4 \equiv$ expected number of people seriously injured or killed per aircraft accident

$X_5 \equiv$ number of people displaced by airport expansion

$X_6 \equiv$ number of people subjected to a high noise level, in this case to 90 CNR or more

The Alternatives

The alternative strategies for the static analysis were defined as the feasible set of all combinations of designs that might be established by different decisions about location, operational configuration, and timing. Two sites were considered, Texcoco and Zumpango. Operations were classed into four categories: domestic, international, general, and military. And three decision points, 1975, 1985, and 1995, defined the time dimension.

The total number of combinations to be considered was thus of the order of $((4)^2)^3$, or 4096. In fact, somewhat less than 4000 alternatives were considered since some initial (1975) choices precluded, in practice, some subsequent choices. One was not likely, for example, to move all operations to Zumpango in 1975 only to switch them all back to Texcoco in 1985. Especially since we were dealing with a six-dimensional set of attributes for each alternative, a computer model was necessary to execute the analysis.

SPECIFYING THE POSSIBLE IMPACTS
OF EACH ALTERNATIVE

To identify the possible impact of each alternative, a_j, it is necessary to assess joint probability distributions over the six measures of effectiveness conditional on each alternative. Thus for each alternative a_j, we wish to assess the joint probability distribution, which we will denote by $P^j (x_1, x_2, x_3, x_4, x_5, x_6)$.

A number of important assumptions were made to simplify the probability assessments. Most importantly, it was assumed that, conditional on a_j, the X_i were mutually probabilistically independent, such that:

$$P^j (x_1, x_2, \ldots, x_6) = P_1^j(x_1)P_2^j(x_2)\ldots P_6^j(x_6) \qquad (24.1)$$

where $P_i^j(x_i)$ is the marginal probability density for X_i conditional on alternative a_j. The reasonableness of the assumption was qualitatively checked by identifying underlying factors contributing to the amounts of each X_i and then examining the degree of overlap. For instance, the size of the airport and quality of facilities would obviously affect both construction costs and capacity. Based on this, it was decided that the probabilistic independence assumption was not "correct" but that it was probably appropriate given the use we had in mind: trying to gather some insights into what were effective developments.

The actual assessments of the marginal probability distributions were done by the leaders of the study in SOP, the Directors of the Departments of Airports and of the Center for Computation and Statistics and their staffs. The fractile technique discussed in Raiffa (8) was used for this purpose.

Thus all the probabilities represented the quantified judgment of a group of people very familiar with the impact of different alternatives. In most cases they had many data to back up their judgment, which, indeed, had formed their judgment.

As an example of the types of data collected for this purpose, consider the probability distribution for access time. First, the Mexico City area was divided into relatively homogeneous districts with the population characteristics of each identified. The airport usage characteristics of these groups were also identified. A central node was designated for each district, and experiments were conducted under many conditions (time of day and week, weather, etc.) to see how long it would take to get to the two airport sites.

Information used for any particular alternative could also be used for evaluating other possibilities. For example, if an alternative had domestic aircraft landing at Texcoco, one could generate the contribution to the access-time distribution from domestic travelers by knowing the percent of domestic travelers from each district.

Using such information, joint probability distributions were generated for each of the 16 alternatives in each of the three years: 1975, 1985, and 1995. The assessments were conducted in group sessions meeting 2 to 3 hours at a time over a period of 1 month. Consistency checks were performed to indicate any "unreasonable" assessments which needed to be altered.

The probability distribution for access time for the "all Texcoco" alternative in 1975 is illustrated in Figure 24.2. The dots represent the fractiles on the curve, which were empirically assessed.

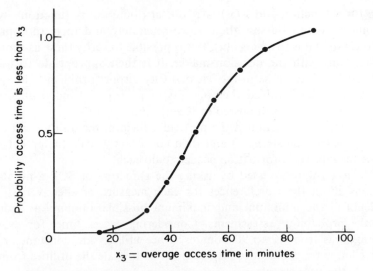

Figure 24.2 Probability distribution for access time from points in Mexico City to Texcoco Airport in 1975.

PREFERENCES FOR THE VARIOUS POSSIBLE IMPACTS

Choosing a "best" alternative requires consideration of both the possible impact of each alternative and the relative preferences of the decision maker for each of these impacts. The probability density functions quantify the possible impacts, and the preferences are quantified by assessing a utility function over the six measures of effectiveness. This utility function $u(x_1, x_2, \ldots, x_6) \equiv u(\mathbf{x})$ will assign a number to each vector of impacts $\mathbf{x}$ such that:

1. The higher of two numbers indicates the preferred set of impacts.
2. In situations involving uncertainty, the alternative with the higher expected utility is preferred.

Given the probabilities for any impact and the utility for every set of impacts, it is straightforward to calculate the expected utility, which is a measure of relative desirability, or preference, for each alternative. This follows from the reasonable axioms of rational behavior (11).

The assessment of a multiattribute utility function is a complex task. It involves interaction with the party or groups involved to quantify their subjective preferences. Details of the assessment are given elsewhere (5), but here we will try to indicate the general idea. In structuring the discussion, we will try to identify assumptions such that:

$$u(\mathbf{x}) = f[u_1(x_1), u_2(x_2), \ldots, u_6(x_6)] \qquad (24.2)$$

where f is a function and $u_i(x_i)$ is a one-attribute utility function over X_i. It is much easier to assess these $u_i(x_i)$ separately and then appropriately scale them than it is to assess $u(\mathbf{x})$. If it is possible to verify these assumptions in discussions with the decision maker, it is then appropriate to use Eq. (24.2). Such was the case in the Mexico City airport problem. The specific assumptions used involved the concepts of preferential independence and utility independence as discussed in Keeney (6).

Once we had ascertained the particular form of the utility function, we needed to assess the six $u_i(x_i)$ and then consistently scale them. Let us illustrate the types of information needed to do each.

The $u_i(x_i)$ were assessed by asking the Directors of SOP a number of questions about their preferences for each measure of effectiveness. First we identified the minimum and maximum possible amounts as indicated from the probability assessments. Considering access time, for example, the range was from 12 to 90 minutes. Since utilities are constant up to a positive linear transformation, it was possible to scale the utilities from zero to one. In this case we set:

$$u_3(12) = 1 \qquad (24.3)$$

and:

$$u_3(90) = 0 \qquad\qquad (24.4)$$

since 12 is preferred to 90. By questioning the group and obtaining a consensus, we found the utility function illustrated in Figure 24.3 using the techniques discussed in Schlaifer (9).

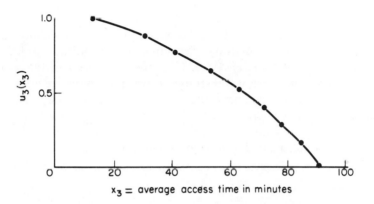

Figure 24.3 Utility function for access time from Mexico City to the Airport.

To scale the $u_i(x_i)$ consistently, we asked the representatives of the SOP to trade off amounts of different attributes. The questions were of the form: "For what probability P_i are you indifferent between: (a) a sure consequence with all attributes but X_i at the least preferred level and X_i at the most preferred level; and (b) an alternative with two consequences, all attributes either at their most preferred level with probability P_i, or at their least preferred level with probability $1 - P_i$?"

To evaluate the impacts that might occur at each of the different points in time considered (1975, 1985, and 1995), it was necessary to determine a means for placing these impacts on a common basis of comparison. For the cost impacts, this was done in the usual way by using discounted present values (1). For all but one of the other impacts, "average" values over years seemed appropriate. Specifically, we used:

1. The average access time weighted by the number of passengers.
2. The expected number of casualties in accidents in the 30-year period.
3. The total number of people displaced.
4. The average number of people subjected to high noise levels.

There was no reasonable way to average capacity over time since the desirability of capacity is highly dependent on the volume of traffic, which is presumed to grow significantly with time. To assess the utility function for the six attributes, over the three decision periods, it was necessary to evaluate an eight-dimensional function in terms of the five average or discounted attributes and three attributes representing capacity at the different times, denoted by $(x_2)_{75}$, $(x_2)_{85}$, and $(x_2)_{95}$. The function to be assessed was then:

$$u(\bar{x}_1, \bar{x}_2, \ldots, \bar{x}_6) = u\{\bar{x}_1, [(x_2)_{75}, (x_2)_{85}, (x_2)_{95}], \bar{x}_3, \ldots, \bar{x}_6\} \quad (24.5)$$

where the $\bar{x}_i$ represent the "averaged" amounts of X_i as defined above. The procedure for doing this using a decomposition as indicated in Eq. (24.2) is described in detail by Keeney (5). Once the assessment of the $u_i(\bar{x}_i)$ and the scaling were completed, it was straightforward to synthesize $u(\mathbf{x})$.

This function $u(\mathbf{x})$ not only allows one to determine the relative preferences for different alternatives but also specifies the value tradeoffs between specific design features; it is possible to specify how much of attribute X_i is worth a specified amount of X_j given levels of each.

The Computer Model

An interactive graphical input–output program was developed to assist in evaluating the alternatives. Computationally, the program was quite simple: given any set of probability distributions and utility functions, it calculated the expected utility for specified alternatives. The interactive capability offered a very flexible and efficient way to do sensitivity analysis of the results. Appropriate graphical terminals for doing this were installed both at the offices of the analytic study team, and in the office of the Secretary and Under-Secretary of SOP, in the Presidencia, and in the President's own offices. This capability was used daily by the SOP and could also be used by the other interested parties to examine the relative merits of alternative developmental policies.

In practice, the probability and utility functions which had been assessed in this study were stored in the computer. Alternative ones could be input by changing the ranges of possible values on the probability distributions, and the single attribute utility functions *could* not be done with the interactive program.

A particularly useful feature of this program was a routine that calculated certainty equivalents. For a given alternative, the certainty equivalent of anyone whose preferences have been assessed for a particular attribute is the amount of that attribute which is indifferent to its probabilistic distribution. Using this routine, the overall possible impact of any alternative could be reduced to an equivalent impact described by a vector of certainty equivalents. This permitted an analysis of dominance, and gave insight into how much of attribute X_i it would be necessary to trade off for a specified amount of attribute X_j for any alternative to be preferred to another.

Results

Two general types of developmental strategies appeared better than all others. These were:

1. "Phased Development at Zumpango," that is, a gradual development of this site over the next 30 years, with some activities remaining at Texcoco.
2. "All Zumpango," implying that all airport activities should be moved to Zumpango by at least 1985.

This conclusion appeared to be fairly insensitive to the specific utilities and probabilities used.

But optimum plans determined by a formal decision analysis, such as those above, do not necessarily translate directly into decisions. The planning horizon, for 30 years in this case, may be entirely different from the decision horizon, which may be only 4 or 5 years.

Indeed, any decision maker does not have to define what he will do for the next 30 years right at the beginning. On the contrary, he would be foolish to adopt a fixed master plan since the future is so uncertain. His better course is probably to make some initial decision and then, based upon subsequent events, to revise his strategies as necessary.

The wise decision maker recognizes that any action taken now can be ambiguous insofar as it is the first step of several strategies, and adapts his decisions to account for significant shifts in political preferences and community priorities. To advise him properly, it is necessary to examine the specific actions that are immediately available to him, taking into account both the utilities that have been formally estimated for the alternatives implied by these actions, and the other considerations which may be or become significant. This was the task undertaken in the dynamic analysis.

THE DYNAMIC MODEL OF GOVERNMENTAL OPTIONS

The purpose of the dynamic model was to decide what action should be taken in 1971 which might best serve the overall objective of providing quality air service to Mexico City for the remainder of the century. This model assumed that the second step in the decision process could be taken in 1975 or 1976, at the end of the current President's 6-year term. The action taken then would depend both on the action taken now and critical events which might occur in the interim. This model was less formal than the static model.

Alternatives for 1971

A very important aspect of the problem involved identifying the alternatives open to the government in 1971. It was decided that the essential differences in 1971 between all the many alternatives involved the degrees of commitment to immediate construction at the two sites. To categorize these into a manageable number, we roughly divided the alternative space into the 16 nominal cases indicated in Figure 24.4. Thus an alternative might be mod-

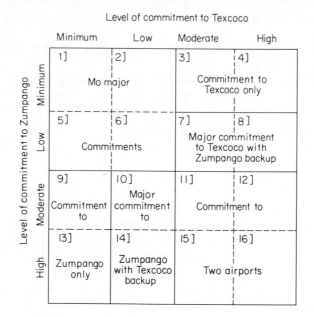

Figure 24.4 The 16 nominal governmental alternatives for 1971.

erate commitment at Zumpango and low commitment at Texcoco. Actually, each nominal case in the figure represents a class of specific alternatives. The idea was to do a first-cut analysis to decide which alternatives were sufficiently viable to be examined in more detail. It should be noted that the two strategies defined by this static analysis could be compatible with all the nominal governmental options except 11, 12, 15, and 16.

The next step involved defining what was meant by the alternatives. Briefly summarized, the alternatives at Texcoco (for the period 1971–1975) were defined as follows:

minimum = maintenance and introduction of safety equipment only.

low   = extend the runways, upgrade support facilities such as terminals, do all routine maintenance, and introduce new safety equipment.

moderate = in addition to that done with a low strategy, buy and prepare land for building a new runway, expand passenger facilities, and so on.

high   = build a new runway and passenger facilities, improve the airport access; in short, build a totally new airport at Texcoco.

Similarly, for Zumpango, we defined the commitment levels:

minimum = at most, buy land at Zumpango.

low   = buy land and build one jet runway and very modest passenger facilities.

moderate = buy land, build a first jet runway and plan others, build major passenger facilities, and build an access road connection to the main Mexico City highway.

high   = build multiple jet runways, major passenger facilities, and access roads, that is, a large new airport at Zumpango.

Critical events that could occur in the period 1971–1975 which were likely to affect the best strategy in 1975 were identified. These involved safety factors and air disasters, shifts in demand in terms of both passengers and aircraft, technological innovations, changes in citizen attitudes toward the environment, and changes in priorities concerning, for example, national willingness to have government funds used for major airport construction.

For each of the 1971 options, the manner in which the strategy should be altered by 1975 in light of each of the events listed above was defined. The purpose of this exercise was to indicate better what the overall impact of a 1971 decision might be. Certain options in 1971 eliminate the possibility of other options in 1975, regardless of the events that occur in the interim.

Objectives

We identified four major objectives that were important in choosing a strategy for airport development. These involved the effectiveness, political consequences, externalities, and flexibility of the various alternatives. The components of the effectiveness are indicated by the six measures of effectiveness covered in the static model. The political consequences of concern were those important to the government: they involved the political effects which would be felt by SOP, by SCT, and by the Presidencia. Flexibility was concerned with the range of options open to the government at the second stage of the decision-making process, with what alterations could be made contingent on events that might occur. Finally, all other considerations that would be important were lumped together as "externalities." These included the number of access roads needed, the distribution of federal expenditures between the Mexico City region and the rest of the country, the distribution of expenditures for airports and other uses, regional development away from central Mexico City, and the national prestige associated with new airport facilities.

First Evaluation of Nominal Alternatives

The 16 alternatives, defined in Figure 24.4, were evaluated in a series of extensive discussions between the Directors of the Department of Airports and of the Center for Computation and Statistics, other staff members of SOP, and ourselves.

A preliminary evaluation indicated that 7 of the 16 alternatives could be discarded. Alternative 1 did not provide for maintaining the present service levels due to anticipated increases in demand. Alternatives 7, 8, 11, 12, 15, and 16 were undesirable since a high level of commitment to Texcoco in 1971 would make it the major airport for the near future and remove the need for simultaneous construction at Zumpango. Finally, since the location of the new runway specified by the moderate Texcoco commitment would require new passenger facilities, there was not much difference between options 3 and 4, so they were coalesced into alternative 3–4.

The next stage of the analysis involved having the members of SOP rank the remaining broadly defined alternatives on the attributes of flexibility, political effects, externalities, and effectiveness, as described before. The particular rankings were arrived at by open discussions, and, as such, they represent the consensus judgment. When some alternatives were "indistinguishable" on a particular attribute, they were assigned the same ranking. For the political considerations and externalities, the assessments on the components were first carried out, and then the overall ranking for these attributes was established. The ranking of alternatives according to effectiveness was provided directly by the results of the static model.

The results of the first ranking effort are shown in Table 24.1, where the smaller numbers represent the better rankings. From this table it can be seen that alternatives 3–4, 9, 13, and 14 are each dominated by others on the basis of their overall rankings for the four measures of effectiveness. Alternative 6, for instance, is better than alternative 14 in terms of all four of the measures. Hence alternative 14, and likewise alternatives 3–4, 9, and 13,

Table 24.1 Preliminary Evaluation of Plausible Governmental Options for 1971 by Rank Order

Alternative	Attributes										
	Political Effects on:					Pres-tige	Externalities Due to:				Effec-tiveness
	Flexibility	Presi-dencia	SOP	SCT	Overall		Reg. Dev.	Bal. of Fed. Exp.	Roads	Overall	
2*	1	1	8	2	3	4	4	1	1	3	7
3–4	7	4	5	1	4	1	4	6	3	7	8
5*	2	3	6	4	3	3	3	2	1	1	3
6*	3	2	7	3	2	3	3	3	1	3	1
9	4	6	3	6	5	2	2	4	2	2	4
10*	5	5	4	5	1	2	2	5	2	4	5
13	6	8	1	8	7	1	1	6	4	5	2
14	8	7	2	7	6	1	1	7	4	6	6

*Dominant on overall ranking of four attributes.

can be dropped from further consideration. The dominant alternatives were those represented by the nominal cases 2, 5, 6, and 10.

Final Analysis of Governmental Options

To refine the analysis of the possible governmental decisions, it was necessary to define the dominant alternatives more precisely. This was done as follows:

2.   At Zumpango, do no more than buy land for an airport. At Texcoco, extend the two main runways and the aircraft apron; construct freight and parking facilities, and a new control tower. Do not build any new passenger terminals.

5A. Build one jet runway, some terminal facilities, and a minor access road connection at Zumpango. Buy enough land for a major international airport. At Texcoco, perform only routine maintenance and make safety improvements.

5B. Same as alternative 5A, only buy just enough land for the current Zumpango construction.

6.   Extend one runway at Texcoco and make other improvements enumerated in alternative 2. Buy land for a major international airport at Zumpango, and construct one runway with some passenger and access facilities.

10.   Same implications for Texcoco as alternative 6. Build two jet runways with major passenger facilities and access roads to Zumpango.

These five alternatives were ranked in the manner previously described. The results are given in Table 24.2. Proceeding as before, we can quickly see that alternative 6 dominates 10; and alternatives 2, 5A, and 6 all dominate 5B. The three remaining viable alternatives are 2, 5A, and 6.

Table 24.2 Final Evaluation of Governmental Options for 1971

| Alternative | Attributes | | | |
	Flexi-bility	Political Effects	Exter-nalities	Effec-tiveness
2*	1	4	4	3
5A*	2	3	3	2
5B	4	5	5	4
6*	3	1	1	1
10	5	2	2	1

*Dominant on overall ranking of four attributes.

The relative advantages of these three options were, finally, subjectively weighed by the SOP personnel as follows. Alternative 6 ranks better on effectiveness, externalities, and political considerations than either 2 or 5A. Although it is worse in terms of relative flexibility, it does allow the government to react effectively to all the critical events that might occur between 1971 and 1975, when the second stage of the airport decision could be made. Hence, in the opinion of the members of SOP working on this problem, alternative 6 was chosen as the best strategy.

OVERALL RESULTS

The final recommendation developed from the static and dynamic analyses was that a phased development should begin at Zumpango but that no final commitment should be made as to the final size or ultimate impact of this program. Specifically, it was suggested that land be acquired at Zumpango, that a major runway and modest terminal facilities be planned for immediate construction. It was also proposed that the government reserve until 1975 or 1976 a more detailed decision on how the airport facilities for Mexico City should be developed.

In a major sense, the process for arriving at this final result was as important as the recommendation itself. Indeed, the analysis revealed many important issues to the staff of the SOP, for example the importance of flexibility, and changed their original intuitions as to the nature of the optimal solution. Whatever the final decision that will be evolved by the Government of Mexico, it can be done with greater awareness of the relative importance of the different attributes and of the dynamic issues.

The analyses described in this report were completed in early September 1971. The staff of the SOP then began a planned series of discussions within the Secretariat and greatly extended the original studies. In November the study was formally presented by SOP to the President. In a sense, this marked the formal end of the analysis.

It is not possible to define precisely what were the exact consequences of the analysis. It would appear, from all the evidence that the participants were able to assemble, that the essential recommendations of the study were followed. Specifically, the SOP was authorized to spend money to prepare formal design plans and specifications, and to lay the legal groundwork for the acquisition of land at Zumpango. On the other hand, they had not, as of early 1973, actually been authorized to buy the land or to begin construction. Money was, however, actively being sought from the World Bank to finance this major effort. It would thus appear that the President's decision, which as of 1973 had never been formally announced as a decision, was actually very close to what the study had recommended: an active pursuit of the strategy of large-scale construction at Zumpango, without a binding irrevocable commitment.

DISCUSSION

The analysis described here did not lead to an immediate public decision about the development of facilities. It may even not have been especially influential. Indeed, as decision-analytic approaches have hardly ever been used for the evaluation of public systems, it is appropriate to consider whether, and to what extent, these approaches are useful in this context. Specifically, it is interesting to ask particular questions about this project, which are addressed in a more general frame by Keeney and Raiffa (7): Why was the decision analysis chosen? What was its contribution? What should its role be in the analysis?

In some respects this decision-analysis technique was chosen for spurious reasons. As indicated previously, leaders of the SOP had essentially made a commitment to the use of computer-based methods in the belief that this was a progressive way to proceed. This kind of situation is a typical problem in practically all situations: it is a really rare occasion when the institutional framework is not predisposed to a particular form of analysis and process of design. The analyst should recognize this fact of life and deal with it in formulating his own special version of the analysis.

As it happens, we believe that the use of computer-based analysis was an appropriate choice for the SOP; it represented the only realistic means to sort out the huge number of combinations of alternatives while taking reasonable account of the six dimensions of effectiveness. The one change that might have been made, had there not been the original mandate to use the computer analysis, would be to start off with the dynamic analysis. This would have reduced the number of alternatives analyzed formally by an order of magnitude, possibly, and made the static computer analysis more efficient.

The contribution of the procedures used should be understood in terms of a method for decision analysis rather than for decision making. Given the full cooperation of a party to a decision, either a person or a homogeneous interest group, the analysis permits one to sort out the complex issues and to provide them with a systematic ranking of their alternatives which is consistent with their real preferences. This process can be, and apparently was in this case, very useful to the persons for whom the analysis is done. It is certainly vastly preferable to the alternatives—no analysis or a haphazard approach.

This decision analysis does not, however, lead directly to a decision unless the party for whom the analysis is done is, in fact, the decision maker and is capable of acting according to its preferences. In this instance the SOP's preferences provided only one element of the decision, although a major one. The contribution of the decision analysis was to define important issues and provide a logical, consistent framework for analysis by interested parties.

In general, decisions about collective facilities are collective choices to some degree. In such a content, decision analysis can be used to define the multiattribute utility functions for the several parties to a decision. It does not, itself, provide the mechanism to resolve the expected, legitimate conflicts between groups with different interests or different perceptions of what they view as the public interest. This is more the domain of negotiation, about which relatively little is known.

ACKNOWLEDGMENTS

We thank Subsecretario R. Felix, F. J. Jauffred, and F. Dovali R., and the members of their staff in the Secretaria de Obras Publicas, Howard Raiffa of Harvard University, and Leonard Blackman of MIT, all of whom we had the pleasure of working with on this study.

REFERENCES

1. DE NEUFVILLE, R., AND STAFFORD, J. H., 1971. *Systems Analysis for Engineers and Managers*, New York: McGraw-Hill Book Company.
2. HOWARD, R. A., 1968. The Foundations of Decision Analysis, *IEEE Transactions on Systems Science and Cybernetics*, Vol. SSC 4, No. 3, Sept., pp. 211–219.
3. *ICAO Bulletin*, 1971. Safety of Scheduled Traffic, May, pp. 42–44.
4. Ipesa Consultores and the Secretaria de Comunicaciones y Transportes, 1970. *Estudio de Ampliacion del Aeropuerto Internacional de la Ciudad de Mexico*, Mexico, D.F., Oct.
5. KEENEY, R. L., 1971. Utility Independence and Preference for Multiattributed Consequences, *Operations Research*, Vol. 19, No. 4, July-Aug. pp. 875–893.
6. KEENEY, R. L., 1972. *Multiplicative Utility Functions*, Technical Report 70, Cambridge, Mass: Operations Research Center, MIT.
7. KEENEY, R. L., AND RAIFFA, H., 1972. A Critique of Formal Analysis in Public Decision Making, Chapter 5 in *Analysis of Public Systems*, A. W. Drake, R. L. Keeney, and P. M. Morse, eds., Cambridge, Mass: MIT Press.
8. RAIFFA, H., 1968. *Decision Analysis: Introductory Lectures on Choices Under Uncertainty*, Reading, Mass: Addison-Wesley Publishing Company, Inc.
9. SCHLAIFER, R. O., 1969. *Analysis of Decisions under Uncertainty*, New York: McGraw-Hill Book Company.
10. Secretaria de Obras Publicas, 1967. *Aeropuerto Internacional de la Ciudad de Mexico*, Mexico, D.F., Sept.
11. VON NEUMANN, J., AND MORGENSTERN, O., 1953. *Theory of Games and Economic Behavior*, 3rd ed., Princeton, N.J.: Princeton University Press, 1st ed., 1944.
12. Wilsey y Ham de Mexico, S.A. de C.V., 1967. *Aeropuerto Internacional de la Ciudad de Mexico*, Mexico, D.F., July.

25

Planning for the IBM System/360*

Ruth P. Mack

Decision itself is only one incident in the life history of an entire decision or, rather, deliberative process. That history starts with the events that bring some problem into focus and ends with the last efforts to correct and carry through on actions chosen. Uncertainty permeates the entire history, since knowledge is incomplete throughout. The task of this case study is to examine the whole time-consuming process to achieve a perspective on the role that uncertainty plays in it, and what this implies about how to view the ongoing deliberative process. The case illustrates one class of decision situation: decisions that are not routine or made repetitively, and in which many people must participate in an open-ended, creative, and adaptive fashion.

Such decisions are characterized by diffuse structure, in contrast to those which are well structured. Thinking about a decision situation in terms of six structural attitudes helps to identify the basic nature of the decision problem and, consequently, the type of approach likely to be useful to its solution. The attributes of well structured decisions are:

1. Homogeneity of the decision agent and focus of the decision in time and space.

*Adapted from *Planning on Uncertainty: Decision Making in Business and Government Administration* by Ruth P. Mack, "Characteristics of the Deliberative Process," Chapter 9 and "Intendedly Rational Conduct of Deliberative Processes," Chapter 11. Copyright © 1971, by John Wiley & Sons, Inc., New York. By permission of the author and John Wiley & Sons, Inc.

370

2. Capacity on the part of the decision-making group to perceive, retain and process complex information in a highly rational fashion.

3. Good quality of the information bearing on expected outcome.

4. The clarity with which goals and utilities can be defined and associated with specific outcomes.

5. Seriability (versus relative uniqueness) of possible outcomes.

6. The potential for innovative action (higher potential implying greater indeterminacy and poorer structure).

Any decision situation can be rated with respect to how well structured are each of its attributes. In the IBM case all attributes bespeak a poorly structured diffuse decision situation.

STAGES OF DECISION PROCESS

Although any deliberative process is continuous, it can sometimes be divided into stages, some of which are clearly distinguishable, some are not. What is important is not the way subdivisions are made so long as the totality of the deliberative process is kept in mind. With that caveat, five stages can be usefully distinguished. First, awareness of a problem must develop to a point where there is some readiness to start to formulate a solution. Second, contemplation must develop specific alternatives, which offer possibilities of improving the situation. Third, a choice among the alternatives must be made. Fourth, the decision must be put into effect, and this is likely to require a number of subsidiary determinations. Fifth, since uncertainty precludes a sure and perfect decision, it is typically necessary to correct and supplement the initial one. These five stages form a decision cycle.

The fourth and particularly the fifth stages imply that decision cycles are recursive: they tend to spiral in developing series of partial approximations of previous and parallel processes. Hence the five stages are themselves often only one phase of a broad deliberative–administrative process.

Much can be learned about decision making by studying the decision process; consequently a great deal of productive work has gone into formal and informal analyses, and a wide variety of formulations has emerged. A large part of the differences in these formulations for the decision process derives from the variations of administrative situations on which they focus, rather than from disagreements as to the character or substance of the deliberative process itself. These descriptions range, to mention just a few, from the rational and orderly deliberative process pictured by managerial economics (8, 10, 14), to the exploratory, remedial, and nibbling process, "disjointed incrementalism," which David Braybrooke and Charles Lindblom see as the nature of day-to-day decisions made by politicians, administrators, and executives (2). Between these views are many others. Systems analysis, for example, features a long and complicated process of circling in on a problem, beginning with study of the broad systemic context of an ongoing program (7). In their models of a process of delineating alternatives and choosing, Herbert Simon and Edward Litchfield accommodate the relationships among individuals and organizations and their environments (9, 12, 13). William Gore, often focusing on a legislative context, highlights the strategic interplay among individuals in a "heuristic" decision process of which the central dynamic is the inducement of shared conceptions of the problem (6). Insofar as these approaches define stages of the decision cycle paralleling those I have formulated they are summarized in Table 25.1. One important question to be asked is whether adequate descriptions of such processes help to arrive at normative conclusions.

THE IBM SYSTEM/360 CASE

It will be useful to translate the five-stage decision process into a concrete example. The literature is lamentably thin in examples of how actual administrative decisions are made, although a few interesting case studies are available (1, 4, 5, 11). However, there is a fortunate exception in *Fortune Magazine*'s account of the genesis of the System/360 computers at International Business Machines Corporation (15). The account has been rearranged in accordance with the stages of the decision process.

Stage 1: Deciding to Decide—Problem Recognition

Prior to early 1961, discontent seemed to be accumulating at IBM with respect to the computer line and the organizational structure for designing new machines. Dissatisfactions with product line centered on lack of an overall concept, overlap on the capabilities of different models, unavailability of some capabilities which competitors had developed or appeared to be on the verge of developing, too much supplementary equipment, and, of particular importance, too great a demand on programming to meet the great variety of equipment and customer needs. Discontent concerning the organizational structure focused on the relationship between the international and the United States divisions of the company, inefficiency resulting from the fact that technological innovation was spread over 15 to 20 engineering units, and difficulties with respect to headquarter's ordering and control of innovation and increased stress on programming.

Two sorts of external changes exacerbated the dissatisfactions. The scientific underpinnings and the basic technology of computer design were undergoing change. Some years earlier this had been recognized in a general way when, in 1956, Emanuel Piore, an outstanding scientist in the field, joined

Table 25.1 Decision Cycle Stages According to Several Authors

Mack	Gore	Simon	Litchfield	Hall	Managerial Economics
1. Problem recognition	Recognition of strain; consideration of how far to open things up	Awareness that present situation is no longer "satisfactory" in view of realistic aspirations	Definition of issues; analysis of existing situation	System studies	Planning
2. Specification of alternatives	Forming a response; delineating criteria	Discovery and design of satisfactory possible courses of action	Calculation and determination of alternatives	Focusing on a problem area; developing alternative systems to satisfy objectives	Search activity
3. Choice (decision proper)	Focusing on something manageable, for which a coalition can be formed	Evaluating alternatives in an "intendedly rational" fashion different for programmed and unprogrammed choices	Deliberation and choice	Detailed enumeration of consequences of each alternative and selection among them	Decision
4. Effectuation	Mounting a response, reforming coalitions, and generating actions		Programming and communicating	Making general objectives operational; communicating task to the engineers	Engineering design
5. Correction and supplementation			Cycles of action: control and reappraisal	Final action phase: evaluation as long as system is in effect	

the company. Nevertheless, the product line was still fundamentally based on "first-generation technology." Customer demand was also shifting. Broadly speaking, business requirements emphasize a large capacity to store information whereas scientific needs are for large computational capacity. Now, however, a greater sophistication on the part of business users resulted in more interest in computational capacity.

Stage 2: Formulating Alternatives and Criteria

The formulation of alternatives and criteria for an acceptable solution was influenced by the history of previous events. One set of events consisted of funeral services for a number of previous efforts. Program "STRETCH," for example, died in May 1961; this was a giant computer that never did better than about 70% of its promised performance and was dropped at a huge loss. A second corpse was a big computer program based on transistor technology, the 8000 series. A third corpse was a machine called SCAMP, designed by the London World Trade Corporation (an IBM subsidiary), which overlapped with the IBM computer 7044.

The decisions to abandon individual products were partly based on interest in a more comprehensive plan, in a computer line that would blanket the market. Accordingly, one division of the company started to work on the "New Product Line." It is possible that the basic idea of the unified line had already been conceived and that some critically placed individuals felt their future to be identified with accomplishing its gestation and birth. If so, in reality the alternatives for the next few years were the unified product line versus the status quo. But this is conjecture.

Discontent culminated in the fall of 1961, in the appointment of a committee called "SPREAD" (an acronym for "systems programming, research, engineering, and development"). The committee included, among others, men who were prime movers in two of the abandoned previous efforts.

The group was closeted together for long intervals during the fall of 1961, and finally, in December, they brought out an 80-page report which made the following recommendations: A new line should be designed aimed at replacing the entire computer line. It should open up a whole new field of computer applications. Compatibility among the various new machines was essential: however, compatibility between a new machine and the comparable members of the old line was not important. The system must be useful for both business and scientific users. To meet this requirement, each new machine in the new line would be made available with core memories of varying size. In addition, the machines would provide a variety of technical features such as "floating-point arithmetic," "variable word length," and a "decimal instruction set" to handle both scientific and commercial assignments. Finally, all peripheral equipment must have standard interfaces so that various types and sizes could be hitched to the main computer.

In this decision, as is so often the case, only one alternative was sufficiently formulated to provide a challenge to the status quo.

Stage 3: Decision Proper

The decision proper was made at a meeting on January 4, 1962, at which the report of the SPREAD committee was presented. The top brass of the corporation was present. In the words of T. Vincent Learson, now president of the corporation (17),

> There were all sorts of people up there and while it wasn't received too well, there were no real objections. So I said to them "All right, we'll do it." Learson continued with some statements that suggested the kind of evaluations that had taken place: The problem was, they thought it was too grandiose. The report said we'd have to spend $125 million on programming the system at a time when we were spending only $10 million a year for programming. Everybody said you just couldn't spend that amount. The job just looked too big to the marketing people, and the engineers. Everyone recognized it was a gigantic task that would mean all our resources were tied up in one project, and we knew that for a long time we wouldn't be getting anything out of it.

In the light of the many statistical techniques for analyzing even complex decisions, it is interesting that the story reports few efforts at numerical evaluation of either the risk or payoff. There is no evidence of foreknowledge that approximately $5 billion would be spent on the program over a period of 4 years. But, of course, the decision of January 4 did not irrevocably commit anything like this sum. Had the next year's work proved clearly disappointing it would no doubt have been possible to retreat from the particulars of the commitment, and perhaps even from the general plan, at a cost, however huge, vastly less than the actual expenditure on the completed venture.

Stage 4: Effectuation

The January 1962 decision was parent to a long line of development work. A number of engineering decisions were required. For example, monolithic circuits were abandoned in favor of a new sort of hybrid circuit. Late in 1962 it was determined that the new line would have operating systems by means of which computers schedule their own operations without manual interruption. A number of primarily managerial decisions were likewise part of the development work. In March the company made a new departure when it determined to construct a plant to manufacture components; this involved the purchase of $100 million of automatic equipment. Late in 1963 it was determined to announce the whole 360 line at once, instead of sequentially as originally intended. Finally, in March, before announcing the model to the public, a "risk-assessment" session was held at which 30 top executives went over all the details of the new program: patent protection, a policy on computer returns, and the company's ability to hire and train an enormous work force in the time allotted. Throughout March and April a sequence of pricing decisions took place. The production people figured a price at which a machine could be produced. The marketing men calculated the volume at the price, and at other prices, and then an additional "loop" was traversed to see whether, if the price were lower and the volume correspondingly higher, profit would increase. "We reviewed the competitive analysis for perhaps the fifteenth time. We had to take into consideration features that could be built in later with the turn of a screwdriver but that were not to be announced formally. We were pulling cost estimates out of a hat" (18).

On April 7, 1964, $2\frac{1}{4}$ years after the "we'll do it" decision, the new line was announced: six separate compatible computer machines, with their memories interchangeable so that a total of 90 different combinations would be available. Peripheral equipment provided 40 different input and output devices. Delivery would start in one year, April 1965.

Stage 5: Correction and Supplementation

Unforeseen events required some change in the IBM line. For example, in the fall of 1964, the Massachusetts Institute of Technology purchased one of the new General Electric 600 line of computers with time-sharing capability. Soon thereafter the Radio Corporation of America announced that it would use the pure monolithic integrated circuitry in some models of its new Spectra 70 line. As a result of these and other challenges, IBM announced some additions to the 360 line in 1964 and 1965, which surrendered some of the insistence on full compatibility. The changes included addition of the 360/90 model, a super computer type designed to be competitive with Control Data's 6800; the 360/44 model designed for special scientific purposes; the 360/67 model, a large time-sharing machine; and the 360/20 computer in the low end of the market.

By the fall of 1966, the new line offered nine central processors (six had been annouced in 1964), and the supplementary devices had increased to over 70.

Recursive Cycles—Subsequent and Parallel

The printer's deadline for the October 1966 issue of *Fortune* may be thought of as the curtain on the first decision cycle. The sequence of decision cycles concerning the computer line included counterparts, some simultaneous and some subsequent, concerning programming and actual production.

Programming appears to have produced mammoth difficulties, far larger than anticipated. The actual construction of the new line to meet delivery schedules strained every nerve in the company. Recursive cycles of related decisions will inevitably continue as long as the new line is on the market.

Reflections on System/360

The startling thing about the IBM case history is that, although all the customary tenets of decision theory are relevant to it, they cover so little of what occurred. The case certainly involved a deliberative, ongoing, staged, recursive, administrative process in which all five stages were traversed. Stages 1 and 2 were important, yet one has the feeling that a very large part of the real stuff of problem recognition and the search among alternatives is invisible. The reason may be that it took place inside the top man's mind. Is it not reasonable to visualize a man, who senses in his blood the long history of the company, saying to himself, "the computer field is reaching the stage of maturation where many able competitors joust for the prize; what can Richard the Lion-Hearted do at his prime that others cannot do and will not be able to do for some time? The answer lies in taking advantage of the extraordinary mystique of the company, the vast marketing organization, the superb staff, and the almost limitless financial resources." Whether or not any such musings ever occurred between dawn and breakfast in this case, they certainly often do. And if they antecede the right person's breakfast, the rest of "the decision process" can be simply a history of giving the idea substance and defending it against the status quo.

The second striking divergence in the IBM case was in stage 3, the calculation of expected payoffs and attendant risks. For one thing, many of the payoffs of the decision were peripheral to designing a product that would sell at a profit, although these have been substantial. The program pushed IBM itself into feats of performance in manufacturing, technology, and communications that its staff did not believe were possible when the project was undertaken. IBM is now a major manufacturer rather than a company specializing in assembly, service, and marketing. It is an integrated international manufacturing and sales company, whereas previously international subsidiaries specialized in marketing; there has been a fundamental shake-up in the major personnel of the company. A transatlantic line has been leased to facilitate constant communication between the United States and foreign offices; international communication reaches into the analyzing, correlating, and sorting of information and this may bypass communications systems by common carriers. It is, in general, a more sophisticated and more thoroughly integrated organization than it was in 1962.

Not only does the gross benefit, and its likelihood, appear to have eluded quantitative estimation, but the cost may have been impossible to prejudge.

The account discusses calculations concerning programming and its large errors but says little concerning the development and manufacture of the line itself. The actual cost of about $5 billion over 4 years appears to have represented on the order of 2 years' gross revenue from the general-purpose computer program prior to the changeover, or about 14 years' gross profits on that business. Of course, this sum was put at risk incrementally as time passed and information accumulated. One view of the character of the gamble was perhaps adequately expressed by Bob Evans, the manager closest to the firing line. "We called this project 'You bet your company,'" but it was a "damn good risk, and a lot less risk than it would have been to do anything else, or to do nothing at all" (16).

The company may have been on the block as Evans said, but not in the usual sense. For the bet was really not *of* but *on* the company, on its power as an organization to bite hold of a gigantic conception and shape it slowly and strongly to its purpose. In any event, the moment of conventional "decision" seems to have been not much more than a day's meetingsmanship and a flamboyant gesture, "all right, we'll do it," by a man with a steel arm.

The IBM case, then, involved all five stages of the decision process, though much of the iceberg was submerged, particularly for stages 2 and 3. It was itself part of ongoing cycles of decision; there had been preceding, parallel, and subsequent decision processes; and there will be more to follow.

It is easy to conjecture other decision situations in which the relative importance of the stages differs widely, or situations that are well structured throughout and consequently for which statistical techniques warrant prominence. In any case, it is instructive to consider, as has been done here, each of the five stages.

THE DELIBERATIVE PROCESS CHARACTERIZED

Decisions exemplified in the design of System/360 are not in their entirety decisions in the usual sense of a choice among largely predelineated alternatives. They are, rather, iterative exercises in the perception, exploration, and solution of problems. A decision is grown, not made. Thoughtful consideration and striving must be involved throughout. Moreover, since interpersonal strategies of a broadly political nature are often part of the process, "deliberative" must be broadly defined to include person-oriented as well as task-oriented dimensions. The process is ongoing in that perception of and effort to solve one problem takes time and, in addition, opens up new problems; the continuity often has a spatial as well as a time dimension. It is staged in that the process runs through the five stages identified. It is recursive since a five-stage sequence often starts over again, after a first go-round, in continuous cycles related to yet different from the first. It encompasses deliberative aspects of administrative activities closely intermeshed with most other aspects of administration.

Needless to say, there is greater freedom in how the ongoing deliberative process can be conducted than is present in its third stage alone, decision among predelineated acts. Indeed, the process as a whole is open at its beginning, middle, and end. Whether a problem is perceived, recognized, and tackled is elective as is its termination. How solutions are conceived and structured is subject to all sorts of determinations, including how much is undertaken in each decision, the length and timing of each cycle, the amount of information to be used or generated, and the sorts and amounts of inputs to be committed to the decision process. The matter of inputs is central since the process should be seeking, innovative, experimental, and dendritic both with respect to improving goals and actions. This demands not only different sorts of capabilities but also a different sort of engagement of staff and organization with the decision process than is essential when a narrower, more structured system framework is appropriate.

REFERENCES

1. BOCK, E. A., 1963. *State and Local Government; A Case Book*, Montgomery, Ala.: University of Alabama.
2. BRAYBROOKE, D., AND LINDBLOM, C. E., 1963. *A Strategy of Decision: Policy Evaluation as a Social Process*, London: Collier-Macmillan.
3. CHAMBERLAIN, N., 1962. *The Firm, Micro-economic Planning and Action*, New York: McGraw-Hill Book Company.
4. CYERT, R. M., AND MARCH, J. G., 1963. *A Behavioral Theory of the Firm*, Englewood Cliffs, N.J.: Prentice-Hall, Inc.
5. CYERT, R. M., SIMON, H. A., AND TROW, D. B., 1956. Observation of a Business Decision, *Journal of Business*, Vol. 29.
6. GORE, W., 1964. *Administrative Decision-Making: A Heuristic Model*, New York: John Wiley & Sons, Inc.
7. HALL, A. D., 1962. *A Methodology for Systems Engineering*, Princeton, N.J.: D. Van Nostrand Co.
8. HARLAND, N. E., CHRISTENSON, C. J., AND VANCIL, R. F., 1962. *Managerial Economics*, Homewood, Ill.: Richard D. Irwin, Inc.
9. LITCHFIELD, E., 1956. Notes on a General Theory of Administration, *Administrative Science Quarterly*, Vol. 1, pp. 3–29.
10. MILLER, D. W., AND STARR, M. K., 1960. *Executive Decision and Operations Research*, Englewood Cliffs, N.J.: Prentice-Hall, Inc.
11. NELSON, R. R., ed., 1962. *The Rate and Direction of Inventive Activity: Economics and Social Factors*, A Conference of the Universities' National Bureau Committee for Economic Research, Princeton, N.J.: Princeton University Press.

12. SIMON, H. A., 1958. The Role of Expectations in an Adaptive or Behavioristic Model, in *Expectations, Uncertainty and Business Behavior*, Mary Jane Bowman, ed., New York: Social Science Research Council.
13. SIMON, H. A., 1960. *The New Science of Management Decision*, Ford Distinguished Lecture Series, Vol. 3, New York University, New York: Harper & Row, Publishers.
14. SPENCER, M. H., AND SIEGELMAN, L., 1964. *Managerial Economics*, Homewood, Ill.: Richard D. Irwin, Inc. (rev. ed.).
15. WISE, T. A., 1966. IBM's $5,000,000,000 Gamble, *Fortune Magazine*, Part I, Sept., p. 118, Part II, Oct., p. 138.
16. WISE, T. A., *ibid.*, Sept. 1966, p. 118.
17. WISE, T. A., *ibid.*, Sept. 1966, p. 228.
18. WISE, T. A., *ibid.*, Oct. 1966, p. 206.

26

Reaching Decisions About Technological Projects with Social Consequences: A Normative Model*

MARVIN L. MANHEIM

In such areas as transportation, water resources, urban development, and others, professionals and the public have in the past seemed to accept a model of the planning process in which the professional was entrusted with very substantial responsibility. The professionals defined the problem; proposed, analyzed, and evaluated the alternatives; and recommended a course of action to higher authority. Until a few years ago, the professionals' recommendation was generally accepted.

This model is no longer accepted by large segments of the public. The technically trained professional can no longer operate in a vacuum, making decisions about large-scale public-works projects in an abstract, supposedly "objective" way. As we shall see shortly, even such powerful, "objective" techniques as benefit–cost analysis can no longer be accepted as the major basis for decisions about public projects. For, in fact, these techniques are not value-free. The public no longer believes in the objectivity of the professional's analysis, and is unwilling to accept his recommendation unquestioningly. The traditional model of the role of the professional in reaching a decision through "objective" analysis is no longer viable.

What is needed is a new model for the process of reaching decisions about large-scale public-works projects. The objective of this chapter is to suggest one possible model for this process, in which the roles of the technical

*Adapted from a paper presented to the Seventeenth North American Meeting, Regional Science Association, Nov. 6–8, 1970. A revised version of this paper was also published in *Transportation*, 2, 1973 pp. 1–24.

professional are different from the more traditional model. Fundamental to the model is a conviction about the role of the technically trained professional in the process of reaching decisions. We suggest that this new definition of role can go a long way toward restoring public confidence in the professionals.

ISSUES IN EVALUATION OF TECHNOLOGICAL PROJECTS WITH SOCIAL CONSEQUENCES

The subject of our concern is those large-scale technological projects which have wide impacts on different groups in the society of a region. Typical of such projects are water resource systems, such as flood control, irrigation and other single or multipurpose projects; transportation projects, such as highways, airports, and transit systems; urban development projects, such as neighborhood development programs, renewal or rehabilitation projects; and so on. Such projects are technological in that there are intricate technical details to be worked out, involving the uses of land, labor, materials, and machines. These projects have social consequences, in that many different groups will be affected by the actions taken.

Incidence of Impacts

A key issue in such projects is the wide variety of groups which typically will be affected. In general, such technological projects will benefit some groups, while others receive no benefits or are, in fact, harmed.

Consider, for example, the construction of a new highway in a densely developed urban area. A highway provides improved service for traffic and relieves to some extent traffic congestion on local streets. A highway may create greater accessibility, thereby stimulating greater development of portions of the area served by a highway. Location of the highway along a particular route may serve well some groups of travelers, while serving others not so well. In another location, the benefits of various groups may be reversed. The choice of a highway rather than a mass transit link may result in higher accessibility for some groups and no improvement, or even reduction of accessibility for others. A highway can also be a disruptive force in the community, by displacing families or jobs; separating people from access to parks, schools, or churches; creating a visual barrier or despoiling a scenic area; or taking parkland or other community facilities out of public use. Construction of highways may change the patterns of air pollution and water flow and may affect land values in a variety of ways. In short, highways cause a wide variety of impacts on many different groups; some groups will benefit, and some will lose.

Highways are not unique in this respect. Major technological projects have potentially far-reaching effects. Each such project generally constitutes a major public intervention in the delicate fabric of society. Many different people and interests can be affected. The total set of these effects, on all groups, must be considered, with particular attention paid to the differential effects—which groups gain, which lose.

It is essential that the process of planning, designing, implementing, and operating large-scale technological projects explicitly recognize and take into account the issues of social equity represented by the differential impacts on various groups. The planning and design of such systems is as much a sociopolitical problem as a technoeconomic problem. For example, a major state highway agency recently declared its policy thus (4):

> A primary goal of the State Highway Program is to provide highway facilities which in their location and design, as well as in their transportation functions, reflect and support the environmental values and community planning objectives of the areas through which they are proposed. These considerations are being weighted increasingly more heavily by the Highway Commission in its route selection decisions, by the Federal Government in its review of our project proposals, and by the general public in its appraisal of the Division's planning efforts. Thus, the accurate assessment of the community and environmental implications of the proposed highway improvements is a major planning responsibility.

Limitations of Present Techniques

The need for weighting the community and environmental effects of technological projects on different groups has been recognized. Unfortunately, techniques for weighting such factors are in their infancy.

To see some of the issues that must be addressed, consider the nature of benefit–cost analysis, the traditional method for evaluating alternative plans and designs. Impacts of a system which can be identified are evaluated in monetary terms, the benefits and the costs are added up separately, and a benefit–cost ratio is computed.

One major problem with benefit–cost analysis, as it is typically used, is that it hides the essential issues. By aggregating all the impacts on different groups through using dollar values for benefits and costs, the differential incidence of alternative systems on different individuals and groups is hidden, when, in fact, it should be brought out clearly.

Such an approach also presupposes that all the benefits and costs can be expressed in quantitative terms. The result is that many impacts which may be very significant, such as social or environmental quality, may be omitted from the analysis altogether, insofar as they do not lend themselves to quantification.

A further characteristic of benefit–cost analysis is that it assumes that weights can be determined, for example that specific monetary values for benefits and costs can be obtained, and that these weights indeed reflect the "value" to society of the different levels of impacts on various groups.

Furthermore, benefit–cost analysis is usually applied "post facto"; that is, it is used to rank alternatives that have already been conceived and designed. What is needed, more appropriately, are techniques which influence the nature of what alternatives are developed and designed, so that the design can more nearly reflect desired mixes of impacts on various groups.

These comments about benefit–cost analysis can apply to many of the formulations of models or evaluation techniques in which there is an attempt to reduce all the impacts to a single overall score: net present value methods, point-rating systems (11), or the use of an objective function plus a set of constraints as in mathematical programming.

The problem with these techniques, and the way they are typically used, is that they ignore these fundamental issues:

Whose goals will be used in choosing some alternative systems?

How is information obtained about the goals of various groups?

What objectives are the alternative designs developed to achieve?

Can a single, consistent statement of the set of goals of a society be obtained which is sufficiently operational to be used to find the most desirable alternative to achieve those goals?

What is the process through which a variety of public and private institutions, interest groups, and individuals will interact to reach a decision?

Interaction of Technical Analysis and Political Process

The questions raised in the last section pose fundamental issues about the interaction of the technical analysis process with the political process.

Instead of benefit–cost analysis, what is needed are techniques for evaluating alternative systems that explicitly identify which groups are benefited and which groups are hurt by each alternative system, and to what extent each is affected; that can deal with impacts which are difficult to quantify; and that promote effective, constructive interaction among the technical team doing a systems study and the various individuals and groups potentially affected.

To see this, we need observe only a few relatively simple facts about the objectives of individuals and groups in society. Individuals are unable to express consistent, operational, fully defined goals in the abstract; they don't know their goals; their values change over time; they clarify their goals by making choices. Individuals are, however, able to make explicit choices among discrete, well-defined alternatives. For example, if we were to ask the man in the street how much dollar value he would place on a certain reduction of fatality rates on highways versus a specified loss of parkland, he would generally be at a loss to identify these relative values, explicitly, in the abstract. However, if we present him with three or four highway alternatives, each with different construction costs, fatality rates, and takings of parkland, he will probably be able to reach a decision about which of the alignments he prefers, provided he is impacted enough personally to become sufficiently involved to take the time to make a choice.

It is, therefore, unrealistic to assume that a technical analysis team can operate effectively in a political context solely via the modes of operation implied by the conventional benefit–cost analysis or optimization formulation. These modes assume that a statement of the values of the community can be made by the technical team and reduced to well-defined consistent numbers; and that, once these numbers have been determined, the job of the analysis team is simply to find some alternatives which are "best" in terms of those simplified statements of goals. In contrast to this simple image, a more comprehensive model for the role of the technical team in the political process is needed.

OUTLINES OF A NORMATIVE MODEL

This section outlines a basic normative model for the interactions between the technical team and the political process. This model prescribes the structure and tone of the activities to be undertaken by the technical team. To best describe this model, two elements are needed: the statements of our conclusions as to the desirable objective of the technical process; and the most effective way of translating this process objective into a practicable method. These are discussed in the following sections.

To assist in understanding this model, it should be pointed out that the objective of the research from which this model was developed was to produce "a practicable method for evaluating the effects of different types of highways, and of various design features, upon environmental values." The presentation is, therefore, specifically couched in terms of the transportation planning process. But the model is suggested to apply more generally to the evaluation of large-scale systems.

Objective of the Technical Process

We begin with this statement of the objective of the process: the objective of the technical team is to achieve substantial, effective, community agreement on a course of action that is feasible, equitable, and desirable. To clarify this statement, we define key clauses below.

The "technical team" is that organization of professionals (engineers, architects, planners, economists, sociologists, community specialists, and so on) which has the task of doing studies of alternative projects. This team may have as few as 2 or 3 professionals, or as many as 100; and may be an element of a federal, state, or local agency, a metropolitan planning council, a consulting firm hired by such agencies, or other groups.

The "courses of action" with which the technical team will deal must coordinate the plan for the "physical" project with other actions. The major public program element of concern generally is a "physical" project—a highway, dam, housing project. However, it will generally be necessary to coordinate the "physical" project with a variety of related public and private actions. For example, a highway plan might be coordinated with plans for construction of replacement housing, multiple uses of rights-of-way, new community facilities, wildlife refuge development and other conservation measures, and rehabilitation of historical sites, to name a few.

"Feasible"—the course of action must be feasible technically, economically, and legally. This may, in some circumstances, require actions by the technical team to stimulate changes in law or administrative interpretation to achieve the basic objective.

"Equitable"—the construction of a major project in a region constitutes a major public intervention in the fabric of the society; some groups may be hurt by this intervention while other groups gain. If there are groups that would receive undue burdens of tangible and intangible impacts, considerations of equity and fairness require that they receive sufficient compensation to correct this imbalance. For example, the traditional concept of paying homeowners "fair market value" for their property is not equitable if equivalent replacement housing cannot be obtained at that price. Conditions such as limited housing supply in a price range, high interest rates, or de facto segregation may require that, to be equitable, additional financial compensation above market value, or even construction of replacement housing, is required.

To achieve equity, the technical team must identify, for the alternative courses of action being considered, any possible inequities. This should guide the team in searching for modifications to the basic design or additional program elements to be included in the course of action, which will redress any undue burdens on any groups.

"Desirable"—after the course of action has been so developed and tailored as to be feasible and equitable, the benefits should still be sufficiently great as to justify the costs incurred, if the action is to be implemented.

"Community"—while "community" is always hard to define in the abstract, a pragmatic definition is applicable in this context: the "community" consists of all those individuals and groups who will potentially be affected, positively or negatively, by any of the courses of action being considered. A basic premise of our approach is that the "community" so defined is composed of diverse groups with very different values. Therefore, it is infeasible to get agreement on a statement of values; it is more feasible to get agreement on a course of action.

"Substantial agreement"—it will never be possible to get total agreement from all the interests affected. However, the technical team should strive for this as an objective. The existence of any sizable group opposed to the course of action should be seen as an indication that there is a legitimate interest which has not been adequately addressed in developing the action. To the maximum extent possible, effort should be devoted to identifying and understanding this interest and developing a component, or modification, of the course of action to be responsive to this interest.

"Effective agreement"—to be "effective," all the interest groups must be involved in the process of reaching agreement. This means that these groups must be confident that their views, needs, and suggestions have been fully considered and taken into account; that the technical team is credible, open, and professionally knowledgeable; that there are no surprises or hidden arrangements; and that the agreed-upon course of action is indeed equitable and desirable from the points of view of the diverse elements of the community.

EFFECTIVE TRANSLATION OF THE PROCESS INTO PRACTICE

To achieve the objective of the technical process, the technical team must engage in extensive community interaction activities as well as more traditional, more "technical," technical/design activities (1).

The "technical/design" activities include the collection and analysis of data; the development of alternative project concepts, sites, and designs, design details, and complementary program elements (relocation assistance, replacement housing, and so on); the prediction of the impacts of each alternative on all the interest groups affected; and the analysis of these impacts.

The "community interaction" activities are all the various ways in which the technical team learns about the community in all its diversity, particularly the needs and values of various groups; the community learns about the technical team and the alternative courses of action and their consequences; and the community and technical team work together to achieve the objective of substantial, effective agreement.

To achieve the objective effectively, it is proposed that a process strategy be followed. A key feature of the method lies in the way this strategy is structured to achieve the objective described above. Although the exact details of what is done must be determined in every specific case, we believe a basic four-stage strategy will be applicable in almost every instance. The four stages are those of:

1. Initial survey.
2. Issue analysis.
3. Design and negotiation.
4. Ratification.

This strategy must be flexible, to respond to changes as new knowledge is developed in the course of the process. Various versions of the strategy can be developed for different contexts.

The four stages of the hypothesized strategy represent the following dynamics. Initially, the technical team has relatively little conception of the issues and alternative actions open to it. As it works with the technical problem in interaction with the community, the issues become clearer. As the issues become defined and a range of meaningful alternatives has been developed, negotiation of an equitable compromise can begin. In this negotiation process, the team may act more or less as a catalyst, as local conditions warrant, while retaining that basic authority over engineering issues which is its legal responsibility. Finally, either substantial, effective agreement is reached or, resources having been expended, the decision is passed to a higher authority.

Within each phase of the hypothesized process strategy, a technical team assigns its resources to the several ongoing activities according to the urgency of the activity and the particular talents and specialties of the team itself. The specific allocation of team resources will depend on the current issues, as well as on the scale of the project and the resources of the location team.

Stage 1: Initial Survey

The objectives of the technical team in the first stage are to acquire basic data and to develop an understanding of the interests, needs, and desires of all potentially affected interest groups. By the end of this stage, the team should have created an initial statement of the issues and goals which define this problem; and should have assembled suitable data for use in generating some initial alternative concepts of the project and related programs such as joint development, relocation, and so on. Further, it has an initial estimate of what the significant technical, social, and political issues are likely to be.

Stage 2: Issue Analysis

The objective of this phase is to develop, for both the team and the interest groups affected, a clear understanding of the issues. It is suggested that this be done by developing a range of alternatives which represent different assumptions about the objectives to be achieved, and which, when presented to various interest groups, help them to clarify their own objectives. Ideally, all parties concerned are seeking to clarify their understanding of the advantages and disadvantages of various alternatives.

In this stage the technical team starts to develop alternative project concepts, locations, and designs. None of these may be finally selected; the purpose is to get a wide range which shows the spectrum of possibilities. The team also engages in a program of direct interaction with the community. The information resulting from these interactions refines and augments the team's perceptions of the interest groups and their values, and feeds back to the technical/design activities, stimulating the search for further alternatives. By presenting the information about alternatives and their impacts to various groups, the team helps them to learn about the issues and demonstrates the tradeoffs which it might be possible to make.

The presentation of information about alternatives and their impacts will occur many times throughout stage 2, to many different groups and individuals. As alternatives become more precisely defined, the presentations will have to be made more carefully to avoid premature polarization of attitudes and positions.

By the end of stage 2, the technical team hopes to have achieved a heightened understanding of the issues in the community without commitments to hardened positions by the groups affected. The understanding of issues, with regard to both the technical possibilities and the value issues of the impacts on different groups, is particularly important to the team's development of its strategy for the design and negotiation activities.

Stage 3: Design and Negotiation

The objective of this phase is to produce substantial agreement on a single alternative. In general, this will probably involve a multifaceted course of action: not only the physical project itself, but also coordinated public and private actions.

As in stage 2, there are both extensive technical/design and community interaction activities. In technical/design, many additional alternatives are developed and their impacts predicted. However, where in stage 2 the emphasis was on a wide range of basically different alternatives, here the emphasis shifts to a focus on variations of several basic alternatives in order to develop potential compromise solutions. For example, application of the criterion of equity will stimulate the search for ways of modifying actions to reduce or eliminate inequities, through redesign, through development of associated nonphysical program elements, or through direct compensation.

Similarly, in community interaction activities, the emphasis shifts from a concern primarily with drawing out information on attitudes and desires, to stimulating constructive negotiation. The technical team hopes to achieve substantial agreement on a single equitable alternative. To effect this, the team attempts to structure a negotiation process, over time, which prevents stalemate and promotes rational bargaining among the affected interests.

The team itself has to consider carefully its role in the negotiations, particularly as a bargaining party. It may have developed its own perception of what an equitable consensus might be through its continuing contact with the community. It has also acquired real bargaining resources in the form of proposals for project design modifications and associated nonphysical program elements. As the representative of the responsible decision maker in a public works project, the team implicitly also represents the interests of voiceless groups or interests which are not active participants in the interaction process. In some situations, these may include the long-term interests of a particular community, national interests, and others for which no representation may be available. Consequently, the team has responsibilities greater than simple mediation.

Stage 3 terminates when substantial agreement has been reached, a complete impasse has developed, or the time and money resources of the technical team are exhausted.

Stage 4: Ratification

Stage 3 has been successful if substantial agreement on a program of action has been reached. Then stage 4 merely formalizes the agreement at the public hearing. In the event that no agreement was reached, the technical team can prepare its recommendation for presentation at a public hearing, together with discussion of the particular advantages and disadvantages of the alternatives and the tradeoffs available. The hearings may serve to catalyze further negotiation as a result of information developed there, and agreement may still be achieved.

Basic Hypotheses

The staged approach described in the preceding section is based upon a particular set of hypotheses about the form of the interaction of technical analysis with the political process (2, 7, 9). The basic premise of this approach is that the role of the technical team is to clarify the issues of choice, to assist the community in determining what is best for itself. We hypothesize that the technical team has a dual role in the sociopolitical context: to develop alternatives and trace out their impacts on various individuals and groups; and to take a positive role in stimulating clarification of goals and the reaching of agreement on a course of action.

The task of developing alternatives and identifying their consequences is, in theory, the relatively traditional role of the technical team. In practice, what is now different from the past is that the alternatives are much more complex, since they include both physical and nonphysical program elements, and that the consequences to be identified are also recognized as more numerous and more complex, as they involve effects on the community, changes in income distribution, and so on (10). Further, we stress identifying consequences in terms of the specific groups on which they fall.

The role of stimulating clarification of goals and negotiation involves a much broader concept of the technical analysis team than is traditional. Given that individuals and groups do not know their objectives, there is a need for the technical team to help them clarify their objectives for themselves. By posing alternatives to individuals and groups, discussion will be stimulated as to what the goals of these individuals and groups might be. People will broaden their perceptions of the impacts of alternatives on themselves and on others. The analysis team can take a positive leadership role in encouraging constructive political negotiation among the interest groups affected in order to reach agreement on a concerted course of action for the overall good of the region.

Let us expand on this second role. The analysis of a complex problem such as planning a transportation network or locating and designing a water project really is a dynamic process. Initially, the technical team is relatively uncertain about what the alternative designs and consequences might be. During the early phases of analysis, the technical team explores the impacts of many different alternatives. It also sharpens its perception of what may be the significant issues of the problem, what goals might be achieved, and what kinds of tradeoffs among different objectives are really crucial. Thus, seen purely as an isolated technical process, the definition of the problem is continuously evolving.

Now let us embed this technical analysis process in its sociopolitical context. Initially, all the relevant interest groups will be relatively unclear about what their objectives and alternatives are, with regard to the specific set of technological designs, such as transit routes or dam sites, under study. As the technical analysis team develops alternative designs and identifies their impacts, this information can be presented to the various interest groups. This will help them understand what objectives they want to achieve, and how much they are willing to sacrifice of X to gain something of Y. Of course, an essential component of this process will be recognition of the differential impacts on various groups; each alternative will help some groups at the cost of hurting others. It is also important to get groups thinking in terms of these tradeoffs of incidence of impacts.

Therefore, the technical analysis team may play a positive role in the sociopolitical process. First the team can help individuals and groups to clarify what their objectives are by developing alternatives and presenting these alternatives and their impacts to them in ways that can be understood (9, 11).

Second, the technical analysis team can attempt to stimulate constructive negotiation among the various interest groups by defining the tradeoffs among these groups, providing factual information as a basis for negotiation, identifying inequities and ways in which groups that are hurt can be compensated, and, itself, being stimulated to search for imaginative, innovative solutions which try to overcome the negative impacts on particular groups. For example, in urban highway location, as long as the engineers are dealing strictly with the highway per se, there is relatively little room for negotiation and for the search for imaginative solutions. However, recently formulated concepts for joint development and corridor development have opened up the idea that the highway can also be associated with various housing, economic, and social programs to achieve a total package of urban development programs in the highway corridor (3, 12–15). With this broadened set of options, there is now flexibility for the technical team to try to stimulate constructive negotiation. Thus, even though some groups may be displaced by a highway, in fact they may become better off, because of the benefits they can gain from a package that combines a relocation program with various joint development and corridor planning options.

Third, the technical team might, or might not, become a spokesman for those interests for which there may not be an effective voice: the interests of the whole, as opposed to special interests; the interests of long-term, as opposed to the short-run; and so on.

In this concept of the process, substantial community involvement plays a central role (6). As one state highway agency has concluded (4):

> There is one main point that we believe is essential in every route study and that is adequate communication between the highway organization and the people involved. These people are all the people, from the single resident, to all the parts of the local governing body. An open door policy with the best understanding possible of just what is being studied helps eliminate many problems that are based on fear of the unknown, and places the honest disagreements on a more factual basis. We never expect to reach utopia and reach one hundred percent acceptance, but that doesn't mean that we should give up trying.

This means much more than simply conducting a "good" public-relations campaign. The whole process must work to create an informed involved community. We believe it is essential that the technical team always operate in a way that is open, that is attentive to the values held by the different groups of the community, and that maintains the community's confidence that the team is indeed open and is searching actively for a good, equitable solution to a difficult problem (8). To maintain this credibility, and to give fair consideration to the full range of choices, the technical team will often want to analyze explicitly the option of "no action," not building any new facility at all, as well as options of different levels of facility. For example, in the highway case, options ranging from arterial street improvements and public transportation improvement to "junior" expressways to full, high-level expressways.

It is important to note two particular features of the process outlined here. First, the public hearing(s) is the end point of the process of community involvement, not the beginning; second, public discussion of alternative courses of action from the very first phases is essential, and these must be meaningful alternatives.

Further, this model does not necessarily force the technical team to focus only on short-run actions. Planners can still deal with long-range, large-scale programs: but the message is, if the community has not been brought into a position of understanding, involvement in the decision, and commitment to the program, long-range planning may be irrelevant and have little or no influence on actual actions.

The primary thrust of this model is oriented to contexts where specific groups can be identified whose interests are directly affected. There are some situations where the relevant interest groups are less articulate, difficult to identify, or even unknown. This is typical of many "environmental" problems. An extreme example would be the projected development of a dam in a remote wilderness area. Here, those interested in the preservation of wilderness quality and the like are less defined. There are certainly some potential interest groups in conservation and similar organizations. Even if such groups do not exist, or are politically ineffective, it is suggested that the process should still be followed. The technical team should strive to bring out the issues of choice, the alternative actions and their consequences, and to create an informed public that can be involved in the process of reaching a decision.

Of course, the way in which the technical team interacts with the sociopolitical process is affected by the structure of that process, particularly the structure of governments (5). This structure can, in some ways, be itself affected by the technical team's actions. Implementation of this positive role for the technical team must reflect understanding of the sociopolitical process and institutions of the particular state and region.

To summarize: We hypothesize that the technical analysis team has two major roles: development of technical alternatives and the identification of their impacts; and constructive participation in a professional manner in the political process. The four phases of the location team strategy reflect this hypothesis.

To fulfill both of these roles in a professional manner is a major challenge to our technical and personal skills. It is an exciting challenge.

MAKING THE APPROACH OPERATIONAL

These notions we have been developing imply some fairly far-reaching changes in the way professionals and their bureaucracies should operate. These implications are relatively subtle and difficult to communicate. Therefore, to get across an understanding of the process as we would like to see it, we have concluded that the results of this phase of our research should be presented as a proposed "procedural guide."

This guide, which will result from this research effort, will be in fulfillment of our commitment to develop procedures for evaluating urban highway impacts. The basic purpose of this document will be to serve as a guide to highway and transportation agencies, location teams, consultants, local governments, and community groups on how to conduct a location and design process to achieve the objective of "substantial, effective agreement on a course of action which is feasible, equitable and desirable."

The guide will indicate the significant issues to be addressed, provide an overall approach, and describe specific techniques for implementing the approach. Space precludes describing the guide in detail here. However, to give a clearer picture about the ramifications of the process model described above, we will discuss briefly the evaluation techniques.

Evaluation Technique

Perhaps the area where the difference in philosophy between the model described here and the traditional models of the past comes through most clearly is in evaluation technique. The reader is surely saying to himself, "But after all is said and done, you still need some technique for taking all the information you have and weighing different factors to come up with a decision." That is, given information about the alternatives and their impacts on different groups, and about the values of groups, how can we operate upon this information to "reach a decision"?

First, let us state carefully what we mean by "reach a decision." We see the objective of an evaluation technique as that of informing decision making by the technical team throughout the process, by assisting in the management of technical design and community interaction activities, the identification of additional potentially desirable elements to be added to a course of action, and the choice of an action which is equitable, feasible, and desirable.

Note that the emphasis in the evaluation technique we suggest is on assisting: evaluation is seen as a process of sharpening the insights of the technical team. The objective is to assist the technical team in understanding the key choice issues, so they can communicate these to the higher authority to whom they report. The objective of evaluation is to bring out the choice issues, not to hide them.

Note also that evaluation is used throughout the process, not just at the end.

The basic steps of our evaluation technique are, first, to analyze the impacts of the actions developed so far. For each action, one should review the information on predicted impacts. For each impact one should then identify who is affected, when, where, how, and, on the basis of information from community interaction, how they are likely to respond to the impact.

The second step is to identify the major differences among projects and make judgments about their relative importance. Alternative actions should be compared to each other and to the status quo, or "do nothing," alternative. One should also identify possible groups for compensation or other actions to ameliorate impacts.

The third step is to investigate new program elements for possible inclusion with particular actions. That is, one should search for potential ways of modifying actions to improve them with respect to feasibility, equity, and desirability. This should result in a list of suggested targets for technical design activities to develop these modifications.

The fourth step is to investigate impact prediction and community interaction. One should specify actions and/or impacts and interest groups and their values on which better information is desired. Establish corresponding priorities for technical design and community interaction activities.

The final step in our evaluation procedure is to make recommendations. This requires that we determine actions which can be dropped from active consideration, elements to be added to actions to modify impacts, and priorities for elements to be explored, for community interaction, and for impact prediction. Based upon differences among present actions, one should establish rankings of actions from points of view of different groups. The degree of consensus, in terms of specific groups, for each action should be recorded. Last, the technical team should establish ranking of actions from the perspective of the technical team.

CONCLUSIONS

We have proposed a basic model for the role of the technical team in the process of reaching decisions about large-scale technological projects with social consequences. We have also described how we are endeavoring to make this model operational. We look forward to comments on this approach and suggestions of alternative approaches. Whatever our differences might be, let us all agree on the challenge we face, however: to define clearly a constructive role for the technical professional in the process of public decision making.

ACKNOWLEDGMENTS

The author gratefully acknowledges the contributions of his colleagues, particularly Hans Bleiker; Lowell K. Bridwell; Harry Cohen (especially in regard to the discussion on evaluation technique); Frank C. Colcord, Jr.; William Porter; Arlee T. Reno; and John H. Suhrbier; Stuart Hill of the California Division of Highways; and is particularly grateful for the thorough and diligent assistance of Carol Walb.

He would also like to acknowledge the research support of the National Cooperative Highway Research Program of the American Association of State Highway Officials, through the Urban Systems Laboratory of MIT, and the California Division of Highways. The views expressed here are those of the author and do not necessarily represent the views of any of the research sponsors.

REFERENCES

1. BLEIKER, H., 1970. *Community Interaction Techniques*, working draft, Transportation and Community Values Project, Cambridge, Mass.: MIT Urban Systems Laboratory.
2. BOYCE, D., DAY, N., AND MCDONALD, C., eds., 1969. *Metropolitan Plan Evaluation Methodology*, Philadelphia: Institute for Environmental Studies, University of Pennsylvania.
3. BRIDWELL, L. K., 1969. Freeways in the Urban Environment, in *Joint Development and Multiple Use of Transportation Rights-of-Way*, Special Report 104, Washington, D.C.: Highway Research Board.
4. California, 1969. Report of the Advisory Committee to the California Highway Commission and the Director of Public Works on Freeway Route Adoption and Design Procedures, Nov. 3.
5. COLCORD, F. C., JR., 1967. Decision-Making and Transportation Policy: A Comparative Analysis, *Southwestern Social Science Quarterly*, Vol. 48, No. 3, Dec., pp. 383–397.
6. HILL, S. L., 1969. Century Freeway (Watts), in *Joint Development and Multiple Use of Transportation Rights-of-Way*, Special Report 104, Washington, D.C.: Highway Research Board.
7. LEGARRA, J. A., AND LAMMERS, T. R., 1969. The Highway Administrator Looks at Values, in *Transportation and Community Values*, Special Report 105, Washington, D.C.: Highway Research Board.
8. MANHEIM, M. L., 1969. Search and Choice in Transport Systems Analysis, Transportation Systems Planning, *Highway Research Record 293*, pp. 54–81.

9. MANHEIM, M. L., BHATT, K. U., AND RUITER, E. R., 1968. *Search and Choice in Transport Systems Planning: Summary Report*, Cambridge, Mass.: Department of Civil Engineering, MIT, Research Report R68–40.

10. MANHEIM, M. L., ET AL., 1969. *The Impacts of Highways Upon Environmental Values, Phase I Report for NCHRP Project 8-8*, Cambridge, Mass.: Urban Systems Laboratory, MIT.

11. RIEDESEL, G. A., AND COOK, J. C., 1970. Desirability Rating and Route Selection, unpublished paper presented to the Highway Research Board, Washington, D.C.

12. TURNER, F. C., 1969. Current Governmental Policies, in *Joint Development and Multiple Use of Transportation Rights-of-Way*, Special Report 104, Washington, D.C.: Highway Research Board.

13. U.S. Federal Highway Administration, 1969. *A Book About Space*, Washington D. C., U.S. Government Printing Office.

14. U.S. Federal Highway Administration, 1969. Instructional Memorandum, 34–50, *Federal Facilities on the Highway Right-of-Way*.

15. U.S. Federal Highway Administration, 1969. Interim Policy and Procedure Memorandum 21–29, *Joint Development of Highway Corridors and Multiple Use of Roadway Properties*.

27

Environmental Control of Construction Project Management*

MICHAEL S. BARAM

This paper develops a framework for understanding how the decision-making process concerning large-scale public projects interacts with the human and natural environments: the available resources, the possible contributions of a new facility, and the desires of diverse interest groups which constitute the public. The model of this process is both dynamic and highly interactive, as intuition suggests it ought to be. It is mainly intended to be suggestive, to illustrate for the reader the kind of interactions that are most likely to occur and should be considered.

This model of how society exercises control over construction projects is particularly directed at recent environmental experience in the United States. As of early 1974, the decision-making procedures for public projects in this country are changing very rapidly. Although the final results of today's ferment are quite uncertain, some of the fundamental forces at work can be identified. There is both a pervasive sense that the public ought to participate more directly and intimately in decisions about systems which affect their future, and there is a fairly substantial shift in the preferences that are expressed by the public.

Previous chapters in this section on evaluation have developed the theoretical basis for multiobjective analyses of projects, have advanced technical means for assessing relative intensities of preferences, and have

*Adapted from an unpublished draft version. Another version of this paper was published as Technology Assessment and Social Control, *Science*, Vol. 180, May 4, 1973, pp. 465–473.

suggested ways to work with community groups to determine their values and desires. This chapter is now concerned with what the preferences of the American public are and how they are expressed with regard to programs for the implementation of constructed facilities.

A review of recent legal history, in the first part of this chapter, indicates that there has indeed been a mounting concern for environmental values, and, presumably, away from the single objective of economic efficiency. This shift in values, coupled with what is often sensed as a reluctance of officials and project management to respond to the new desires of the public, has led to increased conflicts and the gradual emergence of more open decision-making processes for public projects. This trend is exemplified by the process of impact assessment required under the landmark National Environmental Policy Act, discussed in the second section of this chapter. The last section then synthesizes these trends in a descriptive model of how American society now asserts control over the development of large-scale public projects.

MAJOR DEVELOPMENTS IN ENVIRONMENTAL LAW

Traditionally, Americans have relied on various elements of their legal system to prevent, abate, or get compensation for environmental injuries resulting from construction. Common-law concepts of nuisance, negligence, and trespass have been applied by the courts to situations where private rights have been infringed by the construction process and the resultant physical facilities. Constitutional law has provided state and local authorities with "police powers" to protect and enhance public health and welfare, by means of zoning, noise, building, and health ordinances, for example. These legal concepts have provided the major bases for public control over the environmental effects of project decision making (13). Some federal and state authority to protect the environment and quality of life has also been exercised in the form of limited enactments to control the use of specific areas or resources such as navigable rivers, wetlands, historic areas, and wildlife. The government has also used its purchasing power to make contractors comply with design, siting and performance specifications, and numerous other governmental objectives.

These elements of the legal system have, until recently, constituted the framework in which decisions about new systems occur. Since approximately 1970, however, federal and state governments and an environmentally aggressive judiciary have moved beyond these limited traditional approaches and developed major programs for controlling pollution and resource utilization, which will ultimately bring about coherent management of land use. Finally, these approaches are imbued with the desire to make decisions about important projects more responsive to public concerns, and are inevitably bringing about greater citizen roles in all stages of project management. The effects of these developments are now being felt by all parties involved in the funding and management of construction projects. The most significant developments in these areas are summarized below.

Pollution Control

The Federal Water Pollution Control Program was initiated in 1948 and strengthened by major amendments in 1956, 1965, and 1970. All these enactments imposed responsibility on the individual states to establish standards and objectives for the quality of interstate bodies of water, criteria for discharges, implementation schedules, and enforcement proceedings. Results were consequently slow to emerge and meager. As water pollution worsened during the 1960s, environmentalists and the courts increasingly called upon the 1899 Rivers and Harbors Act, which had previously been little used, and which simplistically provided for immediate abatement of all polluting discharges other than domestic sewage (23).

Chaos resulted as the diverse legislative approaches became operative over the same period of time, and the 1972 Water Pollution Control Act was designed, in large measure, to resolve these differences (20). The new law was also intended to establish a coherent, centralized program of action. It provided the federal Environmental Protection Agency with authority to establish national effluent criteria, and a timetable to bring about use of the "best practicable" technology for pollution control by 1977, the "best available" technology by 1983, and to reach a national "no pollution discharge" goal by 1985. The law also authorized the federal government to pay for a larger share of the cost of the facilities for treating wastewater. This new law will certainly affect the planning and siting of constructed facilities, their design when they will produce objectionable effluents, and the construction process itself with its attendant effects of sedimentation and erosion.

The Federal Air Pollution Control Program, now being implemented under the 1970 Clean Air Act, marks a similar approach to the control of activities which impair air quality (15). Once again, a new regulatory framework has been imposed on planning and design of facilities and the construction process. The act also authorizes the government to prohibit the construction of facilities which would add to the serious air quality problems of designated regions. Similarly, the Federal Noise Control Act of 1972 initiates a national effort to control noise emissions (19). The federal government will establish noise standards for a wide variety of products, including those used in construction, and state and local authorities will concurrently, and indeed more aggressively, continue to enforce ordinances controlling noisy activities such as construction and trucking.

A variety of other recent federal and state laws have also added to the new regulatory framework affecting decision making about new systems. For example, the 1970 Occupational Safety and Health Act is now being implemented to safeguard the worker environment, by establishing standards for permissible exposures to noise, asbestos, heat, and other hazards (14). All these new regulatory programs impinge on systems planning and design, and offer new bases for citizens and interest groups to challenge project decision makers through both administrative and judicial procedures.

Management of Land Use

Federal and state authorities in the United States have been, to date, unsystematic in managing land use, because zoning and other land use controls are employed at the local level and reflect local interests. Only certain types of land, such as wetlands, coastal zones, and a few other fragile ecological areas, have been protected by specific state or federal laws designed to regulate construction in those areas. Certain types of publicly authorized facilities have, in addition, been subject, for some time, to siting criteria. The Atomic Energy Commission has, for example, generally discouraged the construction of reactors in densely populated areas, and the Department of Housing and Urban Development has prohibited the construction of subsidized housing in areas with high levels of noise.

More recently, however, there has been a marked trend at both state and federal levels to establish more coherent programs for managing land use; and these programs will directly affect the planning, siting, and design of projects. As of 1972, for example, the state of Vermont has been attempting to control large housing and commercial developments by using new regional authorities, and the states of Rhode Island and Maine have established new regulations for controlling developments in coastal areas, such as power plants and oil refineries. At the federal level, the passage of the Coastal Zone Management Act of 1972 (17) and the possible enactment of a Land Use Management Act will further reinforce the concept and practice of state-wide land use planning. The states will have the complex task of establishing new decision processes to resolve the intensifying and competing social demands on land and other natural resources (22).

These new programs for regulating land use will eventually supersede the existing patchwork of siting criteria and laws which protect wetlands, conservation and historic districts, and other discrete areas. They can be expected to ultimately bring about changes in zoning and other local controls based on traditional "home rule" politics. The future federal and state laws on the management of land use will, therefore, have a significant impact on the stages of project planning, siting, and design. Not only will there be significant restrictions on the use of particular sites, but planners and developers who wish to be eligible for both public and private funding will eventually be forced to comply with public decisions and criteria for land use.

Public Decision Making

In addition to the aggregation of laws and programs which now control the siting and external effects of projects, other laws have been enacted which go to the heart of the process for making decisions about public projects. These laws generally require that the proponents of any public project prepare and use detailed impact assessments as part of their decision-making process.

The most important of these laws which require impact statements is the National Environmental Policy Act of 1970, which is discussed in detail subsequently (16). In addition, the 1970 Airport and Airways Development Act and the 1966 Department of Transportation Act also impose assessment responsibilities on federal and state transportation officials. Legislation and executive orders in a growing number of states mandate similar procedures for state and, in some cases, for local government decision makers.

The development and use of impact assessments in public decision making affects the provision of funds for projects, the authorization of permits, the siting and design of projects, and the implementation of construction programs by private and public groups. Both development and use of impact assessments are subject to judicial review, and citizens and interest groups have therefore been provided with several new bases for litigation, which have been used to delay, redesign, resite, and even block projects and programs. To evaluate the state of public decision making about new systems in the United States, one must focus on the central feature of the legal landscape, the National Environmental Policy Act, and its implementation in the public agencies and the courts.

THE NATIONAL ENVIRONMENTAL POLICY ACT

The National Environmental Policy Act (NEPA) became law on January 1, 1970, and has since surpassed all expectations as to its effects on federal decision making about public projects. Indirectly, NEPA has also had a significant effect on state-level and even private decision making about projects (18).

Requirements of NEPA

NEPA requires federal agencies to assess potential environmental impacts before they take any "major actions." These actions range from the approval by the Atomic Energy Commission of a construction license to a utility to build a nuclear power plant, to the funding of highways by the U.S. Department of Transportation, to the authorization for the use of herbicides and pesticides by the Department of Agriculture. In other words, projects subject to federal permits, funds, or other action are generally subject to NEPA, in addition to projects actually implemented by federal agencies. The assessment of environmental impacts must include full consideration of five issues:

1. Potential environmental impacts.
2. Unavoidable adverse impacts.
3. Irreversible commitments of resources.
4. Tradeoffs between short-term and long-term needs.
5. Alternatives to the proposed action.

Under guidelines established by the Council on Environmental Quality, the federal agencies must make the draft and final assessments available for review by other governmental officials and the public (5, 24). Although NEPA does not provide a veto power to any official even if the project poses real environmental hazards, the act does provide new information to the public by exposing the extent to which an agency has considered environmental effects. Consequently, it also provides the public with more facts on which to ground its opposition, intervention, and litigation and generates an extended record which is subject to review by the courts. On the basis of experience to date, any obvious deficiencies in an agency's procedures, or in the scope or content of its impact assessment, will result in intervention in federal agency proceedings by citizen groups, political opposition, and litigation. Many projects proposed and assessed have been delayed, and in some cases, abandoned. Others have proceeded after having been modified to ameliorate those environmental impacts which have generated controversy (8, 23).

NEPA does not directly require the private sector to prepare impact assessments. Although there have been suggestions that NEPA be extended to the private sector, these do not appear to have been considered seriously at the federal level. However, NEPA does apply to a utility, corporation, or other private institution which is the applicant or intended beneficiary of federal funds, license, or other "major action." Moreover, several states have adopted variants of NEPA which, because of state and local control of land use, have the potential for directly affecting private activities, particularly land development. In California, for example, the state's supreme court ruled that the California Environmental Quality Act required county boards of supervisors to conduct environmental assessments before they could issue building permits for private developments (6). Private developers may be similarly affected in Massachusetts, where recent legislation requires environmental impact statements not only from state agencies but also from all political subdivisions.

Development of Impact Assessments

Most controversy and litigation concerning NEPA has, thus far, been focused on several issues relating to the development of impact assessments:

1. Is the project a "major action" which requires NEPA assessment?
2. At what point in the planning and design process must an impact statement be developed and circulated for comment?
3. Should the scope of the assessment be limited to measurable impacts, or should it also include largely unquantifiable impacts on aesthetics and other aspects of the "quality of life"? Should indirect or secondary impacts on future community development and population migration, for example, also be included?

Is the project a major action? This is the basic issue for planners and designers of projects subject to NEPA. If yes, impact assessments must be developed at a critical point or points in the process of planning, siting, and construction. So the first task for the planner or coordinator of a project now generally is to conduct an informal, preliminary review to determine if the project or its results can be expected to cause either "significant environmental impacts" or significant opposition on environmental grounds from interest groups which could lead to litigation. If either result appears likely, it is advisable for the planner to conduct a formal assessment. If he does not, opponents of the project can be expected to intervene, to bring suit, and to cause greater and costlier delays than if an assessment were conducted in the first place.

Thus far the courts have halted several projects, even when they were well into the construction stage where stoppage is costly, until the NEPA assessments were developed and circulated. However, the courts have sometimes refused to stop construction when they judged the costs of completing the NEPA process to be excessive: they have permitted a highway project to proceed once vegetation had been cleared and delays would have caused erosion and a loss of about 300 jobs (1), and they have refused to hold up construction of a dam on which a 6-month delay would have increased costs by $12.6 million (3). The issue of when a project is a "major action" is far from settled, and the courts will, in general, decide the issue on the facts that surround each particular case.

What constitutes a "major action" under NEPA does not depend upon project size alone. Any number of relatively minor projects, such as a student dormitory, a short stretch of highway, and a drive-in bank, have been held to be "major actions" where the environment has had particularly important aesthetic and ecological qualities. Some agencies have now established internal guidelines for determining whether a proposed project is a "major action." These criteria are, however, subject to review by the courts, which can determine, based upon the facts surrounding a specific project, whether a project is in fact a "major action" likely to have significant environmental effects. It may, therefore, often be more cost-effective for planners and designers to undertake both preliminary and formal NEPA assessments whenever there is any doubt, rather than risk political opposition, court injunctions, a cessation of work, and other results which can greatly increase costs, damage the image of an agency, and bring about a variety of project problems once the construction process has begun.

At what point in the planning and design process must an impact assessment be conducted? This is another important issue that must be faced by the managers of projects subject to NEPA. So far, the courts have held that a formal NEPA assessment should be conducted at least by the time construction contracts are let or the construction itself starts. However, any assessment conducted just before construction of a project minimizes the overall intent of NEPA insofar as it precludes any critical review and feedback from the public during the planning and design stages, when these comments could lead to a more meaningful consideration of alternatives and changes in plans. This is a cause of constant tension: officials and planners wish to plan, site, and design new facilities without public intervention; and concerned groups wish the impact assessments to occur early during the planning process, in order to play significant roles in the decision process.

It now appears that the courts are requiring that impact assessments be conducted quite early where feasible. For example, the U.S. District Court in Hawaii held that the design study and test borings for a highway project be stopped until an impact assessment had been conducted, circulated for review, and used. The argument was that, if the work proceeded without an impact assessment, the work "would increase the stake which . . . agencies already have in . . . (the project)," and reduce any subsequent consideration of alternatives (12). On the other hand, another court held that "NEPA . . . imposes no clear legal duty upon the (agency) to prepare an environmental impact statement prior to an applicant's acquisition of land for a proposed site" (7). It is, thus, still too early to summarize, with certainty, when the courts will require planners to conduct an impact assessment.

Some agencies now conduct an impact assessment at each significant stage of a major project. The Atomic Energy Commission, for example, conducts an assessment both before it grants a license to construct a nuclear power plant, and before it grants an operating permit. Wise project managers may be well advised to allay possible litigation and consequent court injunctions against further work by conducting assessments at significant stages in the planning and design process before the start of construction. Here again, an early responsiveness to NEPA and public concerns may well prove to be cost-effective.

What should the assessment contain? is the third major procedural issue planners should consider. Although economic impacts, consequences for public health, and secondary effects such as population migration and land development caused by a project may be important to a community or to a region, they have frequently been ignored or treated in a cursory fashion. The failure to consider these elements, which are integral to any comprehensive assessment, is partly due to limitations on the time, money, and manpower that have been available to planners. The more fundamental reason for these omissions is that NEPA does not specifically require that these impacts be considered, but is quite open-ended (21).

Recent court decisions, however, have called for fuller consideration of social and secondary impacts. In one case, the court noted that a private housing development would increase traffic and congestion, commercial growth, and the need for facilities to dispose of trash and sewage, and called upon the National Capital Planning Commission to develop an impact assessment that would cover these effects (10). In what is probably the most significant case to date on this issue, a U.S. Court of Appeals has held that the *content* of the impact statement is subject to judicial review (4). This means that NEPA not only forces federal agencies to advance correctly through a set of procedures but also requires that the substantive content of these procedures be sufficiently comprehensive. What this means specifically for the content of an impact assessment is difficult to determine in advance, but it strongly suggests, again, that a wise project manager should not attempt to cut corners in preparing his statements.

Use of Impact Assessments

The development of impact assessments is a meaningless exercise unless they are actually used in decision making. Yet it should be reasonably clear that they are quite difficult to use. They generate information about a variety of factors, often essentially unquantifiable in terms of common units. They embody all the difficulties of a multiobjective evaluation, as discussed in Chapters 20 to 22, in addition to which the information provided is often complicated and unstructured. Understandably, the impact statements are especially difficult to use for agencies which are accustomed to relatively simple variants of benefit–cost evaluations.

A recent court ruling has authoritatively described the balancing process that agencies must undertake in their decision making in order to comply fully with NEPA (6):

The sort of consideration of environmental values that NEPA compels is clarified by (the Act). In general, all agencies must use a "systematic, inter-disciplinary approach" to environmental planning and evaluation "in deci-sion-making which may have an impact on man's environment." In order to include all possible environmental factors in the decisional equation, agencies must "identify and develop methods and procedures . . . which will insure that presently unquantified environmental amenities and values be given appropriate consideration in decision-making along with economic and technical considerations." "Environmental amenities" will often be in conflict with "economic and technical considerations." To "consider" the former "along with" the latter must involve a balancing process. In some instances environmental costs may outweigh economic and technical benefits and in other instances they may not. But NEPA mandates a rather finely tuned and "systematic" balancing analysis in each instance.

In practice, the assignment of values and weights to environmental and social amenities for use in a benefit–cost analysis is a process which is largely quite arbitrary. Sometimes it is even intentionally designed to produce results which had been predetermined by the officials. A potentially useful procedure, a variant of the impact-incidence matrices which are well know to planners, is the Leopold Matrix developed by the U.S. Geological Survey (26). This is a useful mechanism for promoting rational discussion and systemic resolu-tion of project impacts by the proponents and opponents of a project in a nonadversarial setting. The matrix disaggregates impacts, calls for estimates of magnitude and significance of each impact, and can be completed by each of the interested parties in a controversy. Comparative analysis of the results reveals important areas of difference of opinion and enables consideration of a variety of strategies to reduce such differences, such as design change or the need for concurrent projects to offset specific impacts.

Despite these difficulties and the numerous conflicts and increased costs which now attend agency programs, NEPA is slowly forcing the development of wiser environmental practices, more sensitive agency bureaucracies, and more effective roles for the public. It has established a new context for plan-ning and design of projects in the United States. It has legitimized the provi-sion of new information to the public, public review of governmental planning and design, the use of interdisciplinary and unquantifiable inputs, and cohe-rent review of primary and secondary impacts.

Suddenly, the social context for project management contains new laws and regulatory programs, criteria, actors, objectives, and review processes. A fundamental task remains: the development of a framework for bringing together these elements which will integrate information about the substan-tive sectors of concern, the relevant legal and regulatory authorities, and the dynamics of public response to proposed projects.

A FRAMEWORK FOR THE SOCIAL CONTROL
OF PROJECTS

Projects are implemented by activities in several sequential stages: conception, planning, siting, design, construction, and operation. Each stage requires different levels and types of resources or inputs. These inputs include, for example, manpower, funds, time, land, and equipment. Conversely, the facility that emerges from the construction stage, and indeed the construction process itself, brings about social and environmental effects or outputs. These include effects on

1. Ecology, such as sedimentation and erosion, changes in landscape and wildlife.
2. Economy, both public and private, relating to jobs, property values, tax and insurance rates.
3. Community quality of life, including aesthetic and recreational opportunities, congestion and traffic, noise, and odors.
4. Social and political factors, as regards new residents and life styles, new opportunities, and different demographic and political characteristics.

The inputs to the projects and their effects may be considered to be sequentially related, as shown in Figure 27.1.

The implementation of any project depends upon numerous decision makers, both public and private, at national, regional, and local levels. These decision makers control the development of a new system essentially

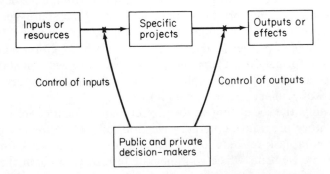

Figure 27.1 A model of the relationship of inputs to a project, the project, its outputs, and the public and private decision makers.

in two ways, as also illustrated in Figure 27.1. First, they control the use of resources. They regulate the availability of manpower and funds for planning and design, they control the use of land through zoning, they may require special procedures for obtaining the right to use certain resources or operate particular equipment. Second, the decision makers can control the effects of the facility. The courts may prevent certain activities, or award damages and allow continuation for others. Federal agencies such as the Department of Transportation and the Environmental Protection Agency and their regional counterparts set standards of performance and regulate many activities. Managers, insurers, building and health authorities, and others can cause a plan or design to be changed to ameliorate specific effects.

To further develop this model of the social control of projects, some of the major influences on planners and designers need to be determined. As shown in Figure 27.2, these influences generally include information about:

1. The availability of land and other resources.
2. The technical and economic feasibility of a project.
3. The actual or potential effects.
4. What may, for convenience, be called the "operational–institutional values."

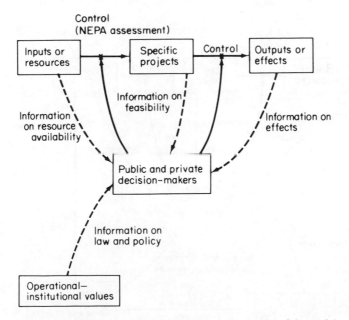

Figure 27.2 Information flows in the descriptive model of the social control of projects.

The operational–institutional values comprise those values which are embedded in common law, legislative and social policy, and which planners must accept as given at any particular time. These values include diverse and sometimes conflicting laws and policies. For example, while NEPA fosters the conservation of resources, programs sponsored by the U.S. Department of Housing and Urban Development encourage expansive use of land by single-unit housing and new communities.

To complete this general model of the social control of projects, the social dynamics brought about by new projects need to be considered. Specifically, we need to examine the responses of individuals and of organized groups to the potential or actual effects they perceive. As suggested by Figure 27.3, these responses can first be manifested through institutional procedures for changing the laws and policies which influence decision makers. This is a lengthy process which requires an extensive coalition of voters, and generally only influences future projects, not the particular project which evoked the response. Alternatively, responses can be manifested through adversarial procedures that challenge particular decisions. Such responses can be formal, as when injured parties go to court for redress, appeal zoning decisions, intervene in agency proceedings, or seek judicial review of governmental decisions. Informal procedures can also be used to feed back community sentiment.

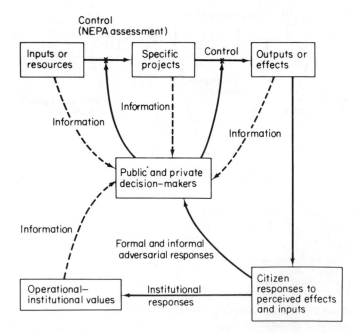

Figure 27.3 Responses to perceived effects of projects.

Demonstrations, raucous town meetings, or quasi-political campaigns can be used. The environmental protection movement serves as a vivid example of how these pressures on decision making can be applied.

Although the sector of society which perceives detrimental effects and protests the specific project does not normally constitute a democratic majority in its early stages, the issues raised by such adverse responses deserve serious consideration. First, the responses represent new perceptions, new "pieces of the truth," which were previously either unknown, ignored, or lightly considered by decision makers. Second, they represent political influence which can easily be magnified by use of the media. Third, although they may be initially ignored, they will continue to reappear in various forms and may later cause extensive delays, as the Atomic Energy Commission, for example, has now discovered. The construction and operation of nuclear power plants are now more than two years behind schedule, with greatly increased costs due to extensive litigation and hearings, because of earlier failures to consider public concern over thermal and radioactive waste disposal, reactor safety, and related ecological and health issues. Fourth, the adverse responses are based on real concerns, will often find large public support, and eventually could result in stringent legislation or judical decisions which decision makers would have to learn to live with. Finally, citizens reflecting a diversity of interests probably represent the most effective mode of promoting the accountability of decision makers to the full social context in which they operate.

The process of listening to community responses and incorporating their ideas into the planning and design of a project certainly complicates decision making. Over the short term, it is less efficient and more costly. Over the longer term, however, significant advantages can be expected to accrue to the public in terms of their larger social interests. In more pragmatic economic and political terms, it has also become increasingly apparent that it is in the long-run self-interest of public officials to be open and responsive to the interests of minority members of the public. Already, in attempts to minimize public opposition to new projects and to restore credibility to their programs, several federal agencies have promoted citizen participation in planning and design beyond the requirements of NEPA. For example, the U.S. Department of Transportation has incorporated new modes of public participation into its policies and procedures for highway planning (25) based to a considerable extent on recent research into community values (9). Similarly, the Corps of Engineers recently initiated its "fishbowl planning" concept, which attempts to bring citizens into the early stages of planning for a new project (11). These efforts represent attempts to provide timely information to interested parties before decisions have been made so that subsequent responses can be constructive and responsible.

Without such attempts, the discussions between interest groups and the decision makers can rapidly degenerate into endless adversarial processes which are both costly and painful. The task facing planners of both public and private projects is to transform this adversarial relationship to one of joint decision making and negotiation of differences in good faith among all interested parties. In short, the task is to establish an ongoing dialogue and joint effort for the planning and design of facilities. This effort will not only require new procedures for incorporating public responses but may, ultimately, entail substantial structural changes in the American political system.

As these changes develop, as both the new public values become better defined and as the new processes of evaluation and decision making evolve, many difficult issues will arise. The descriptive model of the social control of public projects does not provide the answers to these questions. Hopefully, it provides a framework for understanding the issues and how they relate to each other in a dynamic, interactive process.

REFERENCES

1. *Brooks v. Volpe*, Environment Reporter-Cases, 1972, Vol. 4, p. 1532.
2. *Calvert Cliffs Coordinating Committee v. Atomic Energy Commission*, Environment Reporter-Cases, 1971, Vol. 2, p. 1779.
3. *EDF v. Armstrong*, Environment Reporter-Cases, 1972, Vol. 4, p. 1744.
4. *EDF v. Corps of Engineers*, Environment Reporter-Cases, 1972, Vol. 4, p. 1721.
5. Federal Register, 1971. NEPA Guidelines No. 1, Vol. 36, pp. 7724–7729.
6. *Friends of Mammoth v. Mono County*, Environment Reporter-Cases, 1972, Vol. 4, p. 1593.
7. *Gage v. Commonwealth Edison*, Environment Reporter-Cases, 1972, Vol. 4, p. 1767.
8. GREEN, H., 1972. *NEPA in the Courts*, Washington, D.C.: Conservation Foundation.
9. MANHEIM, M. L., ET AL., 1969. *The Impacts of Highways upon Community Values*, Urban Systems Laboratory Report 69-1, MIT, Cambridge, Mass.
10. *McLean Gardens v. National Capital Planning Commission*, Environment Reporter-Cases, 1972, Vol. 4, p. 1708.
11. SARGENT, H., 1972. Fishbowl Planning Immerses Pacific Northwest Citizens in Corps Projects, *Civil Engineering*, Sept., p. 54.
12. *Stop H-3 Association v. Volpe*, Environment Reporter-Cases, 1972, Vol. 4, p. 1684.
13. SWEET, J., 1970. *Legal Aspects of Architecture, Engineering, and the Construction Process*, St. Paul, Minn.: West Publishing Co.
14. U.S. Code, Vol. 29, Section 651 et seq.: Occupational Safety and Health Act (1970).
15. U.S. Code, Vol. 42, Section 1857 et seq.: Clean Air Act (1970).
16. U.S. Code, Vol. 42, Section 4321 et seq.: National Environmental Policy Act (1970).
17. U.S. Congress, 1972. Coastal Zone Management Act, Public Law 92–583.

18. U.S. Congress, House of Representatives, Subcommittee on Fisheries and Wildlife Conservation, Committee on Merchant Marine and Fisheries, 1972. *Administration of the National Environmental Policy Act*, Hearings, Vol. 92–24, Vol. 92–25, Washington, D.C.: U.S. Government Printing Office.
19. U.S. Congress, 1972. Noise Control Act, Public Law 92–574.
20. U.S. Congress, 1972. Water Pollution Control Act, Public Law 92–500.
21. U.S. Corps of Engineers, Institute for Water Resources, 1972. *An Analysis of Environmental Statements for Corps of Engineers Water Projects*, Springfield, Va.: National Technical Information Service.
22. U.S. Council on Environmental Quality, 1971, *Quiet Revolution in Land Use Control*, Washington, D.C.: U.S. Government Printing Office.
23. U.S. Council on Environmental Quality, 1972. *Annual Report for 1972*, Washington, D.C.: U.S. Government Printing Office.
24. U.S. Council on Environmental Quality, 1972. *Memorandum for Agency and General Counsel Liaison on NEPA Matters*, NEPA Guidelines No. 2.
25. U.S. Department of Transportation, 1972: *Policy and Procedure Memorandum 90-4.*
26. U.S. Geological Survey, 1971: *A Procedure for Evaluating Environmental Impacts*, Circular 645, Washington, D.C.: U.S. Government Printing Office.

Biographies of Authors

MICHAEL S. BARAM is an attorney with an LL.B. degree from Columbia University School of Law. He is Associate Professor in the Department of Civil Engineering at MIT, and a special member of the faculty of the Boston University School of Law. His research centers on the legal aspects of new technology and environmental control.

JACK R. BENJAMIN is Professor at Stanford University. He received his Sc.D. from MIT and his special field of interest is structural engineering and probability models. He is coauthor of the book *Probability Statistics and Decision Analysis for Civil Engineers*.

A. BRUCE BISHOP is Associate Professor at the Utah Water Research Laboratory, Utah State University, where he teaches and has research interests in water resources and public works planning. He received his Ph.D. from Stanford University in engineering economic planning. In 1971 he received the Highway Research Board Award for a paper of outstanding merit.

FRANK Y. H. CHIN is a Research Assistant in the Systems Division of the Department of Civil Engineering, MIT where he received his S.M. He has worked for the Port of New York Authority and the Transportation Administration in New York City.

JARED COHON is Assistant Professor of Geography and Environmental Engineering at Johns Hopkins. He received his B.S. from the University of Pennsylvania and an S.M. and PH.D. from MIT. His research interests include large-scale investment decision making in water resources under multiobjective time constraints and chance.

PAUL H. COOTNER is C.O.G. Miller Professor of Finance at the Graduate School of Business, Stanford University. He received his Ph.D. from MIT and did his undergraduate work at the University of Florida. He is the author of several books, including *Water Demand for Steam Electric Generation*, and was the editor of *The Random Character of Stock Market Prices*.

RICHARD DE NEUFVILLE is Associate Professor in the Center for Transportation Studies at MIT. From 1970 to 1973 he was Director of the Civil Engineering Systems Laboratory at MIT. He received his Ph.D. from MIT and was a White House Fellow from 1965 to 1966. He is the author of the book *Systems Analysis for Engineers and Managers*, and has done extensive research in airport planning and design. In 1973 he was awarded a Guggenheim fellowship.

FRANKLIN M. FISHER is Professor of Economics at MIT. He received his Ph.D. in 1960 from Harvard and taught at the University of Chicago before coming to MIT. He is the author of many books, most notably *The Identification Problem in Econometrics*, and numerous articles, and he serves as editor of *Econometrica*. He is an active consultant to governmental and private groups and is a director of Charles River Associates, Inc.

JOSEPH I. FOLAYEN was born in Nigeria and is working in Lagos, where he is a partner in Progress Engineers, a Division of Dames and Moore of San Francisco, where he had worked previously. He received his Ph.D. from Stanford University in civil engineering. His special fields of interest are soil mechanics and foundation engineering and he is the author of several papers on soil settlement.

JOSEPH F. FOLK graduated summa cum laude from the University of Pittsburgh and received his Ph.D. from MIT. He is currently Assistant, in charge of special studies, to the Executive Vice-President of the Penn Central Transportation Company.

DAVID W. HENDRICKS is Associate Professor of Civil Engineering at Colorado State University. He received his Ph.D. in sanitary engineering from the University of Iowa and specializes in water and waste engineering, solid waste management, water chemistry, and water resources. He is the author of many professional papers.

KAARE HÖEG, born in Drammen, Norway, is Associate Professor of Civil Engineering at Stanford. He received his Sc.D. from MIT and is the author of the textbook *Applied Soil Mechanics*, as well as numerous papers.

JOHN F. HOFFMEISTER III received a S.B. degree from MIT and an M.B.A. degree from Columbia University School of Business before returning to MIT to receive an S.M. and Civil Engineer's degree. He is now associated with Peat, Marwick, Mitchell and Co. as a transportation consultant in Washington, D.C.

HENRY JACOBY is Professor of Management at the Sloan School of Management, MIT. He received his Ph.D. from Harvard in economics after extensive experience with their Development Advisory Service, as well as the International Bank for Reconstruction and Development. He serves as a consultant on energy projects to the Rand Corporation.

JOHN H. JAMES is working at the Institute of Transportation and Traffic Engineering at the University of California at Berkeley, where he is a doctoral candidate in industrial engineering and operations research. He has an S.M. in electrical engineering and a B.S. from Cornell. From 1967 to 1971 he worked for TRW Systems, Redondo Beach, California.

RALPH L. KEENEY is Associate Professor of Operations Research and Management at the Sloan School of Management, MIT. He received his Ph.D. in operations research from MIT and is the coeditor (with A. W. Drake and P. M. Morse) of *Analysis of Public Systems*. He received his undergraduate degree from UCLA.

GUY LECLERC is Water Resource Analyst at Hydro-Québec, Montréal. A 1969 graduate of École Polytechnique in Montréal, he received his Ph.D. from MIT. His special field of interest is hydrologic modeling.

JON C. LIEBMAN is Professor of Civil Engineering at the University of Illinois. He received his Ph.D. from Cornell and was Assistant and Associate Professor of Geography and Environmental Engineering at Johns Hopkins from 1965 to 1972. His special field of interest is environmental systems.

GEORGE O. G. LÖF is a Consultant for Resources of the Future, Inc., Washington, D.C.

DANIEL P. LOUCKS is Associate Professor in the Department of Environmental Engineering at Cornell, where he teaches graduate courses in the application of mathematical programming, probability, and economic theory to problems in environmental quality management. He received his Ph.D. from Cornell, S.M. from Yale, and B.S. from Pennsylvania State University. In 1972–1973 he worked as an economist with the International Bank for Reconstruction Development.

RUTH P. MACK is Director of Economic Studies at the Institute of Public Administration, New York. She was a member of the senior staff of the National Bureau of Economic Research and has also served as a consultant in economics and fiscal problems for the city of New York. She is the author of several books, including *Planning on Uncertainty*.

DAVID C. MAJOR is Associate Professor in the Department of Civil Engineering at MIT. He received his Ph.D. in economics from Harvard and did his undergraduate work at Wesleyan. He is also acting as Chief Economist to the North Atlantic Regional Study Group of the U.S. Army Corps of Engineers and has been a consultant to the Water Resources Council on their proposed guidelines for evaluation.

MARVIN L. MANHEIM is Professor in the Transportation Division of the Department of Civil Engineering at MIT. He received his Ph.D. in transportation and urban planning from MIT. His major interest is in community participation in the transportation planning process and he has done extensive work for the National Cooperative Highway Research Planning Program and the California Department of Highways. He is an active consultant to many groups in the United States and abroad.

DAVID H. MARKS is Associate Professor in the Water Resources Division and is currently Director of the Civil Engineering Systems Laboratory in the Department of Civil Engineering, MIT. He received his Ph.D. in geography and environmental engineering from Johns Hopkins and did his undergraduate work at Cornell. His principal research interest is the application of operations research to water resources and environmental planning.

ADOLF D. MAY is Professor of Transportation Engineering and Research Engineer at the University of California at Berkeley. He received his Ph.D. from Purdue and is the author of numerous papers in traffic engineering, two of which have won prizes. He has held several guest professorships in Europe, Mexico, South America, and, most recently, Japan.

NEIL B. MURPHY is Marine Bankers Association Professor of Finance, College of Business Administration at the University of Maine. He received his Ph.D. in economics from the University of Illinois and did his undergraduate work at Bucknell. He has published numerous papers on finance and applied econometrics.

FARROKH NEGHABAT is a member of the Technical Staff at the Bell Telephone Laboratories. A native of Iran, he did his graduate work at the University of Delaware and received his Ph.D. in operations research and systems engineering as applied to civil engineering. He won the 1970 James F. Lincoln Prize in Engineering Design.

MARC NERLOVE is Professor of Economics at the University of Chicago. He received his Ph.D. from Johns Hopkins and has an M.A. (hon.) from Yale. He is the author of *Estimation and Identification of Cobb–Douglas Production Functions*, among other books, and of more than 50 articles in professional journals, conference volumes, and symposia.

MICHAEL RADNOR is Professor of Organization Behavior at Northwestern University. He received his Ph.D. from Northwestern and has also received a diploma in business administration from the London School of Economics and the Diploma of Imperial College in production engineering. He is the author of books and papers on organizational behavior and operations research.

ALBERT H. RUBENSTEIN is Professor of Industrial Engineering and Management Sciences at Northwestern University, where he established and heads the Organizational Theory area and the Program of Research on the Management of Research and Development. He received his Ph.D. in industrial engineering from Columbia and is editor of *Transactions on Engineering Management of the IEEE*.

ROBERT M. STARK is Associate Professor of Civil Engineering and of Statistics and Computer Science at the University of Delaware. He received his Ph.D. in applied mathematics and civil engineering from the University of Delaware. In 1972–1973 he was visiting Associate Professor at the Department of Civil Engineering, MIT. He is the coauthor of *Mathematical Foundations for Design—Civil Engineering Systems*.

JOSEPH M. SUSSMAN is Associate Professor in the Department of Civil Engineering, MIT. He received his Ph.D. from MIT in civil engineering systems and is Associate Director of the MIT Civil Engineering Systems Laboratory. He specializes in the application of simulation to civil engineering problems with focus on railroad transportation. He is a founding member of ECI (Engineering Computer International).

DAVID A. TANSIK is Associate Professor of Management in the College of Business and Public Administration at the University of Arizona. He received his Ph.D. in organizational behavior from Northwestern University and is the author of many papers on operations research.

Name Index

Subject Index